The Asian *Vigna:*
Genus *Vigna* subgenus *Ceratotropis* genetic resources

The Asian *Vigna:*

Genus *Vigna* subgenus *Ceratotropis* genetic resources

by

Norihiko Tomooka

*National Institute of Agrobiological Sciences,
Tsukuba, Japan*

Duncan A. Vaughan

*National Institute of Agrobiological Sciences,
Tsukuba, Japan*

Helen Moss

*West Vancouver,
British Columbia, Canada*

and

Nigel Maxted

*University of Birmingham,
Birmingham, United Kingdom*

SPRINGER SCIENCE+BUSINESS MEDIA, B.V.

A C.I.P. Catalogue record for this book is available from the Library of Congress.

ISBN 978-1-4020-0836-8   ISBN 978-94-010-0314-8 (eBook)
DOI 10.1007/978-94-010-0314-8

Printed on acid-free paper

National Institute of Agrobiological Sciences (NIAS)
The contribution of the National Institute of Agrobiological Resources (NIAS), Japan, towards the costs of publishing this book is acknowledged. NIAS is an independent administrative institution supervised and financially supported by the Ministry of Agriculture, Forestry and Fisheries (MAFF), Japan. NIAS was established in 2001 to conduct basic Life Science research on insects, animals and plants strategically and intensively in order to produce pioneering results. NIAS aims at accelerating genome research and its use for plants (especially rice) and animals (insects and domestic animals) as a leading research institute. Among the several research thrusts of NIAS is to conduct fundamental and technological research on development and use of biological resources in agriculture.

International Plant Genetic Resources Institute (IPGRI)
The contribution of the International Plant Genetic Resources Institute (IPGRI) through the subregional network EA-PGR (EXPAND) to some of the studies that contributed to this publication is acknowledged. IPGRI is an autonomous international scientific organization, supported by the Consultative Group on International Agricultural Research (CGIAR). IPGRI's mandate is to advance the conservation and use of genetic diversity for the well-being of present and future generations. IPGRI's headquarters is based in Rome, Italy, with offices in another 19 countries worldwide. It operates through three programmes: (1) the Plant Genetic Resources Programme, (2) the CGIAR Genetic Resources Support Programme, and (3) the International Network for the Improvement of Banana and Plantain (INIBAP).

CONTENTS

AUTHORS

Norihiko Tomooka, obtained his PhD from the University of Kyoto, Japan. The title of his thesis was "Genetic diversity and landrace differentiation of mungbean, *Vigna radiata* (L.) Wilczek, and evaluation of its wild relatives (the subgenus *Ceratotropis*) as breeding materials". Much of the research for his PhD was conducted during 5 years that he spent based at the Chai Nat Field Crops Research Center, Department of Agriculture, Thailand, as a visiting scientist from the Ministry of Agriculture, Forestry and Fisheries (MAFF), Japan. From 1992 to the present he has been responsible for the *Vigna* grain legumes collection in the MAFF, Japan, Genebank system based in the National Institute of Agrobiological Sciences (NIAS). He has participated in 17 germplasm collecting missions for grain legumes in Japan and 12 abroad. In 1998, he visited Europe as an OECD fellow and visited major herbaria, while based at Kew, U.K., to gather data on *Vigna* subgenus *Ceratotropis* germplasm.

Duncan Vaughan obtained his BSc in Agricultural Botany from the University of Reading, UK, and MSc and PhD from the University of Illinois, Champaign-Urbana, USA. His graduate research was on the wild relatives of soybean in the subgenus *Glycine* and genetics of soybean seed quality. He spent in total 8 years at the International Rice Research Institute, the Philippines during which he wrote a genetic resources handbook for the genus *Oryza*. In 1993, he joined the MAFF, Japan, where he has been working together with Dr. Tomooka in the Crop Evolutionary Dynamics Team of NIAS. This team has focused much of its research effort on the wild relatives of crops including the Asian *Vigna*, particularly the *Vigna angularis* complex, molecular analysis and evaluation of gemplasm for various traits.

Helen Moss has an MSc degree in the Conservation and Use of Plant Genetic Resources from the University of Birmingham, UK. During the 1980's she worked as a consultant for the International Board for Plant Genetic Resources and undertook an extensive field collection and ecogeographic studies of both African and Asian *Vigna*.

Nigel Maxted is Senior Lecturer in Biodiversity Conservation at the University of Birmingham. He has three degrees in biology, taxonomy and conservation, his PhD research focused on the application of database technology to taxonomy and conservation. He has twenty years experience in research, consultancy and teaching biodiversity conservation. He has primarily focused on the taxonomy and ecogeography of legume diversity, as well as the *in situ* and *ex situ* conservation of plant genetic resources. He regularly works as a consultant for the leading international conservation agencies. He has published over 100 research papers and in the last 3 years has written two books on *in situ* and *ex situ* conservation and one on legume genetic diversity in the Mediterranean. He is Chair of the ECP/GR Wild Species Conservation in Genetic Reserves Task Force.

ACKNOWLEDGEMENTS

This book could not have been prepared without the help of many people. We would like to acknowledge the support and encouragement of the present and former management of the National Institute of Agrobiological Sciences (NIAS) particularly: M. Nakagahra, N. Katsura, S. Miyazaki, K. Okuno, and J. Kurisaki.

Legume scientists that have had a major influence on our thinking and have cooperated with us in our studies, Y. Egawa, Japan International Research Center for Agricultural Sciences (JIRCAS), Y. Tateishi, University of the Ryukyus and K. Murata, Hokkaido Prefectural Agricultural Experiment Stations.

We acknowledge the great contribution of present and former staff and visitors within the Crop Evolutionary Dynamics Team that have undertaken much of the research reported in this book and collected germplasm with us: A. Kaga, K. Doi, K. Kashiwaba, S. Tsukamoto, K. Motoyoshi, M. Inoue, R. Q. Xu, X. W. Wang, M. S.Yoon, O. K. Han, B. Chaitieng, A. Konarev, E. Potokina, M. T. Federici Rodriguez, I. Kalubowila, L. E. Nsapato, S. S. Rao, Samsudin, P. Dominguez de Sanctis, Sayafrudin, D. Weerasekera, and X. X. Zong.

Present or former staff of the National Institute of Agrobiological Sciences that have helped us in collecting germplasm in various parts of Japan and our *Vigna* field operations: H. Okano, T. Yoshida, T. Nobori, S. Hirashima, H. Kuwahara, T. Komatsuzaki, M. Akiba, T. Oomizu, S. Yanagiya, T. Taguchi, Y. Tsubokura, H. Tomiyama, T. Chibana, S. Hattori, and T. Yokoyama.

Other visitors that have collected with us and supplied ideas for our work that include K. Hammer, University of Gesamthochschule Kassel, Germany, D. Jarvis, IPGRI, Italy, B. Pickersgill, University of Reading, U. K.

The staff of the Japan International Cooperation Agency (JICA) have helped greatly with our field work, particularly we would like to thank M. Kawase, K. Egara, K. Irie and J. Takahashi.

We would like to thank Imetyzez Khodabux a former student at Birmingham University for help with geographic information system analysis particularly in relation to figures 5.7 and 5.9.

Our collaboration in various countries of Asia would not have been possible without the kind help and support of many people. We would like to acknowledge from Thailand: C. Thavarasook, C. Lairungreang, S. Pichitporn, S. Chotechuen, P. Nakeeraks, P. Poonsavasde and S. Nuplean, Chai Nat Field Crops Research Center N. Boonkerd, Suranaree Industrial University, D. Boonmalison, National Genebank, P. Ornnaichart and B. Taengsan, JIRCAS office, P. Srinives and S. Siripin, Kasetsart University; in Sri Lanka Department of Agriculture, S. D. G. Jayawardena, M. Jayasuriya, A. S. U. Liyanage M. Samarasinghe, S. K. Senevirathne, D. K. Edirisinghe, H. M. Tilakaratne Banda, H. J. Warshakoon, J. Samaranayake, J. Simon, Raja and M. D. Dassanayake; in Malaysia: S. Anthonysamy and Ithin bin Binjang, Universiti Pertanian Malaysia; in Myanmar: U Tin Soe, U Kyaw Soe, U Tan Sein and Daw Oo Mar Awn.

We are grateful for information on germplasm and genebank collections supplied by the following people: L. Engle and S. Shanmugasundaram, AVRDC, Wang Shumin, Institute for Crop Germplasm Resources, China, and S. Dillon, Australian Tropical Crops Genetic Resource

Centre, Australia, N. Q. Ng, IITA, T. Vanderborght, National Botanic Garden of Belgium, C. L. Guevara and D. Debouck, CIAT, and the curator of the seed bank, USDA.

We are grateful to the staff of IPGRI in Asia for their cooperation particularly P. Sajise, V. Ramantha Rao, Z.W. Zhang and M. D. Zhou

We appreciate assistance of taxonomists at the herbarium the Royal Botanic Gardens, Kew, particularly B. Verdcourt, G. P. Lewis, B. Schrire and Rico-Arce M. de Lourdes, during the first authors stay as an OECD fellow in 1998. We thank curators of the following herbaria for permission to consult specimens: R. Vickery, British Museum (Natural History), UK; D. Reué, Paris, Natural History Museum, France; A. Robyns, National Botanic Garden of Belgium; G. Thijsse, Rijksherbarium, Leiden, The Netherlands; H. Ohashi, Tohoku University, Japan; N. Murakami, Kyoto University, Japan.

We thank to the following for permission to use copyright material in this book:
The Botanical Society of Japan (Journal of Plant Research) for Figs. 3.1, 3.2;
American Society of Agronomy (Crop Science) for Figs. 6.7, 6.8;
Springer-Verlag (Theoretical and Applied Genetics) and N. Young for Fig. 6. 10a;
Springer-Verlag (Molecular and General Genetics) and A. Kaga for Fig. 6 10b;
Kluwer Academic publishers (Euphytica) for Figs. 6.9, Table 6.5, (Genetica) for Figs. 2.8.b, (Genetic Resources and Crop Evolution) for Figs. 2.7a and b;

Finally we would like to especially acknowledge the inputs of Kyoko Motoyoshi who not only prepared the figures that appear in chapter 4 but also help very much with preparation of the book in camera ready format.

Note

CHAPTER 1

INTRODUCTION

1. *VIGNA* SUBGENUS *CERATOTROPIS* IN CONTEXT

The genus *Vigna* is a large pan-tropical genus with 82 described species distributed among 7 subgenera (Maréchal *et al.,* 1978). Among the subgenera in the genus *Vigna* only subgenus *Ceratotropis* has its center of species diversity in Asia (Fig. 1.1). However, several species from other subgenera occur in the Asia Pacific region and these are listed (Table 1.1). In this book, we consider the Asian *Vigna* only from the perspective of the species in the subgenus *Ceratotropis* (Table 1.2).

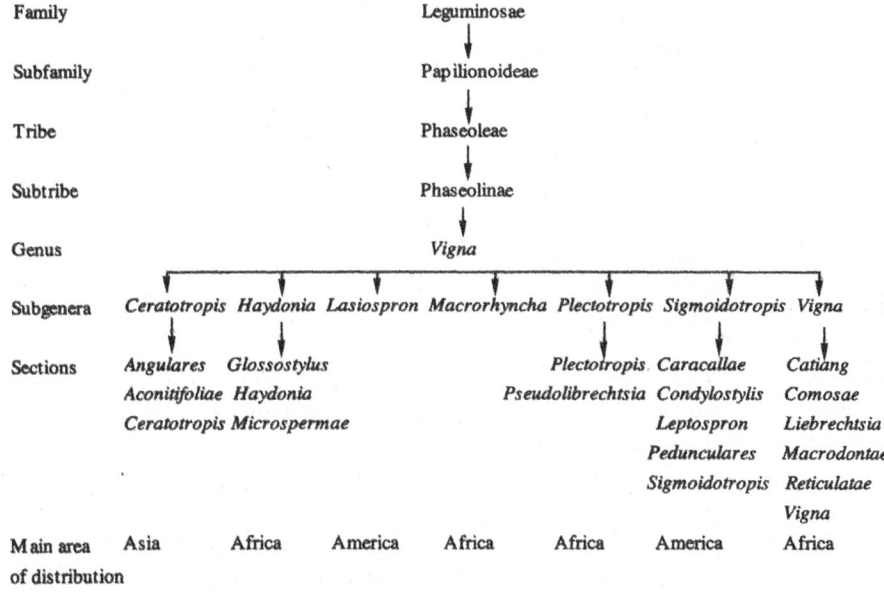

Figure 1.1. Summary of classification scheme and main area of distribution of the genus Vigna *(Updated from Maréchal* et al., *1978).*

The Asian *Vigna* genepool includes cultigens that constitute the primary or secondary legume consumed by many Asian people. Thus, the Asian *Vigna* are an important component of a balanced diet for perhaps one third of humanity.

The Asian *Vigna* consist of 21 species and remarkably, eight of these are used for human or animal food (Table 1.3). This contrasts with the African *Vigna* (the subgenus *Vigna*) in which only 2 species have been domesticated out of 36 species (Maréchal *et al.*, 1978), and the closely related genus *Phaseolus* of the New World that consists of about 50 species and of them only 5 are cultivated (Debouck, 2000).

Table 1.1. Vigna *species not in the subgenus* Ceratotropis *that occur in the Asian Australia Pacific region.*

Subgenus
Section
Species
Sigmoidotropis (Piper) Verdc.
Leptosporon (Benth.)Maréchal, Mascherpa & Stainier
V. adenantha (G.F.Meyer)Maréchal, Mascherpa & Stainier
Macrorhyncha Verdc.
V. grahamiana (Wight & Arn.) Verdc.
Plectotropis (Schumach.)Baker
Plectotropis
V. vexillata (L.)A. Rich.
Vigna Savi
Catiang (DC.)Verdc.
V. unguiculata (L.)Walp. (escape)
Vigna
V. hosei(Craib) Backer
V. luteola(Jacq.) Benth.
V. lanceolata Benth.
V. marina (Burm.) Merr.
V. parkeri Baker

In addition, *V. oahuensis* A. Gray and *V. sandwicensis* A. Gray are reported from the Hawaiian Islands and *V. clarkei* Prain is reported from India. The subgenus and section affinity of these species is not known to us.

Nobel prize winner Norman Borlaug (1973) referred to the domesticated Asian *Vigna* as "slow runners" compared to improvements made in other major crops. Perhaps this reflects the local or national base for the research and crop improvement of most of the cultivated members of the Asian *Vigna*. Only mungbean, *V. radiata*, of the Asian *Vigna* cultigens is a mandate crop of an international center [the Asian Vegetables Research and Development Center (AVRDC), Taiwan, China] that promotes regional

Table 1.2. Taxa of the genus Vigna *subgenus* Ceratotropis

Section	Species	Distribution
Angulares	*V. angularis* var. *angularis*	Japan, Korea, China, Bhutan, Nepal, Vietnam
	V. angularis var. *nipponensis*	Japan, Korea, China, Bhutan, Nepal, India(Himalaya)
	V. dalzelliana	India, Sri Lanka
	V. exilis	Thailand
	V. hirtella	North India, Southeast Asia
	V. minima	Southeast Asia, New Guinea
	V. nakashimae	North China, Korea, Japan (West Kyushu)
	V. nepalensis	Bhutan, East Nepal, Northeast India
	V. reflexo-pilosa var. *glabra*	India, Mauritius, Philippines, Vietnam
	V. reflexo-pilosa var. *reflexo-pilosa*	Japan (Ryukyu), China (Taiwan, Hainan), Southeast Asia, New Guinea, Australia, Oceania
	V. riukiuensis	Japan (Ryukyu), China (Taiwan)
	V. tenuicaulis	Thailand, Myanmar
	V. trinervia var. *trinervia*	Madagascar, South India, Sri Lanka, Myanmar, Thailand, Malaysia, Indonesia, New Guinea
	V. trinervia var. *bourneae*	Southern India
	V. umbellata (cultivated)	Subtropical and warm temperate Asia
	V. umbellata (wild)	Northeast India, Myanmar, Thailand, Indo-China
Ceratotropis	*V. grandiflora*	Thailand, Cambodia
	V. mungo var. *mungo*	South Asia
	V. mungo var. *silvestris*	India, Myanmar, Thailand
	V. radiata var. *radiata*	Asia, Africa and Australia
	V. radiata var. *sublobata*	Asia, Africa and Australia
	V. subramaniana	India
Aconitifoliae	*V. aconitifolia*	South Asia
	V. aridicola	Sri Lanka
	V. khandalensis	India
	V. stipulacea	India, Sri Lanka, Indonesia, New Guinea
	V. trilobata	South India, Sri Lanka

and global research and crop improvement activities. The other Asian *Vigna* domesticated species are the subject of local research and crop improvement initiatives. For example, the main center in Japan for crop improvement related to azuki bean, the second most important legume in Japan is the Hokkaido Prefectural Agricultural Research Station. Thus, the Asian *Vigna* may have greater potential for a major advance in crop improvement than most other major crop groups that have already experienced major production advances.

A recent book devoted to the grain legumes (Smartt, 1990) devotes 54 pages each to peanuts and the cultivated *Phaseolus* species but only 11 pages to the Asian *Vigna*. This, in part reflects the relative lack of knowledge of this group of cultigens and their wild relatives. One reason that may explain the lack of a comprehensive body of work on the Asian legumes and their wild relatives to date may be that this group of species are found in three different eco-geographic regions between which scientific communication tends to hampered by ethno-linguistic and historic factors. Zeven and De Wet (1982) reviewed data related to domestication of crops and suggested the cultigens in the subgenus *Ceratotropis* have been domesticated in three adjacent diversity centers of cultivated plants – the Chinese-Japanese center, the Hindustan center and the Indochinese-Indonesian center.

The subgenus *Ceratotropis* diversified along three principal evolutionary lines that are reflected in the morpho-physiology (taxonomy) and distribution (ecology) of the subgenus. Within each of these evolutionary lines, man has selected and domesticated

Table 1.3. Cultivated species in the genus Vigna *subgenus* Ceratotropis.

Species	Common or local name	Origin
Vigna aconitifolia (Jacq.) Maréchal	Moth bean	South Asia
V. angularis (Willd.) Ohwi & Ohashi	Azuki bean	East Asia
V. mungo (L.) Hepper	Black gram	South Asia
V. radiata (L.) Wilczek	Mung bean	South Asia
V. reflexo-pilosa Hayata var. *glabra* (Maréchal, Mascherpa & Stainier) N.Tomooka & Maxted	Creole bean	Southeast Asia
V. trilobata (L.) Verdc.[1]	Jungli bean	South Asia
V. trinervia (Heyne ex Wall) Tateishi & Maxted	Tooapée (Thai)	South and Southeast Asia
V. umbellata (Thunb.) Ohwi & Ohashi	Rice bean	Southeast Asia

[1] *V. trilobata* is frequently mentioned as used for forage or seeds collected in times of food shortage (Hanelt, 2001). Based on our observations we believe the correct identity for this species that is cultivated and harvested should be *V. stipulacea*.

wild species. One line consists of the species adapted to hot dry, tropical lowland habitats and the group is typified by *V. aconitifolia* (moth bean). A second line consists of the species of the dry to subhumid tropical lowlands typified by *V. mungo* (black gram) and *V. radiata* (green gram), respectively. The third line consists of the species of cool, moist subtropical highlands and warm temperate regions typified by *V. angularis* (Rachie and Roberts, 1974; van der Maesen and Somaatmadja, 1989) (Fig. 1.2).

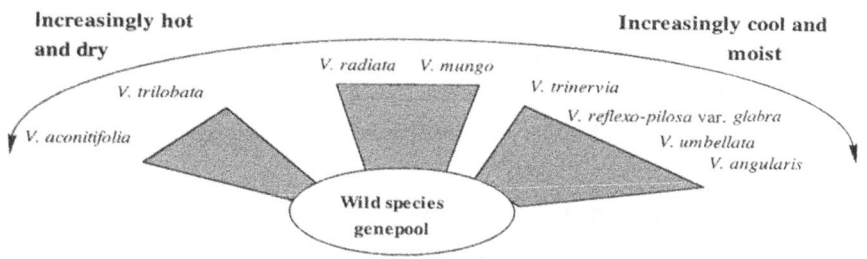

Figure 1.2. Diversity of Vigna *subgenus* Ceratotropis
cultigens in different climatic zones.

The archaeological and historical record for the Asian *Vigna* is sketchy. This is in part related to the lack of familiarity with the key characters of the *Vigna* spp. particularly of the seed where the hilum characteristics can help identify several species. Hence, archaeo-ethnobotanical literature related to the Asian legumes is notoriously inaccurate (Chang, 1983). Southern China is stated as the place of origin of azuki bean (*V. angularis*) since many vegetables were brought into cultivation in that region. However, new research suggests that Northeast Asia may be a more likely region of origin (Mimura *et al.*, 2000). Azuki bean has been referred to as small bean (hsiao-tou) ever since the Eastern Chou period (2770-2255BP) in China (Chang, 1983). In Japan a few plants are believed to have been cultivated by 5500BP including azuki bean (Imamura, 1996).

In South Asia, *Vigna* species have been excavated from a site in western Uttar Pradesh, India. This site is dated at between 3700-3000BP (Meadows, 1996). *Vigna* species were among a wide array of other crops found at the site. *V. radiata* is reported at an archaeological site in central India (Navdatoli) dated at 3500-3000BP (Jain and Mehra, 1980). *V. aconitifolia*, *V. mungo* and *V. radiata* are reportedly mentioned in several ancient Vedic and post Vedic texts (2800BP) of the Hindus (Jain and Mehra, 1980).

Today in several parts of Asia, wild or weedy relatives of the Asian *Vigna* are gathered for their pods and/or seeds for human consumption. In India *V. radiata* var. *sublobata* (wild ancestor of mung bean) is gathered and eaten. Seeds of *V. trilobata* are reported to be gathered for food by some people in India (Jain and Mehra, 1980). In

Japan wild azuki is not used, but weedy azuki bean has been eaten by people as a substitute for the cultivated form (Yamaguchi, 1992).

Unlike many other major crops *Vigna* subgenus *Ceratotropis* consists of cultigens for which the centers of genetic diversity and production overlap. Some Asian *Vigna* cultigens, such as mungbean and azuki bean are grown outside Asia to some extent. However, for the production of the Asian *Vigna* to be grown more widely in areas beyond their current cultivation area recognition of their potential will be important. Below is a list of the various traits that may make this group attractive:·

- Very high nutritional value and high harvest index (*V. umbellata*) (Herklots, 1972);
- Grain legume and cover crop adapted to very dry regions (*V. aconitifolia*) (Duke, 1981);
- Forage legume adapted to dry regions (*V. trilobata*) (Duke, 1981);
- Grain legume for sweet paste production (*V. angularis*) (Lumpkin and McClary, 1994);
- Grain legume for nutritious bean sprouts (*V. radiata*) (Smartt, 1990).

As stated by Smartt (1990) agricultural improvements depend as much on recognizing opportunities and exploiting them effectively as in the execution of complex research programs. We hope that this book will provide insights into the opportunities the Asian *Vigna* afford.

2. SCOPE AND AIMS OF THIS BOOK

It is necessary to explain the perspectives of the authors of this book so that the reader will quickly understand the bias of the presentation in the following chapters. The authors are all genetic resources scientists with a background and interest in crop diversity, wild relatives of crops and taxonomy. Both Dr. Tomooka and Dr. Vaughan work in the Ministry of Agriculture Forestry and Fisheries, Japan where evaluation and use of genetic resources is an important aspect of their research. Dr. Maxted and Ms. Moss have worked primarily on taxonomic and conservation issues in relation to legume genetic resources in various parts of the world. Consequently, the chapters that follow reflect these various genetic resources perspectives. This book details the diversity and conservation of the species in the genus *Vigna* subgenus *Ceratotropis*. However, we do not focus on agricultural production or practice.

The field experiences of the authors also reflect a bias in the knowledge presented. The authors have direct field experience of studying wild and cultivated *Vigna* in Japan, Malaysia, Myanmar, Sri Lanka and Thailand. Their knowledge of other regions comes indirectly from the literature, herbarium specimens and interaction with genetic resources workers in many countries.

This book is written for scientists and advanced students working with legumes. It has been written to provide a sound basis in the current knowledge on taxonomy and genetic resources of the Asian *Vigna*. It is expected that this book will be of broad interest to legume scientists and agriculturalists, particularly those working on the leguminous crops that can be classified as the warm weather group to which the Asian *Vigna* are most closely related – *Cajanus*, *Glycine*, *Phaseolus* and non Asian *Vigna*.

We have deliberately tried to make this book practical and not just academic. Thus, we have included details related to conservation practices including specific details on how to grow the Asian *Vigna*, both cultivated and wild species along with assessment of current conservation status. In addition, practical methodology related to evaluation of *Vigna* germplasm is provided and this includes details of how to extract *Vigna* DNA for molecular studies. In this way, we hope that this book will stimulate further research on genetic resources of the subgenus *Ceratotropis*.

3. TAXA ABBREVIATIONS

Throughout this book, we use a standard set of abbreviations for taxa in the subgenus *Ceratotropis* where abbreviations are sometimes necessary in, for example figures. Numbers that follow abbreviations refer to accessions. The abbreviations for taxa names we use are given below.

V. aconitifolia (aco); *V. angularis* var. *angularis* (an-a); *V. angularis* var. *nipponensis* (an-n); *V. aridicola* (ari); *V. exilis* (exi); *V. dalzelliana* (dal); *V. grandiflora* (gra); *V. hirtella* (hir); *V. khandalensis* (kha); *V. minima* (min); *V. mungo* var. *mungo* (mu-m); *V. mungo* var. *silvestris* (mu-s); *V. nakashimae* (nak); *V. nepalensis* (nep); *V. radiata* var. *radiata* (ra-r); *V. radiata* var. *sublobata* (ra-s); *V. reflexo-pilosa* var. *glabra* (rp-g); *V. reflexo-pilosa* var. *reflexo-pilosa* (rp-r); *V. riukiuensis* (riu); *V. stipulacea* (sti); *V. subramaniana* (sub); *V. tenuicaulis* (ten); *V. trilobata* (tril); *V. trinervia* (trin); *V. umbellata* cultigen (um-c); *V. umbellata* wild form (um-w).

CHAPTER 2

BIOSYSTEMATIC BACKGROUND

1. BASIC STRUCTURE OF *CERATOTROPIS* PLANTS

The flower structure of *Ceratotropis* is rather complex compared to other legumes since it is asymmetrical and has specialized structures. To facilitate ready understanding of subsequent discussion within this book and the key prepared in the last section in this chapter, the following pages illustrate the principal structures of *Vigna* subgenus *Ceratotropis* species (Fig. 2.1, 2.2 and 2.3).

Taxonomic characters of the leaf that are useful in distinguishing Asian *Vigna* species include leaflets lobes and the stipule shape (Fig. 2.1). The stipule in *V. khandalensis* is particularly large and leaf-like compared to other species. The infloresence has several useful taxonomic characters including length and pubescence of the bracteole, primary and secondary bract size and shape, keel pocket size, style beak length and shape, and flower color (Fig. 2.2). Seedling characters, such as cotyledons appearing above the soil and presence of first and second leaf petiole, help to distinguish sections in the subgenus *Ceratotropis* (Fig. 2.3a). Seed characters of taxonomic significance include extent of aril development, hilum length and seed shape (Fig. 2.3b and c).

2. TAXONOMIC HISTORY OF *VIGNA* SUBGENUS *CERATOTROPIS*

2.1. Family Leguminosae (Fabaceae)

The Leguminosae is the third largest plant family after the Orchidaceae and Compositae (Asteraceae) and second most important economically after the Graminae (Poaceae) (Heywood, 1978; Heywood and Chant, 1982). The Leguminosae consists of 650 known genera and about 18,000 species. Current understanding of the broad outline of legume

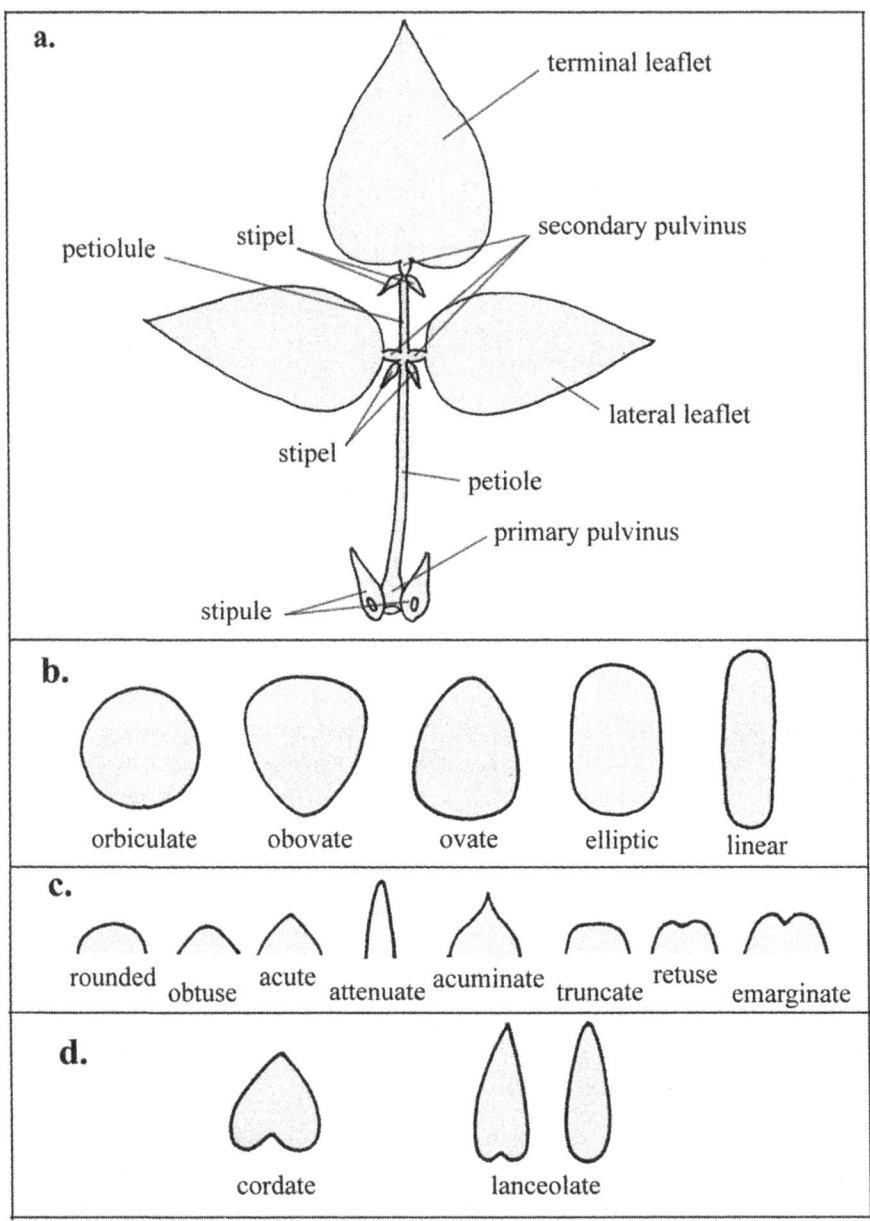

Figure 2.1. a. Leaf and its component parts. b. Main shapes of Vigna *species leaflets.*
c. Leaf apices. d. shape of first and second leaves.

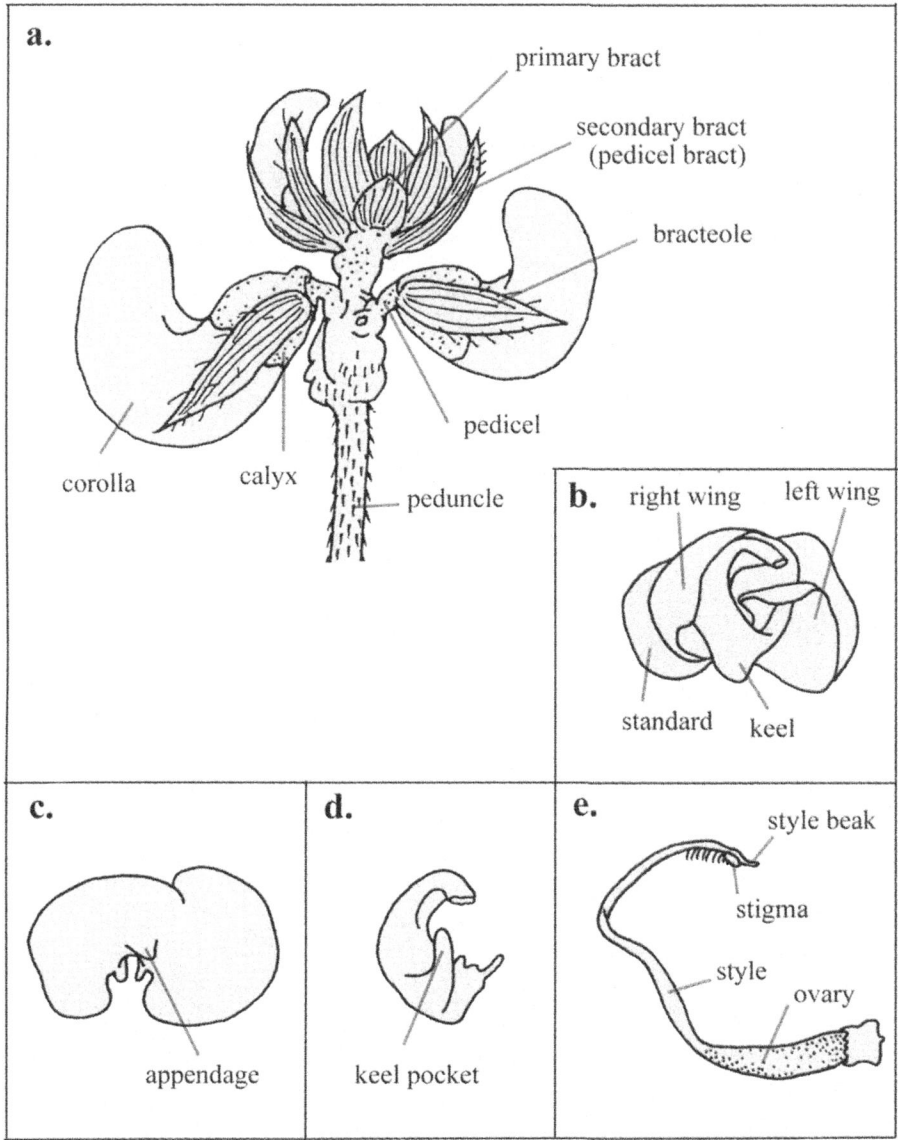

Figure 2.2. a. Inflorescence and its component parts. b. Parts of the flower. c. Standard showing appendage. d. Keel showing shape of the keel pocket found in subgenus Ceratotropis *species. e. Ovary, stigma and style.*

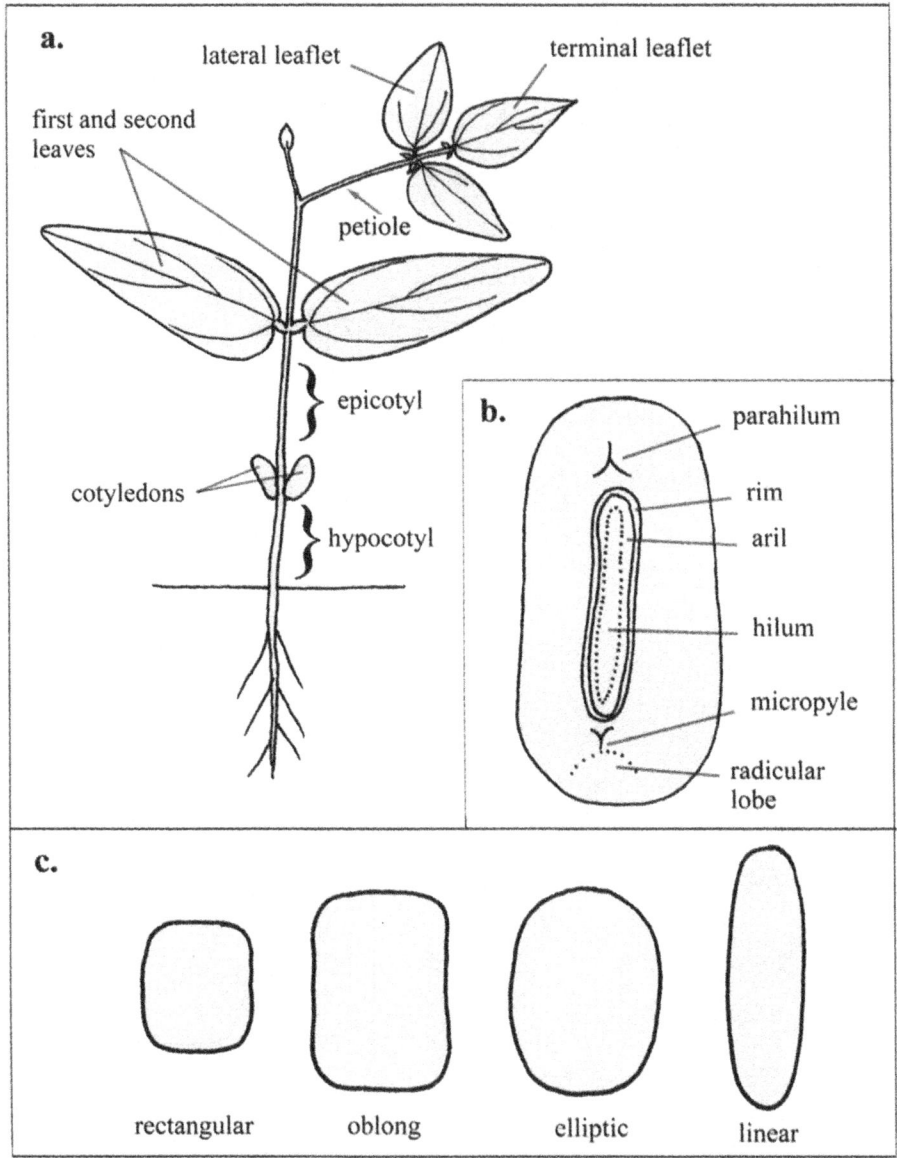

Figure 2.3. a. Seedling and its components. b. Parts of the seed. c. Seed shape.

evolution has resulted from basic analysis of morphological characters to which recently molecular data has been added (Chappill, 1995; Doyle, 1995). Within the Papilionoideae subfamily (440 genera, 12,000 species) the tribe Phaseoleae (84 genera, 1,500 species) includes the main tropical grain legume genera of importance to man, *Cajanus*, *Glycine*, *Phaseolus* and *Vigna*. The subtribe Phaseolinae has 23 genera of which the following genera include cultivated species: *Lablab*, *Macrotyloma*, *Phaseolus*, *Psophocarpus* and *Vigna* (Lackey, 1981).

Table 2.1. History of taxonomic treatment of Asian Vigna *subgenus* Ceratotropis
(after Tateishi and Ohashi, 1990)

References	Genus	Asian *Vigna*	Genus
De Candolle (1825) Bentham (1837, 1865) Baker (1876)	*Phaseolus*	*(Strophostyles)*[1]	*Vigna*
Piper and Morse (1914) Piper (1926)	*Phaseolus*	*(Ceratotropis)*[1]	*Vigna*
Ohwi (1953)	*Phaseolus*	*Azukia*	*Vigna*
Maekawa (1955)	*Phaseolus*	*Azukia & Rudua*	*Vigna*
Verdcourt (1970) Maréchal et al. (1978) Tateishi (1985)	*Phaseolus*	*(Ceratotropis)*[1]	*Vigna*

[1]Names in parenthesis refer to the subgenus or section of the genus in the same box.

2.2. Genus Vigna *Savi*

Savi named the genus *Vigna* in 1824 after a professor of botany at Pisa University, Domenico Vigna (Baudoin and Maréchal, 1988). *Vigna* is a large genus consisting of 7 subgenera that are found in the New and Old World (Maréchal et al., 1981). The history of the taxonomic treatments of the Asian *Vigna* at the genus and subgenus level is summarized (Table 2.1). The Asian *Vigna* were initially classified into the genus *Phaseolus* by De Candolle (1825). A Japanese taxonomist, Ohwi (1953) proposed a new genus *Azukia* for this group. Subsequently, Maekawa (1955) further divided this group into two genera, *Azukia* and *Rudua*, mainly based on seedling characteristics.

Table 2.2. Species and infraspecific taxa of the genus Vigna subgenus Ceratotropis recognized by Verdcourt (1970), Maréchal et al. (1978) and this book.

Verdcourt (1970)	Maréchal et al. (1978)	This book
1 *V. aconitifolia* (Jacq.) Maréchal	1 *V. aconitifolia* (Jacq.) Maréchal	1 *V. aconitifolia* (Jacq.) Maréchal
2 *V. angularis* (Willd.) Ohwi & Ohashi	2 *V. angularis* (Willd.) Ohwi & Ohashi	2 *V. angularis* (Willd.) Ohwi & Ohashi
var. *angularis*	var. *angularis*	var. *angularis*
var. *nipponensis* (Ohwi) Ohwi & Ohashi	var. *nipponensis* (Ohwi) Ohwi & Ohashi	var. *nipponensis* (Ohwi) Ohwi & Ohashi
		3 *V. aridicola* N. Tomooka & Maxted
3 *V. dalzelliana* (Kuntze) Verdc.	3 *V. dalzelliana* (Kuntze) Verdc.	4 *V. dalzelliana* (Kuntze) Verdc.
		5 *V. exilis* Tateishi & Maxted
		6 *V. grandiflora* (Prain) Tateishi & Maxted
4 *V. hirtella* Ridley	4 *V. hirtella* Ridley	7 *V. hirtella* Ridley
5 *V. grandis* (Dalz. & Gibson) Verdc.	5 *V. khandalensis* (Santapau) Raghavan & Wadhwa	8 *V. khandalensis* (Santapau) Raghavan & Wadhwa
6 *V. minima* (Roxb.) Ohwi & Ohashi	6 *V. minima* (Roxb.) Ohwi & Ohashi	9 *V. minima* (Roxb.) Ohwi & Ohashi
7 *V. riukiuensis* (Ohwi) Ohwi & Ohashi	7 *V. riukiuensis* (Ohwi) Ohwi & Ohashi	10 *V. riukiuensis* (Ohwi) Ohwi & Ohashi
8 *V. nakashimae* (Ohwi) Ohwi & Ohashi	8 *V. nakashimae* (Ohwi) Ohwi & Ohashi	11 *V. nakashimae* (Ohwi) Ohwi & Ohashi
9 *V. mungo* (L.) Hepper	9 *V. mungo* (L.) Hepper	12 *V. mungo* (L.) Hepper
	var. *mungo*	var. *mungo*
	var. *silvestris* Lukoki, Maréchal & Otoul	var. *silvestris* Lukoki, Maréchal & Otoul
		13 *V. nepalensis* Tateishi & Maxted
10 *V. radiata* (L.) Wilczek	10 *V. radiata* (L.) Wilczek	14 *V. radiata* (L.) Wilczek
var. *radiata*	var. *radiata*	var. *radiata*
var. *sublobata* (Roxb.) Verdc.	var. *sublobata* (Roxb.) Verdc.	var. *sublobata* (Roxb.) Verdc.
	var. *setulosa* (Dalz.) Ohwi & Ohashi	

Table 2.2.cont.

Verdcourt (1970)	Maréchal et al. (1978)	This book
		15 *V. reflexo-pilosa* Hayata
11 *V. radiata* var. *glabra* (Roxb.) Verdc.	11 *V. glabrescens* Maréchal, Mascherpa & Stainier	var. *glabra* (Maréchal, Mascherpa & Stainier) N. Tomooka & Maxted
12 *V. reflexo-pilosa* Hayata	12 *V. reflexo-pilosa* Hayata	var. *reflexo-pilosa*
		16 *V. stipulacea* Kuntze
		17 *V. subramaniana* (Babu ex Raizada) M. Sharma
		18 *V. tenuicaulis* N.Tomooka & Maxted
13 *V. trilobata* (L.) Verdc.	13 *V. trilobata* (L.) Verdc.	19 *V. trilobata* (L.) Verdc.
		20 *V. trinervia* (Heyne ex Wall.) Tateishi & Maxted
		var. *trinervia*
14 *V. bourneae* Gamble	14 *V. bourneae* Gamble	var. *bourneae* (Gamble) Tateishi & Maxted
15 *V. umbellata* (Thunb.) Ohwi & Ohashi	15 *V. umbellata* (Thunb.) Ohwi & Ohashi	21 *V. umbellata* (Thunb.) Ohwi & Ohashi
	var. *umbellata*	
	var. *gracilis* (Prain) Maréchal, Mascherpa & Stainier	
16 *V. malayana* M.R.Henderson	16 *V. malayana* M.R.Henderson	
17 *V. papuana* Baker f.	17 *V. papuana* Baker f.	

Synonyms between two systems are listed in the same row.

However, these treatments, that consider the Asian *Vigna* constitute a distinct genus or genera, have not gained general acceptance.

Verdcourt (1970) proposed a restricted concept for *Phaseolus*, limiting it to those American species with a tightly coiled style and pollen grains lacking coarse reticulation. The concept of *Vigna* was revised to include the Asian *Vigna* as subgenus *Ceratotropis*. Verdcourt listed 17 species in the subgenus *Ceratotropis* (Table 2.2).

Maréchal *et al.* (1978) in their monograph on the *Phaseolus-Vigna* complex, which has become the standard taxonomic system for this group, recognized seven subgenera in the genus *Vigna*, namely, *Ceratotropis, Haydonia, Lasiospron, Macrorhyncha, Plectotropis, Sigmoidotropis* and *Vigna*. The relationships and geographical distribution of these subgenera are summarized (Fig. 2.4). The African subgenus *Vigna* has a simple floral morphology and is species rich (Maréchal *et al.*,1978). While in Asia the subgenus *Ceratotropis* has differentiated towards a homogeneous group with very specialized floral morphology (Maréchal *et al.*, 1978). Maréchal *et al.* (1978) recognized 17 species in the subgenus *Ceratotropis* (Table 2.2).

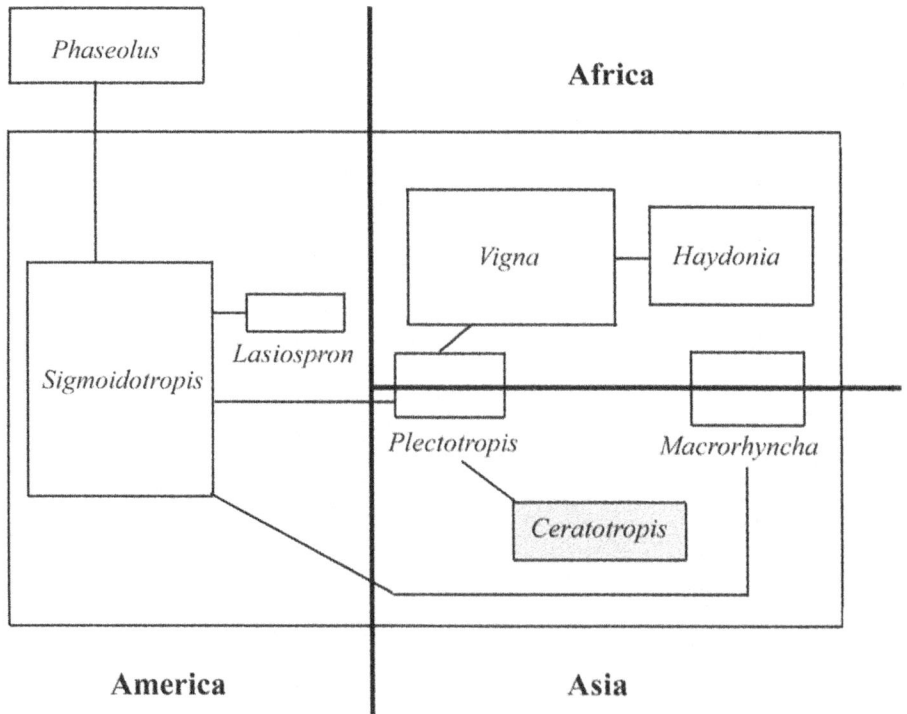

Figure 2.4. The relationships and geographical distribution of seven subgenera in the genus Vigna *and closely related genus* Phaseolus *(after Maréchal* et al., *1978).*

2.3. *Subgenus* Ceratotropis (Piper) Verdcourt: the Asian *Vigna*

Tateishi (1985) proposed a revision of the Asian *Vigna* based both on examining specimens in European and Asian herbaria and intensive field studies in many Asian countries. Some of these proposed revisions have been published (Tateishi, 1984; Tateishi and Maxted, 2002). Recently, studies using a range of techniques have resulted in a revised nomenclature for the subgenus *Ceratotropis* that is presented in this book and this is compared to the taxonomic treatments of Verdcourt (1970) and Maréchal *et al.* (1978)(Table 2.2).

2.3.1. Key morphological characteristics of the Asian Vigna

Maréchal *et al.* (1978) described the useful discriminating characters between the genus *Vigna* and the genus *Phaseolus* as follows (Fig. 2.5):

- Stipule: In the genus *Phaseolus*, the stipule is attached to the stem by its basal part (Fig. 2.5a S-b 0) and does not spread underneath the attachment point. By contrast, in the genus *Vigna* the stipule is attached in several ways but in the subgenus *Ceratotropis*, the stipule is always attached to the stem by its central basal part (Fig. 2.5a S-b 3);
- Tubercle (with extra-floral nectaries): Tubercles are reduced from the inflorescence branch in the genus *Vigna* (Fig. 2.5b tb+), but tubercles are not seen in the genus *Phaseolus* (Fig. 2.5b tb-);
- Keel: In *Phaseolus* the upper part of the keel is narrowly curled around the style. In *Vigna* the keel is variously shaped and the upper part may be narrowly curled to broad and uncurled. In *Ceratotropis* the upper part of the keel is a sickle shaped cylinder curled to the left through between 160°- 360° (Fig. 2.5c).
- Style: The end of the style in *Phaseolus* is curled through more than 360° whereas in *Vigna* it may not be curled or curled through more than 360° (Fig. 2.5d).
- Style-beak: In the genus *Phaseolus*, the stigma is positioned at the end of the style (Fig. 2.5e Bk 0). In the genus *Vigna*, the tip of the style sometimes elongates to form a style beak, so that the stigma is situated somewhat laterally on the style especially in the subgenus *Ceratotropis* (Fig. 2.5e Bk 2).

These *Vigna* characters are represented in the subgenus *Ceratotropis* with a high degree of specialization (Tateishi and Ohashi, 1990). Subgenus *Ceratotropis* species always have flowers colored various shades of yellow, but are never purple, violet, blue or white as is often found in other *Vigna* subgenera (Baudoin and Maréchal, 1988).

The Asian *Vigna* are considered to be a morphologically homogeneous group that have specialized and complex floral organs. The key morphological characters recognized for the subgenus *Ceratotropis* were summarized by Tateishi and Ohashi (1990) as follows:

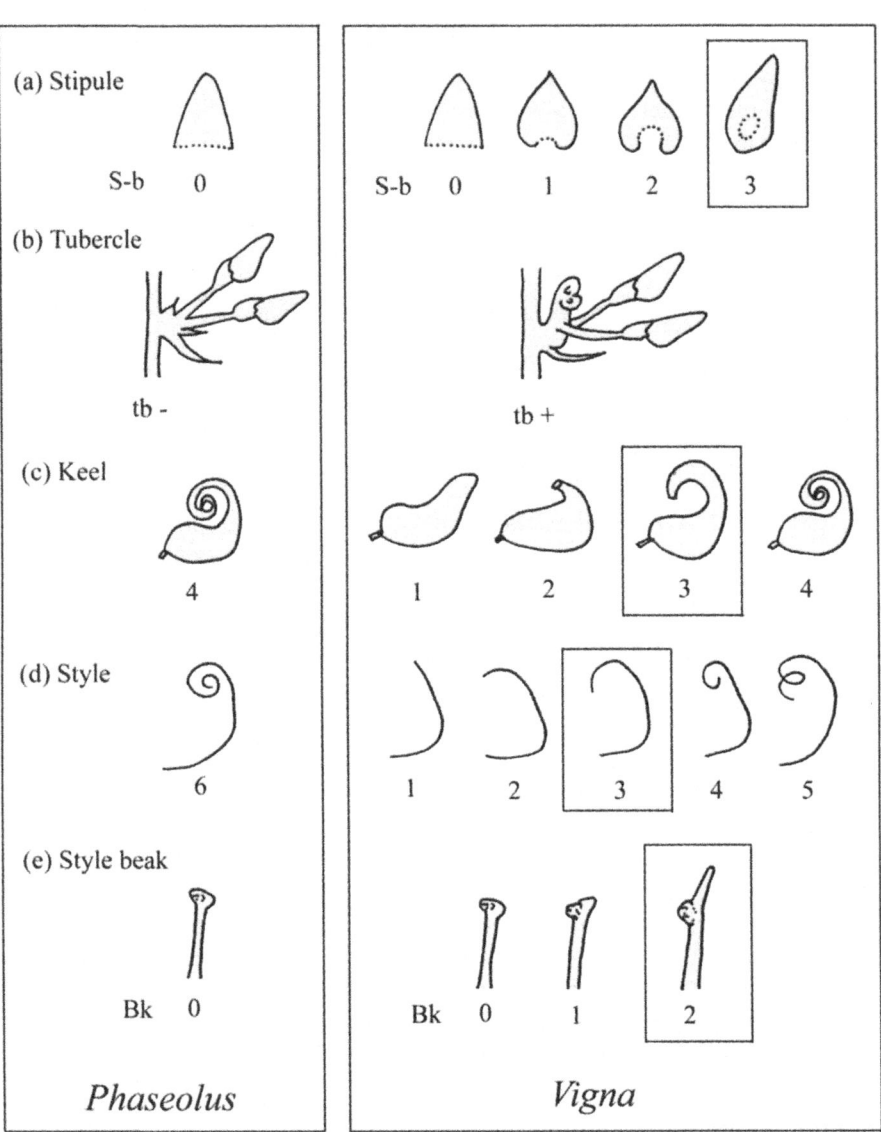

Figure 2.5. Key traits that distinguish Phaseolus and Vigna.
(Modified from Maréchal et al., 1978)
Trait of Ceratotropis in the Vigna box is surrounded by a box.

- peltate stipule (Fig. 2.5a S-b 3);
- standard with an appendage near the center of the inner surface of the lamina (for sections *Ceratotropis* and *Angulares*) (Fig. 2.2c);
- keel petals curved to the left in the upper part (Fig. 2.2b);
- pocket on the left keel petal (Fig. 2.2d);
- style extending beyond the stigma as a beak (Fig. 2.2e, Fig. 2.5e Bk 2);
- pollen grains with a coarse reticulate sculpture.

The morphological similarity of these key characters among subgenus *Ceratotropis* species has lead to the view that the subgenus *Ceratotropis* is monophyletic (Tateishi and Ohashi, 1990).

Although the Asian *Vigna* have rather homogeneous morphology, Maekawa (1955) pointed out that the first and second leaves are sessile in *V. radiata* and *V. mungo*, while those of *V. angularis* and *V. umbellata* are petiolate. He used this character to divide these species into two genera *Rudua* and *Azukia*. These two groups of species differ in the position of the cotyledon on germination. *Rudua* species have epigeal cotyledons and *Azukia* have hypogeal. Baudet (1974) reported that *V. aconitifolia* has epigeal cotyledons and petiolate first and second leaves. This seedling type is intermediate between *Rudua* type and *Azukia* type. Based on these results and detailed morphological studies three subgroups were recognised in the subgenus *Ceratotropis*, the Azuki bean group *s. str.*, the Mungbean group *s. str.* and an Intermediate group (Tateishi, 1996).

Based on several traits, seedling characteristics, size of floral parts, habit and habitat, the three groups within *Ceratotropis* have been recognized as sections, section *Angulares* (Azuki bean group), section *Ceratotropis* (Mungbean group) and section *Aconitifoliae* (Intermediate group) (Tomooka *et al.*, 2002a) (Table 2.3). Many studies (for example Doi *et al.*, 2002; Kaga, 1996; Konarev *et al.*, 2002; Tomooka *et al.*, 2002b) have confirmed the validity of the 3 sections of species within the subgenus *Ceratotropis*. The variation in seedling characteristics in each section is shown in Fig. 2.6. Seedling characteristics are classified into three types based on 1) petiolate or sessile first and second leaves and 2) epigeal or hypogeal germination.

Type A: petiolate first and second leaves + hypogeal germination

Type B: sessile first and second leaves + epigeal germination

Type C: petiolate first and second leaves + epigeal germination

Species in the section *Aconitifoliae* exhibit all of the three seedling types. In contrast, species in the section *Ceratotropis* show seedling type B while species in the section *Angulares* show seedling type A.

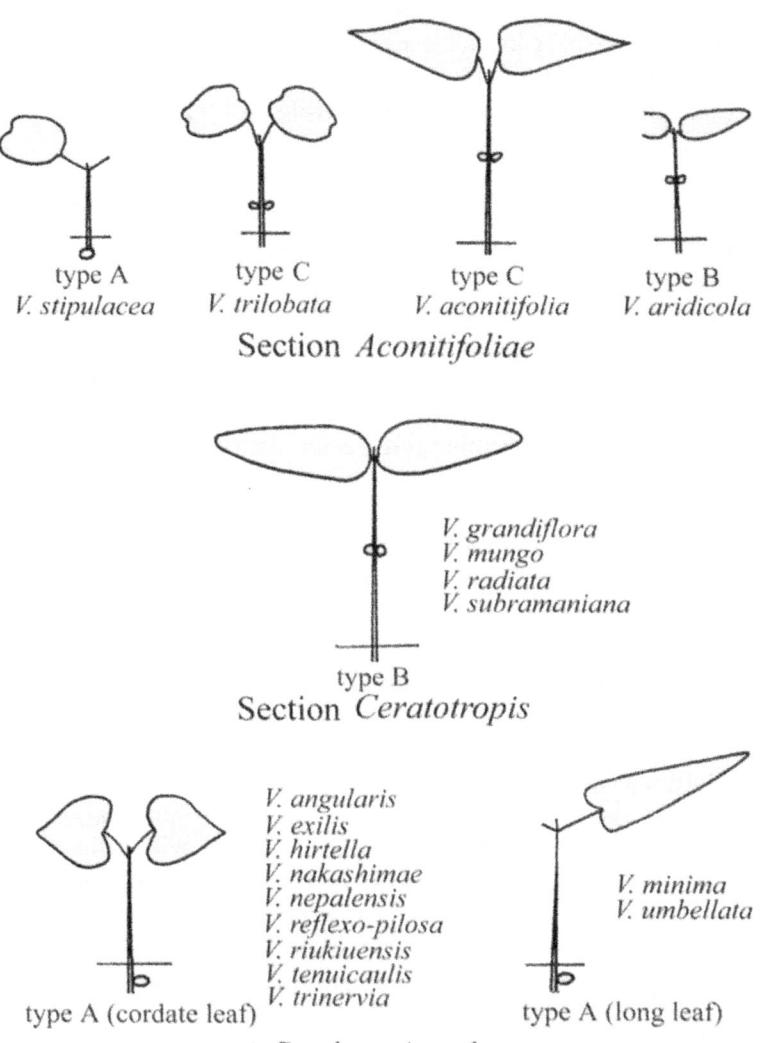

type A
V. stipulacea

type C
V. trilobata

type C
V. aconitifolia

type B
V. aridicola

Section *Aconitifoliae*

V. grandiflora
V. mungo
V. radiata
V. subramaniana

type B
Section *Ceratotropis*

V. angularis
V. exilis
V. hirtella
V. nakashimae
V. nepalensis
V. reflexo-pilosa
V. riukiuensis
V. tenuicaulis
V. trinervia

V. minima
V. umbellata

type A (cordate leaf)

type A (long leaf)

Section *Angulares*

Figure 2.6. Variation in seedling characters among the three sections of the subgenus Ceratotropis. For the definition of each seedling type (type A, B & C) see text (p.19).

Table 2.3. A comparison of selected characters that distinguish sections Aconitifoliae,
Ceratotropis *and* Angulares

Character	*Aconitifoliae*	*Ceratotropis*	*Angulares*
Habit	Trailing (erect for *V. khandalensis*)	Twining (except cultigens)	Twining (except cultigens)
Root	Thick tap root with short lateral roots	Relatively thin tap root with many lateral roots	Relatively thin tap root with many lateral roots
Leaflet	Deeply lobed (rarely entire in *V. trilobata* and *V. aridicola*)	Entire to shallowly lobed	Entire to shallowly lobed
Flower	Small, 5.7-12.3 (mean 9.5) mm	Medium, 11.5-16.1 (14.2) mm	Small to large, 9.8-25.3 (15.2) mm
Keel pocket	Short, 0.5-1.2 (0.8) mm	Intermediate, 1.6-5.4 (3.2) mm	Long, 3.2-5.6 (4.2) mm
Style beak	Short, 0.1-0.4 (0.2) mm	Long, 0.5-1.4 (0.8) mm	Long, 0.3-1.3 (0.9) mm
Standard face appendage	Absent	Present	Present and large
Seed coat lustre	Shiny or dull	Dull (except some cultivated varieties)	Shiny (except *V. trinervia*)
Germination	Epigeal or hypogeal	Epigeal	Hypogeal
First and second leaf	Petiolate or sessile	Sessile	Petiolate
Pod	Pubescent or glabrous	Pubescent	Glabrous

2.3.2. Species relationships in the subgenus Ceratotropis

Analyses of subgenus *Ceratotropis* germplasm have, until recently, either focused on a limited number of species in the subgenus (e.g. Fatokun *et al.*, 1993; Jaaska & Jaaska, 1990; Kajiwara & Tomooka, 1998; Vaillancourt and Weeden, 1993) or mainly one section of the subgenus, section *Angulares* (Yee *et al.*, 1999; Yoon *et al.*, 2000) or section *Ceratotropis* (Kaga *et al.*, 1996b; Tomooka *et al.*; 1996). The molecular techniques that have been used to date to analyze subgenus *Ceratotropis* taxa include RFLP (Fatokun *et al.*, 1993; Zink *et al.*, 1994; Kaga, 1996), ribosomal and chloroplast DNA (Doi *et al.*, 2002; Vaillancourt & Weeden, 1993), RAPD (Kaga *et al.*, 1996; Mimura *et al.*, 2000; Tomooka *et al.*, 1996; Yee *et al.* 1999) and AFLP (Yee *et al.*, 1999; Yoon *et al.*, 2000; Tomooka *et al.*, 2002b).

Relationships among species revealed by molecular studies are discussed based on AFLP and nuclear ribosomal DNA internal transcribed spacer (rDNA-ITS) region sequence analyses (Doi *et al.*, 2002; Tomooka *et al.*, 2002b). These two molecular studies both used a set of accessions that were selected based on inter- and intra-specific morphological variation and geographic origin to represent the whole subgenus currently conserved *ex situ* (discussed further in Chapter 6). This representative species collection (species level core collection) includes all known species except *V. khandalensis* and *V. dalzelliana* whose relationship to other species is deduced from morphological information only.

AFLP analysis revealed 863 polymorphic fragments that were used to determine genetic diversity based on inferred nucleotide diversity. For rDNA-ITS analysis direct sequencing of a region of about 670 base pairs was analyzed that revealed 280 polymorphic sites (111 parsimony informative sites). AFLP analysis included 18 diploid species consisting of 50 accessions and rDNA-ITS analysis included 19 species (25 accessions) that included the wild and cultivated forms of the tetraploid species *V. reflexo-pilosa*. Both sets of results showed consistency when analyzed by different methods.

The phylogenetic trees based on AFLP (Fig. 2.7a and b) and rDNA-ITS region sequence analysis (Fig. 2.8a and b) reveals three groups of species that correspond to sections in the subgenus *Ceratotropis*. *V. trinervia* has a central position among the three sections. Flower and seedling morphology of *V. trinervia* is most similar to section *Angulares* thus it has been placed in this section. However, *V. trinervia* has dull seed coat and mature pod with brown hairs that are common characteristics of species in section *Ceratotropis* and are not seen in other species of section *Angulares*. Based on morphological data it has been suggested that section *Angulares* was derived from section *Ceratotropis* through *V. trinervia* (Tateishi, 1985).

Based on the average DNA base pair (bp) differences of the rDNA-ITS region between *V. trinervia* and species in the three sections, *V. trinervia* is 37bp different from section *Ceratotropis* species, 43bp different from section *Angulares* species and 47bp different from section *Aconitifoliae* species (Doi *et al.*, 2002). The intermediate position of *V. trinervia* suggests this may be a useful species to facilitate gene transfer among sections of the subgenus *Ceratotropis*.

Previous studies, including morphological and isozyme variation, have suggested that *V. hirtella* and *V. trinervia* were the genome donors to *V. reflexo-pilosa* (Tateishi, 1985; Egawa *et al.*, 1996). Proteinase inhibitor variation also strongly suggests that *V. trinervia* was one of the genome donors to *V. reflexo-pilosa* (Konarev *et al.*, 2002) (chapter 6). On the other hand, the rDNA-ITS region sequenced from *V. reflexo-pilosa* is similar to several species, *V. exilis*, *V. hirtella* and *V. umbellata* (Fig. 2.8). Thus, one of these species may be another one of the genome donors to *V. reflexo-pilosa*.

Based on AFLP and rDNA-ITS, all the species in section *Aconitifoliae* and section *Ceratotropis* are genetically distinct from each other. No groups are apparent within section *Aconitifoliae*. Previous morphological and molecular studies have suggested that section *Ceratotropis* consists of two evolutionary branches one including *V. radiata* and the other including *V. mungo* (Kaga *et al.*, 1996b). AFLP and rDNA-ITS results, also shows that *V. mungo* and *V. radiata* are highly diverged and represent two distinct phylogenetic lineages.

In contrast to section *Aconitifoliae* and *Ceratotropis*, most species in section

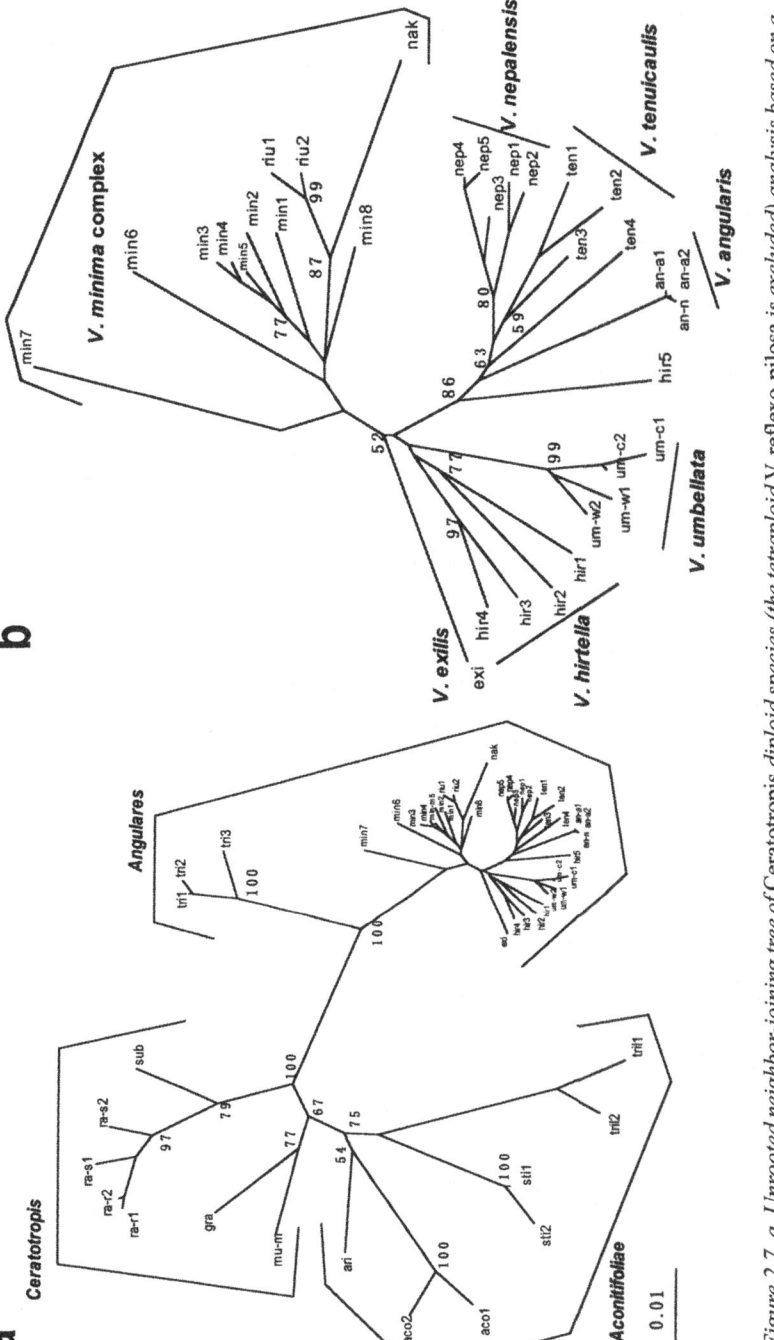

Figure 2.7. a. Unrooted neighbor-joining tree of Ceratotropis diploid species (the tetraploid V. reflexo-pilosa is excluded) analysis based on a matrix of inferred nucleotide diversity using AFLP variation to prepare the matrix. Branch length corresponds to inferred nucleotide diversity. Number at branches indicates probability supporting that branch (modified from Tomooka et al., 2002b). b. Enlargement of part of section Angulares. For taxa abbreviations, see chapter 1 section 3.

a

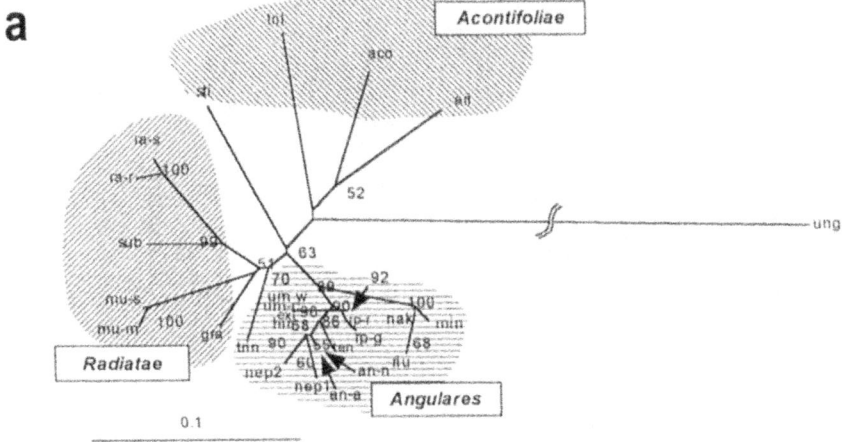

b

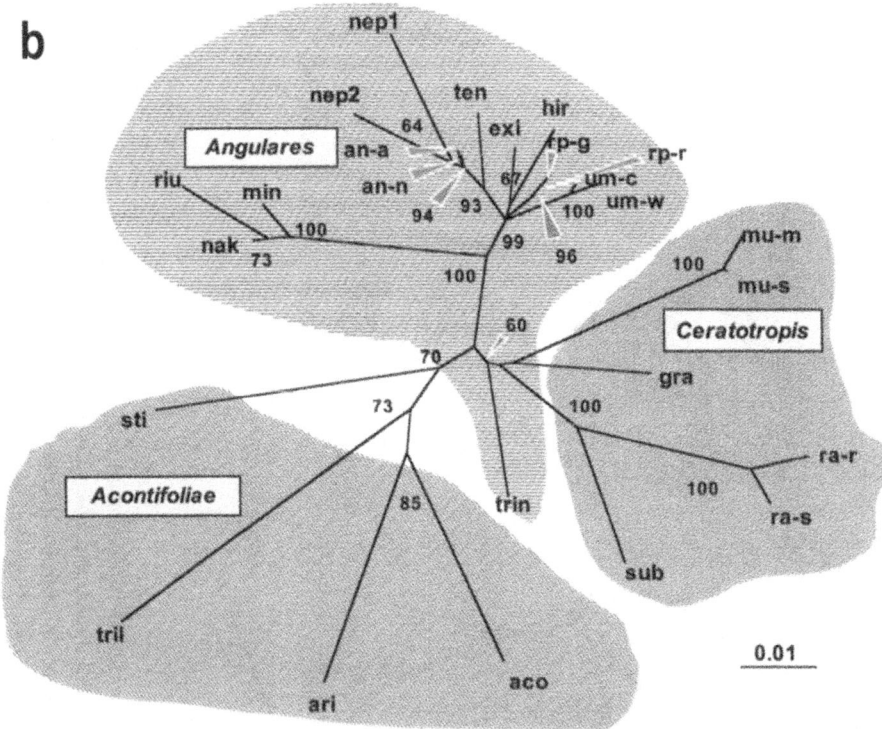

Figure 2.8. Neighbor-joining tree based on sequence data of rDNA-ITS. Numbers beside branches represent bootstrap values (>50%) based on 1000 replicates. a. With sub-genus Vigna *species V.* unguiculata *included as an outgroup. b With V.* unguiculata *excluded (b from Doi et al., 2002). For taxa abbreviations, see chapter 1 section 3.*

Angulares are closely related and have lower inter-species genetic distances than other sections (Table 2.4). The intra-sectional genetic distances suggest, if the rate of evolution is the same for each section, that section *Angulares* is the most recently evolved and section *Aconitifoliae* the ancestral section of the subgenus (Table 2.4). This is supported by analysis of rDNA-ITS region sequence data with subgenus *Vigna* species *V. unguiculata* used as an outgroup species (Fig. 2.8a).

AFLP analysis reveals three main evolutionary branches within section *Angulares* (Fig. 2.7a and b). One branch consists of *V. trinervia*. The second branch consists of the three closely related species *V. minima*, *V. riukiuensis* and *V. nakashimae*. The third branch consists of seven closely related species including 2 cultivated species, *V. angularis* and *V. umbellata*.

The low level of genetic differentiation observed in section *Angulares*, particularly among the species *V. angularis*, *V. hirtella*, *V. nepalensis* and *V. tenuicaulis*, suggest this group of species may currently be in a state of dynamic evolution. Population and geneflow studies of section *Angulares* in the Himalayan region and northern Thailand, both centers of species diversity (chapter 5), may be rewarding. Recent progress in identifying co-dominant (SSR or microsatellite) markers in *V. angularis* may assist in such studies (Wang *et al.*, 2002).

To summarize, taxonomic and genetic studies of the subgenus *Ceratotropis* show an evolutionary trend from small flowered species with absence of standard appendage

Table 2.4. Genetic distances within and between sections in the subgenus Ceratotropis

	AFLP data[1,2] rDNA-ITS data		
Section	*Aconitifoliae*	*Ceratotropis*	*Angulares*
Aconitifoliae	0.03953 (21)[3] *0.099 (6)*		
Ceratotropis	0.05020 (49) *0.108 (24)*	0.02685 (21) *0.064 (15)*	
Angulares	0.07593 (252) *0.104 (60)*	0.06535 (252) *0.087 (90)*	0.02107 (630) *0.044 (105)*

[1] Genetic distance is based on nucleotide diversity (π) that was estimated from the proportion of shared AFLP bands following the method of Innan *et al.* (1999).

[2] For diploid species only

[3] Number in parenthesis is the number of comparisons on which genetic distance is based.

and poorly developed keel pocket and style beak in section *Aconitifoliae* to the large flowered species with well developed keel pocket and long style beak in section *Angulares*. Genetic analyses show interspecies divergence in section *Aconitifoliae* is greatest. Comparison of DNA sequence data between subgenera *Ceratotropis* and *Vigna* suggests section *Aconitifoliae* is the ancestral section in subgenus *Ceratotropis*. Species in section *Angulares* are least diverged and probably derived from species in section *Aconitifoliae* via section *Ceratotropis*. Section *Ceratotropis* is intermediate both morphologically and in terms of inter-species diversity and has two distinct phylogenetic lineages each containing one cultigen, *V. radiata* and *V. mungo*.

3. TAXONOMIC KEY TO TAXA IN SUBGENUS *CERATOTROPIS*

1. Stipules 1.5- 4 cm, foliaceous; bracteole foliaceous, 11-12 mm, concealing the flower bud; primary bract 5-6 mm; secondary bract 9-10 mm; plant erect; distribution India .... *V. khandalensis*
1. Stipules < 1.5 cm, not foliaceous; bracteole not foliaceous, < 11 mm, not concealing flower bud; primary bract < 5 mm; secondary bract < 9 mm; plants trailing or twining (except for cultigens); distribution Asia, Africa, northern Australia and Oceania 2
   2. Plants trailing; leaflets deeply lobed (rarely shallowly lobed in *V. aridicola* and entire on beaches for *V. trilobata*); standard small 5.7-12.3 mm; keel pocket short 0.5-1.2 mm 3
      3. Mature pod with brown hairs 4
         4. Stipule narrowly elliptic, 3.5 – 5 mm long; flowers pale yellow; bracteole hairy; germination epigeal; first and second leaves without petiole .... *V. aridicola*
         4. Stipule ovate, 11-16mm long; flowers bright yellow, bracteole glabrous; germination hypogeal; first and second leaves with petiole .... *V. stipulacea*
      3. Mature pod glabrous 5
         5. Stipule ovate, 4-6 mm; bracteole as long as the calyx; hilum ovate; aril developed .... *V. trilobata*
         5. Stipule narrowly elliptic, 8-10 mm; bracteole much longer than the calyx; hilum linear; aril not developed .... *V. aconitifolia*
   2. Plants twining (or erect for cultigens); leaflets entire to shallowly lobed; standard > 13 mm; keel pocket > 1.2 mm 6
      6. Mature pod hairy; seed coat dull; germination epigeal and first and second leaves without petioles (except for *V. trinervia*) 7
         7. Aril developed; pods ascending 8
            8. Hilum linear, 2.3-2.7 mm; flowers creamy yellow; keel pocket short ca. 3 mm; bracteole as long as calyx; style beak > 1.5 mm .... *V. grandiflora*
            8. Hilum oblong, 1.6-2.4 mm; flowers bight yellow; keel pocket 4-5 mm; bracteole much longer than calyx; style beak < 1 mm.. 9
               9. Plants erect; pods 4-6 cm; seed 0.04-0.06 g; cultivated .... *V. mungo* var. *mungo*
               9. Plant twining; pods 3-4 cm; seed 0.01-0.02 g; wild .... *V. mungo* var. *silvestris*

7. Aril not developed or very thin; pods spreading 10

   10. Flowers bright yellow; seed rectangular; hilum linear, 1.5-2 mm 11

      11. Stem densely covered with long (1.7-2.1 mm) brown hairs, standard smaller 11.0 x 17.5 mm .... *V. trinervia* var. *trinervia*

      11. Stem heavily covered with shorter (0.7-1.0mm) brown villose hairs, standard larger 13.0 x 21.0 mm .... *V. trinervia* var. *bourneae*

   10. Flowers pale greenish yellow; seed elliptic to oblong; hilum oblong < 1.5 mm 12

      12. Standard diameter ca. 10 mm; hilum short, 0.5-1 mm .... *V. subramaniana*

      12. Standard diameter 13-15 mm; hilum long, 1-1.3 mm 13

         13. Plants erect; seed 0.03-0.07 g; cultivated .... *V. radiata* var. *radiata*

         13. Plants twining; seed 0.013-0.15 g; wild .... *V. radiata* var. *sublobata*

6. Mature pod glabrous; seed coat shiny; germination hypogeal and first and second leaves with petiole 14

14. Aril developed 15

   15. Bracteole longer than calyx 16

      16. Style beak flat; distribution mainly in India and Sri Lanka .... *V. dalzelliana*

      16. Style beak linear; distribution mainly in Southeast Asia 17

         17. Stem nearly glabrescent; seeds linear; standard diameter ca. 10mm; pod 2-3 cm .... *V. exilis*

         17. Stems densely hairy; seeds not linear; standard diameter 14-20 mm; pods 5-10 cm 18

            18. Aril thin; seed 2.4-4 mm; hilum oblong or linear, 1-1.8 mm; first and second leaf cordate .... *V. hirtella*

            18. Aril thick; seed 4-7 mm; hilum linear, 2.5-3.5 mm; first and second leaves narrowly elliptic 19

               19. Seed 0.06-0.11 g; cultivated .... *V. umbellata* (cultivated)

               19. Seed 0.016-0.027 g; wild .... *V. umbellata* (wild)

   15. Bracteole shorter than calyx 20

      20. Standard diameter 11-12 mm; flowers pale yellow; distribution northern China, Korean peninsula and western Japan .... *V. nakashimae*

      20. Standard diameter 15-19 mm; flowers bright yellow; distribution Southeast Asia and tropical and subtropical East Asia, New Guinea 21

         21. Seeds per pod 9-12; terminal leaflet linear to ovate with acute apex, membranous; aril thick; distribution Taiwan, Southeast Asia and New Guinea .... *V. minima*

         21. Seeds per pod 7-9; terminal leaflet orbiculate to obovate with rounded or obtuse apex, glossy; aril thin; distribution Okinawa prefecture, Japan and Taiwan .... *V. riukiuensis*

14. Aril not developed 22

22. Standard diameter 12-13 mm; seed length 2.4-2.8 mm; hilum ca. 1 mm .... *V. tenuicaulis*

22. Standard diameter > 13 mm; seed length > 3 mm; hilum > 1mm 23

   23. Seeds oblong, truncate at both ends; primary bract 1-1.5 mm; secondary bract, ovate, ca. 1.5 mm 24

      24. Twining; seed 0.014-0.019 g; wild .... *V. reflexo-pilosa* var. *reflexo-pilosa*

      24. Erect; seed 0.04-0.06 g; cultivated .... *V. reflexo-pilosa* var. *glabra*

23. Seeds elliptic; primary bract 2-6 mm; secondary bract elliptic or cymbiform, 4-10 mm
....25

25. Hilum 1.1-1.8 mm; bracteole glabrous, as long as calyx; bracts glabrous ....
.... *V. nepalensis*

25. Hilum 2.2-4 mm; bracteole pubescent, twice as long as calyx; bracts pubescent .....
.... 26

26. Plants erect; seed 0.03-0.25 g; cultivated .... *V. angularis* var. *angularis*

26. Plant stem slender and twining; seed 0.017-0.03 g; wild ....
.... *V. angularis* var. *nipponensis*

CHAPTER 3

GENETIC RESOURCES

1. CONSERVED GERMPLASM

1.1. Ex situ *conservation*

There are many collections of *Vigna* subgenus *Ceratotropis* germplasm. Most of these collections consist primarily of accessions of the cultigens in the subgenus. The main collections are to be found at the Asian Vegetables Research and Development Center, Taiwan, National Board for Plant Genetic Resources, India, Institute of Crop Germplasm Resources (ICGR), Beijing, China, and Plant Genetic Resources Conservation Unit, Georgia, USA. Large collections of *Vigna angularis* germplasm are maintained by the Hokkaido Prefectural Agricultural Research Centers and Ministry of Agriculture, Forestry and Fisheries Genebank, Japan, ICGR, China and University of Washington, USA. Wild *Vigna* subgenus *Ceratotropis* species are poorly represented in the worlds genebanks. The Ministry of Agriculture, Forestry and Fisheries Genebank, has a collection consisting of 19 out of the 21 species in the subgenus. The National Botanical Garden of Belgium wild *Phaseoleae-Phaseolinae* collection is remarkably rich in species, over 130 species are in the collection of these 9 species are of subgenus *Ceratotropis* (Vanderborght, nd). Some countries have comprehensive collections of their own indigenous *Vigna* genetic resources, such as Australian *Vigna radiata* var. *sublobata* in the CSIRO collections (Lawn and Cottrell, 1988).

A table giving a list of germplasm conserved *ex situ* by various genebank systems around the world is given in Table 3.1.

1.2. In situ *conservation*

In situ conservation is a complementary approach to *ex situ* conservation (Convention on Biological Diversity, 1992; Maxted *et al.*, 1997). We are not aware of any direct action to specifically protect the Asian *Vigna in situ*. However, here we present information relevant to *in situ* conservation of Asian *Vigna*.

Table 3.1. Some of the main genebanks conserving Vigna subgenus Ceratotropis

Species	No. Acc.	Institution
V. angularis (wild)[1]	136	Asian Vegetable Research and Development Center(AVRDC), Shanhua, Taiwan
	339	Australian Plant Genetic Resources System, Australia
	3933 (60)	Institute of Crop Germplasm Resources (ICGR), CAAS, Beijing, China
	3150(72)	Genetic Resources Division, Rural Development Admin. (GRD-RDA), Rep. Korea
	3604(25)	Hokkaido Prefectural Agricultural Experiment Stations, Japan
	753	East Asian Crop Development Program, Washington State Uni., Pullman, USA
	1198(134)	MAFF Genebank, National Institute of Agrobiological Sciences (NIAS), Tsukuba, Japan[2]
	298	Plant Genetic Resources Conservation Unit, USDA (PGRCU-USDA) Georgia, USA
	8	National Botanical Garden, Meise, Belgium
V. aconitifolia	1100	National Bureau of Plant Genetic Resources (NBPGR), New Dehli, India
	32	Australian Plant Genetic Resources System, Australia
	66	Plant Genetic Resources Institute, Islamabad, Pakistan
	57	SRPIS-USDA, Georgia, USA
	4	NIAS, Tsukuba, Japan
	7	National Botanic Garden, Miese, Belgium
V. aridicola	1	NIAS, Tsukuba, Japan
V. dalzelliana		
V. exilis	1	NIAS, Tsukuba, Japan
V. grandiflora	2	NIAS, Tsukuba, Japan
V. hirtella	3	NIAS, Tsukuba, Japan
V. khamdalensis		
V. minima	1	NIAS, Tsukuba, Japan
	9	National Botanic Garden, Meise, Belgium
	1	PGRCU-USDA, Georgia, USA

Table 3.1. cont.

Species	No. Acc.	Institution
V. mungo (wild)	339	Bangladesh Agricultural Research Institute (BARI), Joydebpur, Bangladesh
	82	Australian Plant Genetic Resources System, Australia
	481	AVRDC, Shanhua, Taiwan
	2000	NBPGR, New Dehli, India
	646	Plant Genetic Resources Institute, Islamabad, Pakistan
	300	PGRCU-USDA, Georgia, USA
	6(6)	National Botanic Garden, Meise, Belgium
	105(1)	NIAS, Tsukuba, Japan
V. nakashimae	2	NIAS, Tsukuba, Japan
	40	GRD-RDA, Rep. Korea
V. nepalensis	1	NIAS, Tsukuba, Japan
V. radiata (wild)	5612(4)	AVRDC, Shanhua, Taiwan
	609(34)	Australian Plant Genetic Resources System, Australia
	936	Chai Nat Field Crops Research Center, Chai Nat, Thailand
	3889(1)	PGRCU-USDA, Georgia, USA
	930	Malang Research Institute for Food Crops (MARIF), Indonesia
	498	BARI, Joydebpur, Bangladesh
	4720	ICGR, CAAS, Beijing
	1068(14)	NIAS, Tsukuba, Japan
	135	Instituto Colombiano Agropecuario (ICA), Palmira Valley, Colombia
	3,000	Punjab Agricultural University (PAU), Ludhiana, India
	2220	NBPGR, New Dehli, India
	2172	Central Research Institute for Food Crops (CRIFC), Sukamandi, Indonesia
	626	Plant Genetic Resources Institute, Islamabad, Pakistan
	5736	Inst. Plant Breeding, Uni. of the Philippines, Los Banos (IPB-UPLB), Philippines
	3889	SRPIS-USDA, Georgia, USA
	2100	University of Missouri (UM), Colombia, Missouri, USA

Table 3.1. cont.

Species	No. Acc.	Institution
V. reflexo-pilosa (wild)	1(1)	NIAS, Tsukuba, Japan
	(2)	National Botanic Garden, Meise, Belgium
	1	AVRDC, Shanhua, Taiwan, China
	1	PGRCU-USDA, Georgia, USA
V. riukiuensis	1	NIAS, Tsukuba, Japan
V. stipulacea	1	NIAS, Tsukuba, Japan
V. subramaniana	1	NIAS, Tsukuba, Japan
V. tenuicaulis	1	NIAS, Tsukuba, Japan
V. trilobata[3]	1	NIAS, Tsukuba, Japan
	49	Australian Plant Genetic Resources System, Australia
	8	National Botanic Garden, Meise, Belgium
	1	PGRCU-USDA, Georgia, USA
V. trinervia	1	NIAS, Tsukuba, Japan
V. umbellata (wild)	121	AVRDC, Shanhua, Taiwan, China
	1432	ICGR, CAAS, Beijing, China
	10	Australian Plant Genetic Resources System, Australia
	41	PGRCU-USDA, Geogia, USA
	11(2)	National Botanic Garden, Meise, Belgium
	12	Hokkaido Prefectural Agricultural Experiment Stations
	26(1)	NIAS, Tsukuba, Japan
Vigna spp.	100	NBPGR, New Dehli, India
	200	Chai Nat Field Crops Experiment Station, Thailand
	204	AVRDC, Shanhua, Taiwan, China

[1] Number of accessions for wild form shown in parenthesis

[2] Refers only to the active collection

[3] This species name is often used for germplasm that is *V. stipulacea*.

The ecological conditions into which the cultivated Asian *Vigna* best grow can be divided into 5 categories (based in part on Rachie and Roberts, 1974). These are:

1. Semi-arid - annual precipitation less than 600mm
    V. aconitifolia (moth bean)
2. Semi-arid- subhumid regions 600-900mm precipitation
    V. radiata (mungbean) - short duration
    V. mungo (black gram) - short duration
3. Subhumid - humid 900-1500 mm
    V. radiata (mungbean) - medium to long duration
    V. mungo (black gram) - medium to long duration
4. Humid and very humid regions - above 1500mm precipitation
    V. umbellata (rice bean)
    V. reflexo-pilosa var. *glabra*
5. Temperate/ cool tropics (highland)
    V. angularis (azuki bean)

The wild relatives of these crops and closely related species grow in similar ecological conditions. The centers of diversity of these cultigens reflect the different ecologies to which these crops are adapted. Thus, to comprehensively conserve subgenus *Ceratotropis* species, *in situ* conservation areas should be located within each of these ecological zones.

In situ conservation allows for continued evolution of genetic resources in their natural environment. *In situ* conservation is also important for species that are particularly difficult to conserve *ex situ*. Most of the species in the subgenus *Ceratotropis* readily produce seeds when conserved *ex situ*. Of the species in the subgenus *Ceratotropis, V. trinervia* is the most problematic for seed production *ex situ* at Tsukuba, Japan (36°N) and should therefore be a target for *in situ* conservation. Seed production for *V. trinervia* is problematic since it does not always respond to short day treatment and the factors that initiate flowering are not known. In addition, both *in situ* and *ex situ* the seeds of this species, rather like artificial interspecies hybrids, do not fill the seed chamber a difference from other species in the subgenus.

1.2.1. Asian Vigna *in protected areas*

Based on surveys of *Vigna* subgenus *Ceratotropis* herbarium specimens, passport data and field observations, nine species have been found growing in National Parks or local protected areas. Thus, these habitats are protected from destruction by development activities to some extent (Table 3.2). This list is undoubtedly incomplete and specific surveys of protected areas would provide valuable additional information on taxa in the subgenus *Ceratotropis* within protected areas. Among poorly known taxa in the subgenus *Ceratotropis, V. exilis* represents one species that grows in habitats in Thailand that can be easily identified. It grows on limestone outcrops that are frequently the location of Buddhist temples in Thailand. In these habitats, populations of this species are relatively safe from destruction. At a site seeing location maintained as a park on Ishigaki island, Okinawa prefecture, Japan, local officials annually cut the vegetation to maintain views of the sea. This annual cutting of the grass probably helps prevent succession that would eliminate the population of *V. riukiuensis* that grows in the grass there.

On the other hand, many populations of *Vigna* subgenus *Ceratotropis* grow in lowland areas that are vulnerable to habitat destruction. Among these are some populations of *V. riukiuensis* and *V. reflexo-pilosa* that grow on the islands of southern Okinawa, Japan. These islands have large areas that are being converted to grazing land and sugar cane fields thus *Vigna* populations are vulnerable to destruction. In addition, the wild and weedy relatives of subgenus *Ceratotropis* cultigens, such as azuki bean (*V. angularis*) in Japan and rice bean (*V. umbellata*) in Thailand, grow in disturbed habitats, such as roadsides, that by their very nature are vulnerable to development projects. Several other wild *Vigna* species appear to be well adapted to disturbed habitats such as *V. stipulacea* and *V. tenuicaulis*. A roadside population of *V. tenuicaulis* found in 1989 was revisited 7 years later and the location was buried under a widened road (author N.T., personal observations).

V. minima grows in dry deciduous forest that used to be widespread across northeast Thailand. However, since these forests have now been cut the remaining areas of this type of forest are in protected areas (Table 3.2).

Table 3.2. National Parks or other protected or semi-protected areas where wild Vigna *subgenus* Ceratotropis *species have been reported to grow.*

Species	Location[1]	Ref.[2]
V. exilis	Wat Tum Ma Dear, A. Sai Yok, Kanchanaburi, Thailand	3
V. exilis	Wat Thum Kao Chaank, Nong Pai, Kanchanburi, Thailand	3
V. exilis	Dept. of Forestry, Thum Kao Bin, Ratchaburi, Thailand	3
V. exilis	Wat Koh Thum Talu, Don Sai, Pak Thua, Ratchaburi, Thailand	3
V. exilis	Wat Khao Ban Dai, Petchaburi, Thailand	3
V. exilis	Wat Khao Yoi, Khao Yoi, Petchaburi, Thailand	3
V. exilis	Wat Banpotavas, Tayang, Petchaburi, Thailand	3
V. exilis	Wat Thum Pra Non, Khao Yai Cha Am, Petchaburi, Thailand	3
V. exilis	Wat Kaowon, T. Kaowon, Praputabat, Saraburi, Thailand	3
V. exilis	Put Tanimet Temple, Mittapap, Muak lek, Saraburi, Thailand	3
V. hirtella	National Park Khao Yai, Pakchong, Nakorn Ratchasima, Thailand	3
V. minima	Phu Phan National Park, Sakhon Nakhon, Thailand	H

Table 3.2. cont.

Species	Location[1]	Ref.[2]
V. minima	Nam Nao National Park, Loei, Thailand	H
V. minima	Phu Khieo Game Reserve, Khon Kaen, Thailand	H
V. minima	Temple, Khao Khat Nong Nak, Nong Kae, Saraburi, Thailand	H
V. minima	Mae Yom Nat. Park, Dow Boon, W.Yom River at Baritemine, Song Dist. Prae, Thailand	H
V. radiata	Khaw Khieo National Park, Si Racha, Chonburi, Thailand	H
V. reflexo-pilosa	Banna Koen (Park), Ishigaki, Okinawa, Japan	2
V. riukiuensis	Agarihennazaki, Miyako island, Okinawa, Japan	1
V. riukiuensis	Uganzaki, Ishigaki, Okinawa, Japan	1
V. riukiuensis	Hirakubosaki, Ishigaki, Okinawa, Japan	1
V. riukiuensis	Banna Koen (Park), Ishigaki, Okinawa, Japan	2
V. aridicola	Gongala wewa, Yala Nat. Park, Hambantota, Sri Lanka	4
V. aridicola	Parakramabahan, Polonnaruwa, Sri Lanka	4
V. stipulacea	Patanangala, Yala Nat. Park., Hambantota, Sri Lanka	4
V. stipulacea	Komawa Wewa, Yala Park, Hambantota, Sri Lanka	4
V. stipulacea	Rahuna Nat. Park, Komawa Wewa plot R17, Hambantota, Southern District, Sri Lanka	H
V. trilobata	Kari Villu, Plot W.9, Wilpattu Nat. Park, Sri Lanka	H
V. trilobata	Rahuna Yala Nat. Park, E. Butawa Modera, Sri Lanka	H, 4
V. trilobata	Yala Nat. Park, Uraniya Pond, Hambantota, Sri Lanka	H
V. trilobata	Pallugaturai, West of Wilapattu, West Sanctuary, Sri Lanka	H
V. trilobata	Patanangala, Yala Nat. Park, Hambantota, Sri Lanka	4
V. trilobata	Jamburagala, Yala Nat. Park, Hambantota, Sri Lanka	4

[1] We have included in this list a variety of areas that may be classified as protected or semi protected. Some such as the Yala National Park of Sri Lanka have a high level of official protection. Others such as Buddhist temple gardens have protection from development and in relation to the religious beliefs governing the site.

[2] H refers to data from herbarium specimens, Numbers correspond to 1 - Kaga *et al.*, 2000b; 2 - Tomooka *et al.*, 2000d; 3 - Tomooka *et al.*, 2000e; 4 - Tomooka *et al.*, 2000c.

1.2.2. Species "hot spots" for in situ *conservation*

Sympatric species

In many locations, several species in subgenus *Ceratotropis* are sympatric and geneflow between cross compatible diploid taxa may occur (Table 3.3). The complexity of section *Angulares* and relatively little genetic differentiation among several species within the section suggests this is a rather recently evolved branch of the subgenus *Ceratotropis* (see Chapter 2). Several species in section *Angulares* are cross compatible, at least in one direction (see Chapter 6). This suggests that interspecific geneflow may occur in the wild. This may explain the difficulty some workers have had in assigning species names to newly collected accessions (Tomooka *et al.*, 1996). This situation is reported in *Vicia* where interspecific gene flow has been demonstrated between *Vicia sativa* and *V. segetalis* (Potokina *et al.*, 2000). Areas where more than one *Vigna* species grow in close proximity are logical areas to select for *in situ* conservation. The more species of potential value there are in an area can strengthen the case for designation of *in situ* conservation reserves.

Species distribution

The diversity of species in subgenus *Ceratotropis* in different parts of Asia is not evenly distributed. On the Korea peninsula there are only two wild taxa *V. angularis* var.

Table 3.3. *Wild* Vigna *subgenus* Ceratotropis *species that have been found sympatric based on direct observations of the authors.*

Species	Location	Ref.[1]
V. reflexo-pilosa and *V. riukiuensis*	Komi, Iriomote, Okinawa, Japan	1
V. riukiuensis and *V. reflexo-pilosa*	Banna Koen (park), Ishigaki, Okinawa, Japan	4
V. angularis and *V. nakashimae*[2]	Yangpeong, Kyunggido, Republic of Korea	
V. angularis and *V. nakashimae*[2]	Namyangju, Kyunggido, Republic of Korea	
V. angularis and *V. nakashimae*[2]	Hwacheon, Kangwondo, Republic of Korea	
V. radiata, V. mungo, and *V. umbellata* (all escaped)	Route 117, 58km N. of Nakhon Sawan, Phichit, Thailand	3
V. mungo (escaped) and *V. umbellata* (wild)	Route 12, 22km W. of Phitsanulok, Thailand	3
V. exilis and *V. minima*	Wat Kao Won, T. Kaowon, Phra Phutthabat Saraburi, Thailand	5

Table 3.3. cont.

Species	Location	Ref.[1]
V. exilis and *V. umbellata*	Kao Ngu, T. Pub Pra, A. Ratchaburi, Thailand	5
V. hirtella and *V. minima*[2]	Bok Lua, Nan, Thailand	5
V. minima and *V. umbellata*[2]	Bok Lua, Nan, Thailand	5
V. exilis and *V. umbellata*	Wat Ratsingh Kon, Kao Nor, T. Pub Pra, A. Ratchaburi, Ratchaburi Prov. Thailand	5
V. trinervia and *V. reflexo-pilosa*	Bentong, Pahang, Malaysia	2
V. trilobata and *V. aridicola*	Buttawa modera, Yala National Park, Sri Lanka	6
V. aridicola, *V. stipulacea* and *V. trilobata*	1 km from Patanangala Bungalow, Yala National Park, Sri Lanka	6, 7
V. aridcola and *V. radiata* var. *sublobata*	Wellpallewela, Mahiyangana, Sri Lanka	7
V. dalzelliana and *V. trinervia*	Ramboda Mission and waterfall, Ramboda, Nuwara Eliya Dist. Sri Lanka	7
V. aridicola and *V. trilobata*	Debaragaswala, Yala sanctuary, Hambantota, Sri Lanka	7
V. stipulacea and *V. trilobata*	Komawewa, Yala sanctuary, Hambantota, Sri Lanka	7
V. aridicola and *V. trilobata*	Kotabandi Wewa, Yala sanctuary, Hambantota, Sri Lanka	7
V. aridicola and *V. trilobata*	Near Mahaseelawa Wewa, Yala sanctuary, Hambantota, Sri Lanka	7
V. aridicola and *V. trilobata*	Between Jamuragala and Gonagala, Yala sanctuary, Hambantota, Sri Lanka	7
V. aridicola and *V. trilobata*	Situlpawwa road, Yala sanctuary, Hambantota, Sri Lanka	7

[1] Reference numbers correspond to 1 - Kaga *et al.*, 2000b; 2 - Tomooka *et al.*, 1993; 3 - Tomooka *et al.*, 1997; 4 - Tomooka *et al.*, 2000d; 5 - Tomooka *et al.*, 2000e; 6 - Tomooka *et al.*, 2000c; 7 - Tomooka *et al.*, 2001b. Information from Korea was provided by Dr. Mun Sup Yoon based on his observations in autumn 2000.
[2] Species known to be cross compatible. For most species in this table that are sympatric compatibility relationships are not known.

nipponensis and *V. nakashimae*. Whereas Sri Lanka, an area of comparable size to the Korean peninsula, has six wild taxa, *V. aridicola*, *V. dalzelliana*, *V. radiata* var. *sublobata*, *V. stipulacea*, *V. trilobata* and *V. trinervia*. Focusing *in situ* conservation reserves in areas of high species diversity will likely capture greater genetic diversity. This topic is discussed in further detail in chapter 5 section 3.

1.2.3. Population types for in situ conservation

Island populations - *Vigna riukiuensis* and *V. reflexo-pilosa* in southern Okinawa
The biology of populations that occur on islands has generated a lot of research interest, in part, because islands can enable processes of evolution to be readily measured (Vitousek *et al.*, 1995; Baldwin *et al.*, 1998). Studies of island populations of *Vigna riukiuensis* and *V. reflexo-pilosa* have measured the genetic differentiation between populations of these species on different islands of southern Okinawa, Japan (Fig. 3.1).

In Japan *V. riukiuensis* grows in two neighboring groups of islands that are considered geologically distinct, the Miyako and Yaeyama archepelago's (Maekawa and Shidei, 1974). *Vigna riukiuensis* is a perennial species often growing in grass, on cliff tops overlooking the sea. *Vigna riukiuensis* has a restricted distribution in southern Okinawa and Taiwan. DNA analyses of populations of *V. riukiuensis* from a wide area of southern Okinawa including islands of the Yaeyama and Miyako island archipelagos (Fig.3.1) revealed island to island variation (Fig.3.2). Populations from the Miyako island archipelago (Irabu, Miyako and Tarama islands) were distinct from those of the Yaeyama archipelago (Hateruma, Iriomote, Ishigaki and Yonaguni islands). The results suggest that there is a floristic barrier between the Miyako and Yaeyama archipelagos and it is possible that *V. riukiuensis* is an ancient element of the flora of these islands.

Vigna reflexo-pilosa is also found throughout the islands of southern Okinawa occasionally it is sympatric with *V. riukiuensis*. Okinawa, Japan, represents the northern distribution limit of this species. DNA analyses of 5 populations of *V. reflexo-pilosa* from 5 islands of southern Okinawa, Ishigaki, Irabu, Iriomote, Tarama and Yonaguni (Fig. 3.1) revealed very little variation. The intra-specific genetic diversity measured by Shannon's diversity index (Ha) was 0.006 compared to 0.039 for *V. riukiuensis* (Yoon *et al.*, 2000). This suggests that in southern Okinawa *V. reflexo-pilosa* may have been a relatively recent introduction in the area and populations are derived from a limited number of founder plants since there is little intra-specific variation.

These results suggest different strategies are needed for *in situ* conservation of the island populations of *V. reflexo-pilosa* and *V. riukiuensis*. For *V. reflexo-pilosa* a few populations may capture the majority of the genetic variation of this species across southern Okinawa. On the other hand, since *V. riukiuensis* shows genetic differentiation specific populations in both the Mikako and Yaeyama archipelagos would need to be conserved to ensure the genetic resources of this species are adequately protected.

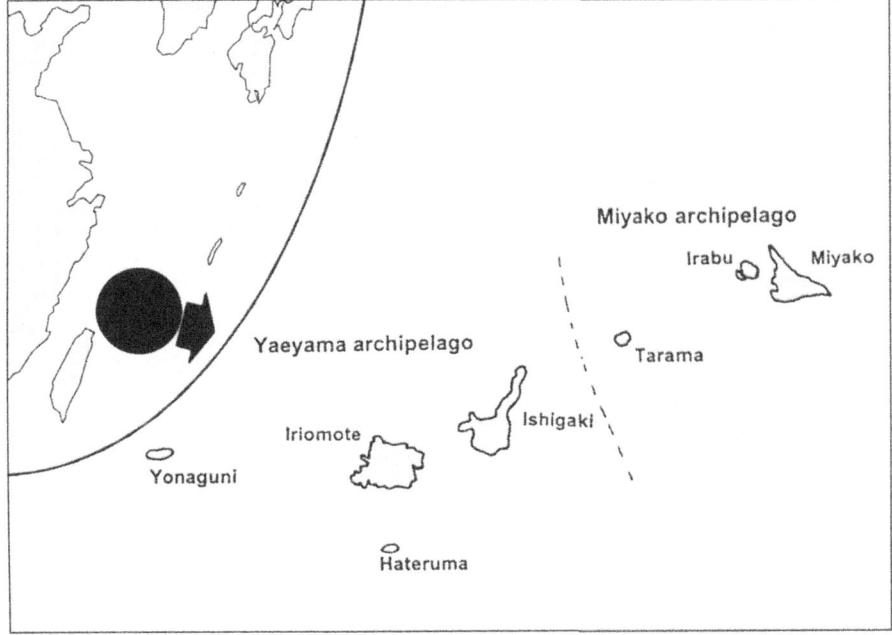

Figure 3.1. Islands of southern Okinawa, Japan.

Geographic disjunction
V. nakashimae has a disjunct distribution in northeast Asia. Most populations are found on the continental mainland. However, a few populations occur in western Japan. DNA analysis of Korean and Japanese populations has shown that Korean populations are genetically distinct from populations in Japan (Fig. 3.2). Thus, to conserve the genetic variation of this species *in situ* conservation should focus on selected populations from both countries. The genetic distance between populations of *V. nakashimae* from Korea and Japan was almost the same as the genetic distance between populations of *V. riukiuensis* in eastern and western islands of southern Okinawa (0.0036 and 0.0039, respectively). This may reflect the fact that these two species have evolved over a similar time in the area studied (Yoon *et al.*, 2000).

The azuki bean (*Vigna angularis*) crop complex
Populations of the relatives of azuki bean may consist of wild or weedy plants or be a mixture of both plant types. Mixed wild and weedy populations have been called complex populations (Xu *et al.*, 2000a, b). Complex populations tend to cover a large area and seem to represent particularly dynamic populations where gene flow among different types of azuki occurs. The azuki bean complex is discussed in detail in chapter 6. Comments here are restricted to information relevant to *in situ* conservation.

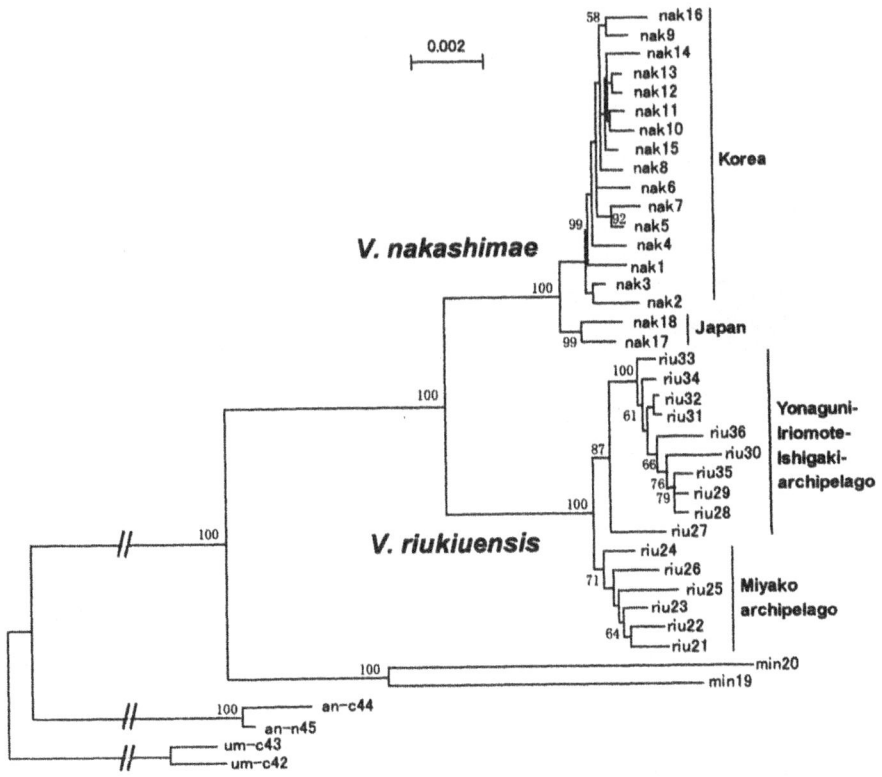

Figure 3.2. Neighbor-joining tree of Vigna nakashimae *and* V. riukiuensis *based on AFLP variation using a matrix of inferred nucleotide diversity.* V. angularis, V. umbellata *and* V. minima *are outgroup species. Number at a branch indicates the probability supporting that branch. Branch length unless broken corresponds to nucleotide diversity (modified from Yoon* et al., *2000).*

Genetic analysis of the azuki bean complex population types suggests that particular attention should be placed on *in situ* conservation of complex populations. Complex populations have a higher level of genetic variation than populations of wild or weedy azuki bean. To understand the changes occurring in populations scientific monitoring both at different times during the year and from year to year is necessary. Monitoring the immediate environment of the population and its general vicinity are important components to observe.

Considering strategies for *in situ* conservation of crop complex gene pools based on studying the azuki bean complex two factors appear to be critical for maintaining the population structure:

a. *Disturbance*, mainly by farmers, and their knowledge of the importance of this disturbance in preventing succession. In several areas of Japan it has been reported that in earlier times when food was in short supply presumed weedy types of azuki bean were gathered and eaten (Yamaguchi, 1992). Indigenous knowledge of the value of wild and weedy forms of azuki may be one means of protecting populations of these plants.

b. *Pollinators*. Carpenter bees, particularly the species *Xylocapa appendiculata* Smith, frequently visit flowers of *V. angularis* for nectar in Japan. These and other bees may be instrumental in pollen transfer between different members of the azuki bean complex growing in close proximity. There are still many unanswered questions on the topic of gene flow within and between sympatric wild, weedy and cultivated azuki including how do the numbers and species of pollinators fluctuate from year to year? The floral features and activity of insects on the flower suggests cross-pollination by insects occurs even if only occasionally.

2. CONSERVATION PRACTICES

2.1. Collection

A thorough compilation of many aspects of collecting plant genetic resources can be found in Guarino *et al.* (1995). Here we discuss collection specifically in relation to the Asian *Vigna*. Collecting consists of three phases, planning (this may be very long if, for example, special grants must be applied for and visits to herbaria to collect distribution and phenology data etc. are required), the collecting phase (time is always a premium during daylight hours) and post-collecting phase (data analysis, report writing, seed and herbarium specimen processing). Essential collecting equipment is shown (Table 3.4).

Passport data

The germplasm collector has a responsibility to collect as much pertinent information as possible on collected samples while in the field. Passport data for wild species includes many items not generally necessary when collecting cultigens (Table 3.5). Collecting one sample of a wild species usually takes much longer than for a cultigen since usually only one sample can be collected from each location whereas one farmer may be able to provide several different varieties of a cultigen. The most critical passport data is related to exact location of the sample – map reference, latitude and longitude (from GPS), altitude and hand drawn sketch map (particularly for wild species). A generalized set of data that may be used for collecting wild and cultivated *Ceratotropis* species is shown in Table 3.5.

Table 3.4. Essential collecting equipment

Herbarium press / newspaper
Trowel - to obtain roots for herbarium specimens and root nodules.
Net bags (made from mosquito net is ideal for collecting seeds of cultivated *Vigna*, from muslin cloth for the small seeded wild *Vigna* species)
Coin envelopes
Pre prepared passport data sheets
Secateurs
Global positioning system (GPS) (extra batteries)
Altimeter
Tubes containing indicator (blue) silica gel - for collecting root nodules
Camera and film and camera batteries
Hand lens
Notebooks, pencils
Local maps

Materials to collect

a. Seeds

b. Herbarium specimens. Germplasm collectors focusing attention on crop germplasm tend not to collect herbarium specimens in part because collecting time does not coincide with flowering time when the best herbarium specimens can be made. However, the genebank of the N.I. Vavilov Research Institute for Plant Industry has an extensive collection of herbarium specimens of land race varieties and these can helpfully supplement genebank seed files.

When collecting germplasm of wild species herbarium specimens are an essential part of the overall collection and usually flowering and fruiting are asynchronous. Procedures for this have been well documented in the literature (e.g. Miller and Nyberg, 1995). Since flower structure of *Vigna* subgenus *Ceratotropis* species is complex, flowers and inflorescence can be preserved by immersing them in small tubes of alcohol to retain 3 dimensional form and enable flower structure to be examined in detail in the laboratory.

c. Samples of *Bradyrhizobium* bearing root nodules. Many scientists conducting research on *Vigna* species are working in research institutes with microbiologists or know microbiologists who are interested in the symbiotic microorganisms associated with legumes. It is therefore worthwhile when field collecting seeds and herbarium specimens of *Vigna* to also collect root nodules from which associated *Bradyrhizobium* can later be isolated (Yokoyama *et al.*, 1999).

The procedure for collecting root nodules is to prepare prior to collecting small vials with airtight caps such as 15ml tubes half filled with indicator silica gel on top of which

Table 3.5. Passport data that can be taken for Vigna *germplasm while in the field and after returning from the field. Items in italics are important for wild species.*

Date:
Collectors:
Site number: Collection number:
Plant codes (if individual plant samples taken):
Scientific name:
Local name: Meaning of local name:
Location:
       GIS coordinates: Map reference:
       Land holders name:
       Village: Nearest town (direction):
       District/Prefecture: Country:
Sketch map
Site characteristics:
       Topography – mountains, hills, plains, other specify
       Altitude: (m)
       Slope:
       Land use
       Soil type
Habitat: Associated vegetation: Forest, bushes, cultivated, grassland,
       others specify
       Associated species: Dominant, others
       Shading: heavy, medium, light, none
Population:
       Size:
       State: vegetative, flowering, mature, past maturity
       Status: wild, weedy, cultivated, mixed
       Population variation: yes (describe) , no
       Disease assessment: leaf.......pods.....
       Pest assessment
       Population variation:
Field check list:
       Photo: site   ; *habitat*   ; *plant habit*
       Herbarium specimen: yes   number of sheets   ; no
       Root nodules: yes/ no;
       Collection method: Bulk, individual (number of individuals);
Data to obtain during the post collection phase:
Soil type from map;
Geology from map:
Climate rainfall, maximum and minimum temperature of collection
location.

a small cotton wool plug is placed. In the field with a strong trowel, dig up roots with fresh young nodules and place the nodules into the vial. Further details of collecting root nodules can be found in Date (1995). Later in the laboratory, the *Bradyrhizobium* can be isolated. Isolation procedures can be found in Date and Halliday (1987).

Handling samples in the field
a. Separate seeds from the pod and maintain in dry conditions. Placing seeds after drying in an airtight box with silica gel can be beneficial.
b. Herbarium specimens – rapid thorough drying of herbarium specimens is essential if the colors of the specimens are to be retained. If drying facilities are not available at the end of each day, frequent changing of the newspaper between specimens in the plant press can accelerate the drying process.

Handling samples after the collecting phase
a. Thoroughly dry the seeds and herbarium specimens
b. Prepare trip report/ herbarium labels and distribute the specimens and seeds to those responsible for long-term conservation, characterization and use.

2.2. Characterization

We provide a list of characterization descriptors that can be used for the wild and cultivated Asian *Vigna* (Table 3.6).

2.2.1. Germination and cultural practices

Pot planting. Most seeds of wild *Vigna* have dormancy so dormancy breaking is necessary for seeds that are not old. This is usually easily done by cutting away a small part of the seed coat on the opposite side of the seed to the hilum.

Seeds of exotic species may then be placed on filter paper in petri dishes for germination in an incubator at 30°C before transplanting into pots. Alternatively, the seeds may be planted directly into 20cm diameter pots. Finely ground but non-sterilize soil is used since sterilization tends to lead to imbalances in the natural soil ecosystem that can adversely affect legume seedlings growth. The outlet for water at the bottom of the pot can be covered with a paper towel to prevent soil loss. Water pots very well initially.

Field planting. Seeds of native species adapted to the growing location may be planted directly in the field. A single row 3m long and seeds spaced 20cm apart will generally provide sufficient space for plants to produce abundant seeds of locally adapted accessions. Spacing between rows of 70cm and alternating cross incompatible species to prevent cross-pollination is recommended. For example, a row of wild azuki beans (*Vigna angularis* var. *nipponensis*) planted next to a row of cowpea (*V. unguiculata*). For large twining species such as *V. umbellata* plant in rows 1 meter apart and plant spacing of 60cm is recommended.

Table 3.6. Characterization data for the Asian Vigna *(modified from IBPGR, 1980)*

Agronomic evaluation identifier
Site
Planting date and year
Population density
Growth stage and growth characteristics
Days to emergence (From planting to 50% seedling emergence)
Days to 50% flowering (Days from planting to 50% of plants with first flower
     open)
Seedling vigor (15 days after emergence – 1 poor; 5 medium; 9-vigourous)
Hypocotyl color [1 green;2 greenish purple; 3 purple; 4 mixed; 5 other
     (specify)]
Growth habit [When first pod changes color. 1 erect; 2 semi erect; 3 spreading]
Growth pattern (1 indeterminate; 2 determinate)
Plant height (in cm average)
Twining tendency (1 none; 5 moderate; 9 pronounced)
Number of pod bearing branches
Length of longest branch
Leafiness (at 50% flowering) (1-sparse; 5 medium; 9 abundant)
Flower period (1 asynchronous greater than 30 days; 2 intermediate 16-30 days;
     3 synchronous less than 16 days)
Flower drop (%)
Number of pod clusters (having at least one fully grown pod)
Days to 50% ripe pods
Lodging (1 none; 5 intermediate; 9 heavy)
Nodulation (1 none; 2 poor; 3 heavy; 4 excessive)
Leaves
Terminal leaflet shape (1 deltoid; 2 ovate; 3 acute; 4 ovate-lanceolate; 5
     cuneate; 6 lobed; 7 other (specify)
Terminal leaflet length for leaf at the 4th node (1 small less than 10 cm;
     medium 10-13cm; 3 large greater than 13 cm)
Leaf pubescence (1 glabrous; 2 sparse short pubescence; 3 sparse long
     pubescence; 4 heavy short pubescence; 5 heavy long pubescence)
Leaf color [at 50% flowering 1 light green; 2 green; 3 dark green; 4 others
     (specify)]
Petiole color (1 green; 2 greenish purple; 3 purple; 4 dark purple)
Color of petiole/leaf blade joint (1 green; 2 purple; 3 dark purple)
Color of basal petioles (1 green; 2 purple; 3 dark purple)
Petiole length (recorded for the leaf at the 4th node)(1 short <12 cm; 2 12-18cm;
     3 long >18cm)
Leaf senescence (recorded when 50% of pods mature) (1 not visibly senescent; 5
     intermediate; 9 conspicuously concurrent)

Table 3.6. cont.

Inflorescence

Nodes to first pod bearing node (node number starting from unifoliate node).

Length of the peduncle (length of the longest peduncle when the first pod changes color) (1 short < 14cm; 2 medium 14-18cm; 3 long >18 cm).

Raceme position (recorded when the first pod changes color) (1 mostly above canopy; 2 intermediate; 3 no pods visible above canopy)

Calyx color [1 green; 2 green purple; 3 other (specify)]

Corolla color (wing and standard) recorded at full bloom between 8 and 10am [1 light yellow; 2 deep yellow; 3 greenish yellow; 4 yellow with reddish tinge; 5 other (specify)]

Pods

Pod color at the immature stage [1 light green; 2 deep green; 3 other(specify)]

Color of the ventral suture of immature pod [1 light green; 2 deep green; 3 purple; 4 other (specify)

Pod color at maturity [1 straw; 2 tan; 3 brown; 4 black; 5 other (specify)]

Shape of ripe pod (1 semi flat; 2 round)

Attachment of mature pod to peduncle (1 pendant; 2 intermediate; 3 ascending)

Pod length mm

Pod pubescent (1 glabrous; 5 intermediate; 9 heavily pubescent)

Constriction of pods between seeds (0 absent; 1 present)

Pod curvature (1 least curved; 2 medium; 3 most curved)

Shattering in the field (1 none; 3 partial; 5 heavy)

Pods per plant (10 plants)

Seeds per pod (10 plants)

1000 seed weight

Yield per plant (10 plants)

Percent of 1st harvest over the total harvest

Shelling % (seed weight divided by pod weight as % from 100 pods)

Seeds

Seed colour [1-yellow, 2-greenish yellow, 3-light green, 4-dark green, 5-brown, 6-grey, 7-black, 8-mixed, 9-others (specify)]

Mottling on the seed

Lustre on the seed surface (1-dull; 2-shiny; 3 - powdery)

Seed shape [1 round; 2-oval; 3-drum shaped; 4-others (specify)]

Hilum (1-concave; 2-slightly protruding; 3-heavily protruding)

Ratio of seed to seed coat weight

Immediately after planting, the field is covered with a coarse grain net to prevent bird and animal damage.

Fertiliser

Top dressing prior to seeding (in Japan).

a. 1000kg/ha of limestone;

b. 1000kg/ha of fused magnesium phosphate;

c. 650kg/ha of 3 parts nitrogen, 10 parts phosphate and 10 parts potassium.

Herbicides

Prior to planting herbicide is applied.

Insecticides

Chemicals are applied (in Japan) to control insects (primarily leaf cutters) prior to planting. After this, insecticides are applied every other week to control insects such as thrips. At flowering, additional protection from insects is required when pests such as stink bug that can affect seeds are prevalent.

Photoperiod sensitivity. Some wild *Vigna* show seasonality in flowering and short day treatment can initiate flowering (see chapter 5.3). Where these conditions do not occur naturally place plants in a photoperiod chamber or cover daily until flower initiation providing an 8 hours light/16 hour dark treatment about 6 weeks after planting.

Since wild species have a twining habit, stakes are necessary about a month after germination. For field grown wild species two stakes at either end of the row with a net strung between the stakes can be effective support for twining species.

2.2.2. DNA extraction from Vigna *for molecular characterization*

Molecular characterization is an increasingly routine operation for germplasm collections. One key to successful molecular characterization is obtaining sufficient quantities of very clean DNA. Since, the method for extracting DNA from *Vigna* species differs from some other plant groups we describe in detail a method we have employed successfully.

Leaf sampling - It is essential that leaves selected for DNA extraction are small unfolded young leaves about 1 cm long.

A. Destructive sampling

Collect leaf blades of the first true leaf before it is open when it is about 1cm long. Place in aluminum foil on ice, then (within 1 hour) immerse in liquid nitrogen. Store at minus 80°C until extraction of DNA.

B. Non destructive

For plants about one month to 6 weeks old gather very young leaves prior to opening (if the species is a cultigen leaves will be about 1cm long, if they are wild *Vigna* leaves will be a little shorter). Do not take the hard parts of small leaves such as petiole. Place small leaves in aluminum foil with label and put on ice in an ice chest. Immerse in liquid nitrogen as soon as possible after harvesting (within one hour). Store at minus 80°C. Sampling 2 or 3 times may be necessary to accumulate sufficient leaves.

DNA extraction

Step 1. Leaf grinding

1. 0.3g-0.5g of small unopened leaves per accession;

2. Prepare 50ml screw top tube in ice and liquid nitrogen mixture;

3. In beaker put liquid nitrogen and immerse spatula;

4. Take mortar and pestle from minus 80°C freezer;

5. Pour liquid nitrogen into mortar and pestle;

6. Leaves must be ground to a fine powder without melting to room temperature. Fine grinding of the young leaves is essential;

7. After grinding, immediately and carefully scrap all the fine powder into the 50ml tube and immediately place in the minus 80°C freezer.

8. It is advisable that after grinding several samples to quickly rinse the dirty mortars and pestles to facilitate cleaning later.

The aim should be to have between 0.3 and 0.5g of very young leaf blades. Stalks and leaves more than about 2cm and unfolded should be avoided. Don't increase weight with hard tissues such as petiole. When grinding ensure all powder is collected. If leaf weight is 0.3-0.5g use 50ml tube and use quantities of solutions as indicated here. If leaf weight is less than 0.3g reduce quantity of solutions accordingly. However, for AFLP analysis 0.3g of powdered leaf is the minimum weight that will give sufficient DNA.

Preparation of solutions used in the description below can be found in Birren et al. (1997).

Step 2.

Put powdered leaf (that has been continuously maintained at minus 80°C) into a pre-warmed 50ml tube (1st tube 50 ml) containing 12µl of α-mercaptoethanol, and 6ml of 1.5X CTAB. Mix completely with vortex before putting into the 65°C water shaker (do not vortex after shaking in water bath incubator since the DNA may break). Every 5 minutes mix gently by inverting the tubes.

Step 3.

After 20 minutes transfer to 15 ml tube (2nd tube 15ml) and cool down for a few minutes. Add 6ml chloroform (24):isopropylalcohol(1). Gently rotate on rotator for 20 minutes. Centrifuge at 3500rpm (or 1300G) for 20 minutes. Transfer supernatant to a new tube (3rd tube 15ml) and add 0.6ml of 10% CTAB (1/10v) at 65°C. Mix completely. Incubate the tube for 5 minutes at 65°C. Add 6ml of chloroform: isopropylalcohol mix for 20 minutes on a rotator. Centrifuge at 3500rpm for 20 minutes. Put supernatant in new tube (4th tube 15ml) using sterile cut pipette tip add 9 ml CTAB precipitation buffer and mix. You should be able to see DNA if you started with 0.3g of leaf. Centrifuge at 500-1000rpm (or 25-100G) for 5 minutes. Discard supernatant.

Step 4.

Add 5ml (1M NaCl, 10mM Tris, 1mM EDTA) dissolve pellet - don't break DNA at this stage by mixing or shaking vigorously. Pellet may dissolve more readily if kept at 65°C for 30 minutes. Add 3µl of fresh RNAase (10mg/ml). Incubate at 37°C for 30 minutes. Add 10ml 100% ethanol and mix gently. Spool the DNA with a cut pipette tip into 70%

ethanol (5th tube 1.5ml). If DNA cannot be spooled, wash twice more in 70% ethanol. Spin down the DNA decant 70% ethanol, add 300-500μl of normal TE (10, 1) and store in refrigerator or freezer at minus 20ºC.

2.3. Evaluation

The range of traits that the Asian *Vigna* genetic resources can be evaluated for is shown in Table 3.7.

2.3.1. Evaluation for resistance to microorganisms (Reddy et al., 1987; Schulze-Lefert and Vogel, 2000; Young et al., 1993)

Screening for powdery mildew (*Erysiphae polygoni* DC.) resistance in mungbean [*Vigna radiata* (L.) Wilczek].
Powdery mildew is one of the most common plant diseases infecting over 650 monocots species and 9000 dicot species (Schulze-Lefert and Vogel, 2000). While powdery mildew is an obligate biotropic fungus it generally does not kill its host but severely reduces yield. Resistance to this disease is one of the important targets for breeders of *Vigna radiata* in the tropics so we describe field and greenhouse/laboratory approaches for evaluating this disease as a model for evaluation of other *Vigna* pathogens.

Field (Young et al., 1993)

Time of planting will depend on the location of testing in northern (southern) hot summer temperate areas planting would be May or early June (November or early December) when temperatures reach 21ºC. In tropical areas, screening can be conducted more than once in a year, heaviest infection occurs during the wet season. A randomised block design is best even when general screening is being undertaken. Replication is necessary since uniformity of inoculation may be difficult to ensure in the field. Susceptible varieties are planted around materials to be tested. Plots are single rows with plant spacing 25cm and 75cm spacing between rows. Inoculation can be natural in areas where powdery mildew is an annual occurrence. Natural infection can be enhanced by supply of overhead irrigation. Plots can also be laid out in a covered vinyl house where higher humidity can provide an environment more conducive to powdery mildew development than open fields. The best temperature for powdery mildew development is between 20 and 25ºC.

Chances of heavy field infection can be improved by shaking susceptible infected plants over young plants in susceptible border rows. Susceptible infected plants are prepared prior to the experiment using local strains of powdery mildew.

Plants are scored for response to powdery mildew infection at 45, 55, 65 and 75 days after germination using a scoring system of:
1= no visible mycelial growth;
2=0-25% foliage area covered by fungus;
3=25-50% foliage area covered by fungus;

Table 3.7. Evaluation of Vigna *genetic resources (IBPGR, 1980; Nagamine and Takeya, 1999; AVRDC, 1988)*

Reaction to fungal diseases

Described under natural or artificial conditions, on scale of 1-9 where 1 is resistant, 3 is moderately resistant, 5 is tolerant, 7 is moderately susceptible and 9 is susceptible.

Damping off [*Corticum* (=*Sclerotium*) *rolfsii*, *Rhizoctonia solani* and *Pythium aphanidermatum*]

Powdery mildew (*Erysiphe polygoni*)

Root rot (*Pythium* spp., *Rhizoctonia solani* and *Fusarium* spp.)

Collar rot

Phytophthora stem rot (*Phytophthora vignae*)

Brown stem rot (*Phialophora gretata*)

Leaf rot

Pod rot (*Fusarium* spp., *Choanephora cucurbitarum*)

Other fungal leaf spots

Wilt

Anthracnose (*Colletotrichum lindemuthianum*)

Scab

Seedling blight

Leaf blight

Web blight (*Rhizoctonia solani*)

Rust

Cercospora leaf spot (*Cercospora canescens, C. cruenta*)

Others (specify)

Reaction to bacterial diseases

Bacterial leaf blight (*Xanthomonas phaseoli*)

Bacterial leaf spot (*Xanthomonas phaseoli*)

Halo blight (*Pseudomonas phaseolicola*)

Bacterial wilt (*Fusarium oxysporum* f. spp.)

Reaction to viruses

Mungbean Yellow Mosaic Virus

Mungbean Yellow Mottle Virus

Mungbean Mosaic Virus

Mungbean Mottle Virus

Leaf Curl Virus

Cucumber Mosaic Virus

Tobacco Ring Spot Virus

Blackgram Sterility Mosaic Virus

Mungbean Witches Broom

Table 3.7. cont.

Other (Specify)

Reaction to Nematodes
Root-knot nematode (*Meloidogyne javanica*)
Soyabean cyst nematode (*Heterodera glycines*)
Reniform nematode (*Rotylenchulus reniformis*)
Others (specify)

Reaction to insect pests (Sehgal and Ujagir, 1988)
Jassids (*Empoasca* spp.)
White fly (*Bemisia tabaci*)
Pod borers (*Lampides boeticus, Heliothis armigera*)
Bruchids (*Callosobruchus* spp.)
Thrips (*Megalurothrips usitatas*)
Cutworm (*Sporoptera litura*)
Leaf caterpillar (*Agrius convolvuli*)
Aphids (*Aphis craccivora*)
Leaf miner (*Acrocercops phaseospora, Phytomyza atricornis*)
Beetles (*Callosobruchus* spp., *Anomala* spp.)
Bean fly (*Ophiomyia phaseoli*)
Others (specify)

Environmental adaptability
(Evaluated under defined conditions)
Photoperiod sensitivity (1- none, 2-slight, 3-sensitive, 4-variable)
Drought tolerance [tolerance to soil moisture deficit coded 1-9 measured as a
        reduction in yield. 1 least tolerant, 5 intermediate, 9 most tolerant]
Tolerance to salinity (measured as reduction in plant height 30 days after
        sowing. 1 least tolerant, 5 intermediate, 9 most tolerant)
Tolerance to acid soils (measured as reduction in plant height 30 days after
        sowing. 1 least tolerant, 5 intermediate, 9 most tolerant)
Cold tolerance (Measured as reduction in general vigor and productivity after
        being continuously exposed to an average temperature of 15°C for at
        least 15 days)
Heat tolerance (Measured as yield reduction when continuously exposed to
        average of 30-35°C during the flowering period. 1 least tolerant, 5
        intermediate, 9 most tolerant). Heat tolerance can also be tested at the
        seedling stage in an incubator.

4=50-75% foliage area covered by fungus;

5=75-100% foliage area covered by fungus.

Average plant or plot score indicates 1 resistant, 2 moderately resistant, 3 intermediate, 4 moderately susceptible and 5 susceptible.

Greenhouse/laboratory (Reddy et al., 1987)

1. Fully expanded third trifoliate leaves are excised from 10 individuals, 21-25 day old plants per accession just above the pulvinus.

2. Leaves are washed in tap water and petioles trimmed to a uniform length.

3. Leaves are then inserted into holes in an opaque plastic sheet and held in place with cotton plug.

4. Plastic sheets with inserted leaves are placed over a tray with water such that the leaf petioles are well immersed in the water.

5. Trays are kept in a growth chamber maintained at 21°C ($\pm$ 1°C) and 12h photoperiod with 4136 lux/m^2 illumination.

6. After 24 hours the excised leaves are sprayed with conidial suspension (3.5×10^6 conidia/ml) in water containing 0.001% Tween 80. Spray volume should be adjusted to 10ml/900cm^2 using a hemocytometer.

7. Disease symptoms are scored at 10 and 20 days after inoculation using a scale from 0 no lesions to 9 completely covered with lesions.

8. Germplasm showing no symptoms can be re-inoculated at 11 and 15 days after first inoculation.

9. The screening is repeated.

2.3.2. Evaluation for resistance to insects (Tomooka et al., 2000b)

Vigna species are affected by many insect pests. One of the most damaging and widespread are the bruchid weevils (*Callosobruchus* spp.) that do much damage to stored seeds. Here, a method for evaluating *Vigna* germplasm for resistance to bruchid beetles is described.

Populations of the insect are maintained in an incubator (30°C, 70% relative humidity) on commercial azuki or mungbean seeds. For the evaluation test, ten seeds are placed in a glass petri dish (9cm diameter) with three replications. The test seeds are infested with about 5 pairs of freshly emerged bruchid adults for one day. The adults are allowed to mate and lay eggs on the seeds. At the end of the 24 hours the 5 pairs of adults are removed from the petri dish. The petri dishes are kept in incubators that are maintained at 30°C and 70% relative humidity. When adults start emerging, this may be about 3 weeks later, the number of emerged male and female adults are counted daily until 50 days after infestation. Counting emerged adults is discontinued thereafter to avoid counting second generation adults. The counted adults are removed from the petri dish daily. To examine the extent of damage, seeds can be observed by X-ray analysis 50 days after eggs are laid on the seeds (e.g. model TV-PBO-C, SOFTEX, Japan). X-ray analysis also enables larvae that fail to emerge to be identified and prevents such seeds as being counted as 'resistant' (Fig. 3.3a).

Preparation of artificial beans (Fig. 3.3b)

To test the potency of resistance factor(s) and to exclude factors such as seed shape, hardness, size and color, from influencing results, artificial beans can be made. Ground powder of the test accessions is mixed in various proportions with ground powder of a susceptible azuki bean cultivar, e.g. cv. Dainagon. Artificial beans, weighing about 0.25g each, are made into a legume seed shape by hand the shape is maintained by adding distilled water to the bean flour. The artificial seeds are freeze-dried. Four seeds are made for each accession. Three of the seeds are used for the bioassay and one for X-ray analysis.

Artificial beans test for reaction of bruchid beetles

4-5 pairs of adult bruchid beetles are added to each of 4, 9 cm petri dishes containing one artificial bean each. After 24 hours, by which time eggs have been laid on the artificial beans, the beetles are removed from the petri dishes. The egg number on the surface of the artificial beans is counted. Petri dishes are kept in an incubator maintained at 30°C and 70% RH. The fourth artificial bean is observed by X-ray analysis on days 10, 20, and 30 after eggs were laid and bruchid development, if any, is recorded. After 40 days three artificial beans are crushed with forceps and the number of larvae, pupae and adults are recorded. These numbers are considered as indices of resistance. However, the fourth artificial seed is not destroyed and X-ray observation continued until 60 days after eggs were laid.

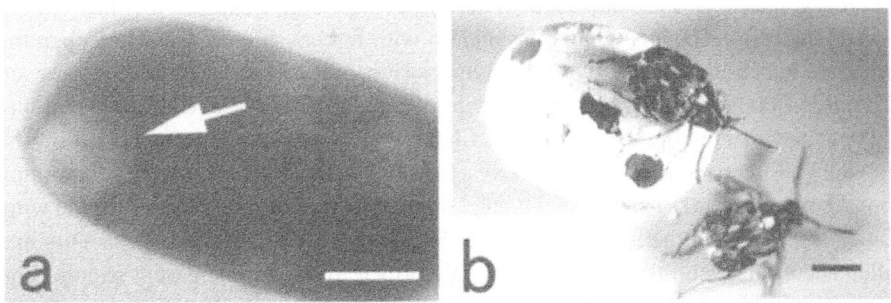

Figure 3.3. (a) X-ray image showing larve in a seed of a susceptible V. umbellata *accession. (b) Artificial seed made from flour of a susceptible azuki accession at the end of the screening test showing emerged adults of azuki bean weevil* (Callosobruchus chinensis).

2. 4. Use

In this section, the first step in using *Vigna* genetic resources is described, a method for making crosses between different accessions of *Vigna* species. Details of a method to hybridize cowpea (*Vigna unguiculata*) that are similar to the method describe here can be found in Blackhurst and Creighton (1980).

Floral characteristics
    While it is generally believed that most cultigens of the subgenus *Ceratotropis* are predominantly inbreeding, the literature lacks reliable estimates of outcrossing. The floral characteristics, such as large size, bright color, production of nectar and keel pocket suggests that species of this subgenus are adapted to cross-pollination by insects.
    Racemes arise in leaf axils or sometimes from leafless nodes (*V. exilis* and *V. umbellata*). The flower number ranges from 2 to about 30 with most species having between 5 and 15 flowers per raceme. The number of pods that develop and mature per raceme is generally from 1 to 4.

Items required for hybridization
1. Fine forceps;
2. Small vial of alcohol;
3. Indelible marker pen;
4. Small (25mm X 25mm) glycine bags;
5. Stapler.

Preparation of the female and pollination (Fig. 3.4)
    In the early morning, select flowers that are one day prior to opening with corolla extended about 2-2.5 times the length of the calyx (Fig. 3.4a). Remove other flowers and pods on the inflorescence. Open the standard with forceps (Fig. 3.4b). Then open the wings and keel with forceps to expose stamens (Fig. 3.4c). Anthers should be removed by holding filaments firmly and with forceps remove all 10 unopened anthers, avoiding all contact with the stigma (Fig. 3.4d). Check that there is no sign of pollen on the stigma. Dip forceps into alcohol. Remove a flower at anthesis that occurs in the morning from the male parent. Push the keel pocket down, like a pollinator landing on the wing (Fig. 3.4e), exposing the top of the stigma (Fig. 3.4f). The top of the style is hairy and pollen accumulates on these hairs. Apply the pollen to the female parent stigma (Fig. 3.4g). Write cross combination and date of crossing on a small glycine bag. Cover pollinated flower with glycine bag, fold the base of the bag and secure with staple (Fig. 3.4h). After one week, cut the top of the glycine bag with scissors to allow the developing pod to grow.

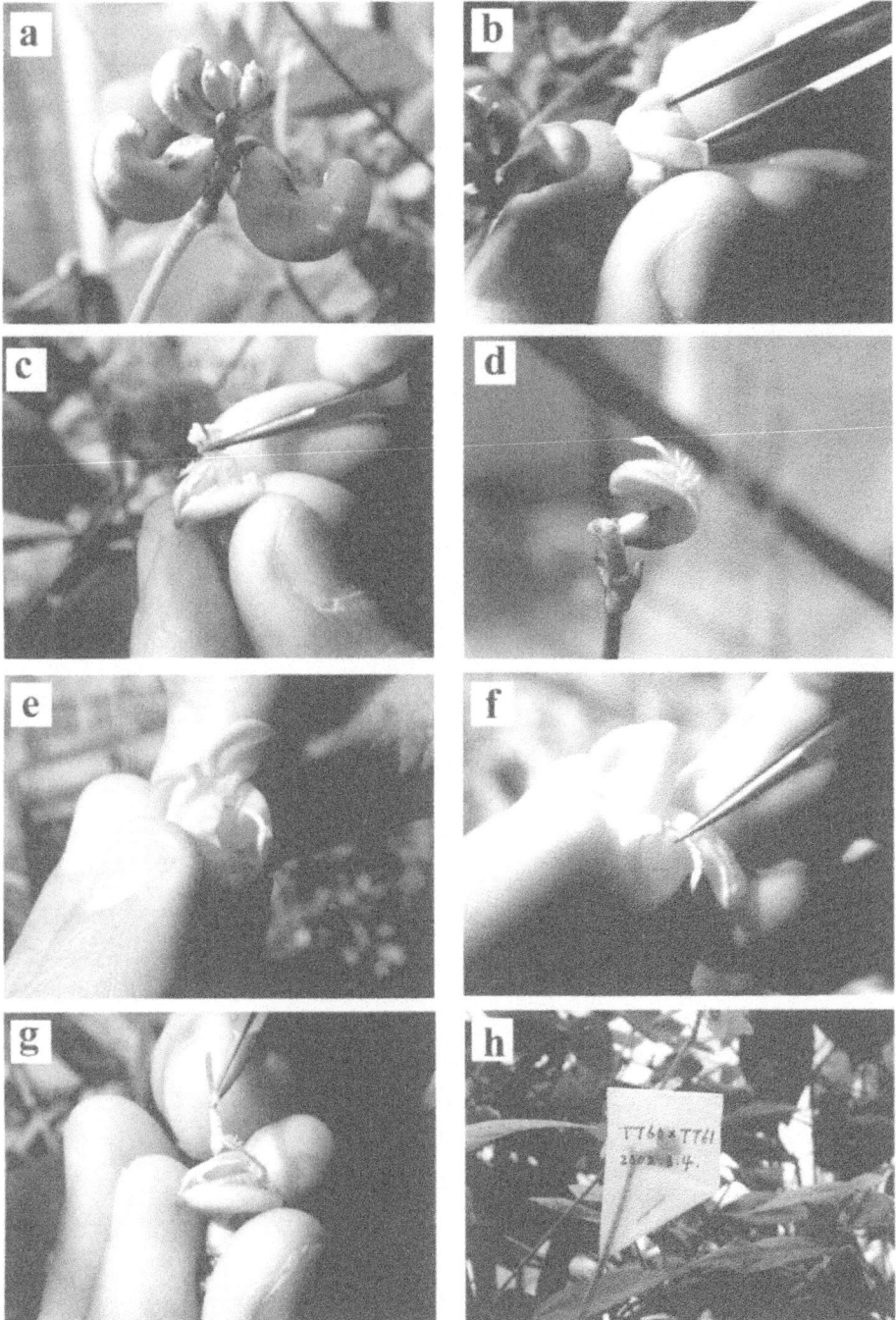

Figure 3.4. Procedure for artificial cross pollination in Vigna.

CHAPTER 4

SPECIES CONSPECTUS

1. DATA COLLECTION

Sources of data that form the basis for the descriptions and analysis that follow are:
(a) Information gathered directly by the authors (NT, DAV, HM) from the following herbaria- (See Appendix II for list of specimens):

BM - British Museum (Natural History), UK;

BO - Herbarium Bogoriense,Bogor, Indonesia;

BR - Herbarium, National Botanic Garden of Belgium, Meise, Belgium;

BSIS - Herbarium, Botanical Survey of India, Calcutta, India;

CAL - Central National Herbarium, Calcutta, India;

DD - Herbarium, Dehra Dun, India;

DUH - Herbarium, Botany Department, University of Delhi, New Delhi, India;

FRI - Australian National Herbarium, CSIRO, Canberra, Australia;

K - Herbarium, Royal Botanic Gardens, Kew, UK;

KYO - Kyoto University, Japan;

L - Rijksherbarium, Leiden, the Netherlands;

NBPGR - National Bureau of Plant Genetic Resources, New Dehli, India;

P - Natural History Museum, Paris, France;

PDA - Botanic Gardens, Peradeniya, Sri Lanka;

TUS - Herbarium, Tohoku University, Miyagi, Japan;

(b) Direct collections by the authors (N.T., D.A.V.) in Japan, Malaysia, Myanmar, Sri Lanka, Thailand. The information from these collections are all published by the NIAS, a recent listing of these reports can be found in Tomooka *et al.*, 2000a.

(c) The figures and species descriptions are based on one or two accessions chosen from the living conserved collections in the MAFF, Japan, Genebank collection unless otherwise stated. These accessions are listed in Table 4.1.

(d) The principal sources in the literature to vernacular names reported here were:
Hanelt, 2001; Maxwell, 1991; Neupane, 1999; Ohwi and Ohashi, 1969; Piper and Morse, 1914; Purseglove, 1974; Quattrocchi, 2000; Skerman *et al.,* 1988; Takeya and Tomooka, 1997; Tateishi and Maxted, 2002; Thuan, 1979; Tomooka *et al.,* 2000e; van der Maesen and Somaatmadja, 1989; Wiersema and León, 1999.

Table 4.1. List of accessions used for figures in Chapter 4.

Species	Drawing of plant / shoot		Drawing of plant parts	
	JP No.	Origin	JP No.	Origin
Section *Angulares*				
V. angularis var. *nipponensis*	87905	Japan	107861	Japan
V. dalzelliana[1]	-	Sri Lanka	-	Sri Lanka
V. exilis	207983	Thailand	205884	Thailand
V. hirtella	108515	Thailand	108851	Malaysia
V. minima	205886	Thailand	205886	Thailand
V. nakashimae	110356	Japan	107879	Japan
V. nepalensis	107881	Nepal	107881	Nepal
V. reflexo-pilosa var. *reflexo-pilosa*	108815	Japan	108815	Japan
V. riukiuensis	108810	Japan	108810	Japan
V. tenuicaulis	108552	Thailand	109682	Thailand
V. trinervia var. *trinervia*	207980	Sri Lanka	108840	Malaysia
V. trinervia var. *bourneae*	207981	India	207981	India
V. umbellata (wild form)	207982	Thailand	109675	Thailand
Section *Ceratotropis*				
V. grandiflora	207984	Thailand	107862	Thailand
V. mungo var. *silvestris*	107874	India	107874	India
V. radiata var. *sublobata*	107877	Madagascar	107877	Madagascar
V. subramaniana	110836	India	110836	India
Section *Aconitifoliae*				
V. aconitifolia	105629	India	105629	India
V. aridicola	207977	Sri Lanka	205896	Sri Lanka
V. khandalensis[2]	-	India	-	-
V. stipulacea	207976	Sri Lanka	205892	Sri Lanka
V. trilobata	207979	Sri Lanka	205895	Sri Lanka

[1]From passport number (2001S33) collected by author (N.T.) herbarium specimen of which is deposited at K.

[2]From type specimen collected by Sedgwick & Bell (7953) deposited at K.

2. CONSPECTUS

2.1. Section Angulares *N. Tomooka & Maxted, Kew Bull. 57(3):613-624 (2002a)*

2.1.1. Vigna angularis *(Willd.) Ohwi & Ohashi, J. Jap. Bot. 44:29 (1969)*

Chromosome number: 2n=2x=22

Taxonomic affinities and diagnosis: *V. angularis* belongs to a group of closely related species that includes *V. tenuicaulis* and *V. nepalensis* within section *Angulares*. Useful

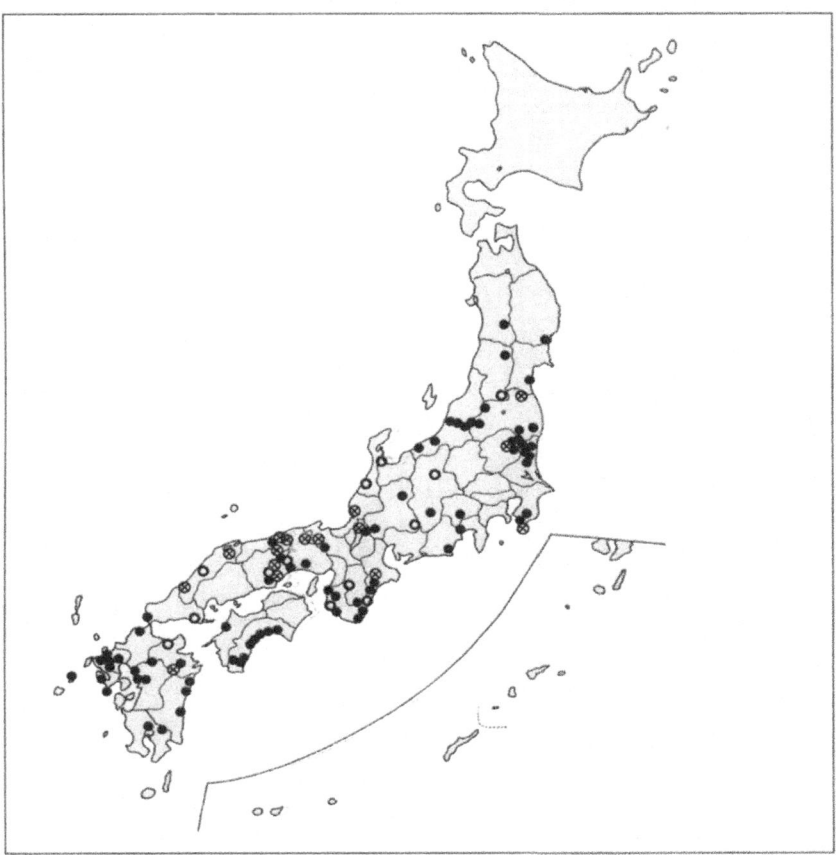

Figure 4.1. Distribution of wild (●), weedy (○) and mixed (⊗) population of the Vigna angularis *complex in Japan based on direct collection by the authors (N.T., D.A.V.).*

characters to identify *V. angularis* are large secondary bract (pedicel bract) concealing flower bud; pendulous glabrous mature pod; smooth seed with long linear hilum without aril. Diagnostic characters which distinguish it from *V. nepalensis* are much more pubescent primary and secondary bracts and bracteole; bracteole much longer than calyx; longer hilum in relation to seed size. Differs from *V. tenuicaulis* by larger overall plant parts, especially flower and seed; large secondary bract concealing flower bud.

Description: Based on wild form var. *nipponensis* JP107861 from Japan

A twining herb. *Stems* slender, sparsely pubescent with long (1.0 - 1.8 mm) white hairs. *Stipules* peltate, narrowly elliptic, acute at the apex, 7.6 x 1.6 mm, pubescent outside with white hairs (0.4 - 0.7 mm). *Leaf petioles* about 10 cm long, sparsely covered with white hairs (0.5 - 0.7 mm). *Leaflets* membranous, sparsely covered with white hairs (0.4 - 1.3 mm) on both surfaces; *terminal leaflet* ovate, entire to narrowly lobed, acuminate at the apex, obtuse at the base, 9.0 x 5.2 cm; *lateral leaflets* obliquely ovate, entire or sometimes faintly lobed, acuminate at the apex, rounded at the base, 7.1 x 5.2 cm; stipels narrowly elliptic, acuminate at the apex, 2.9 x 0.7 mm; pulvinus covered with white hairs.

Inflorescence axillary, 2 - 16 flowered; peduncles 5 - 10 cm long, sparsely covered

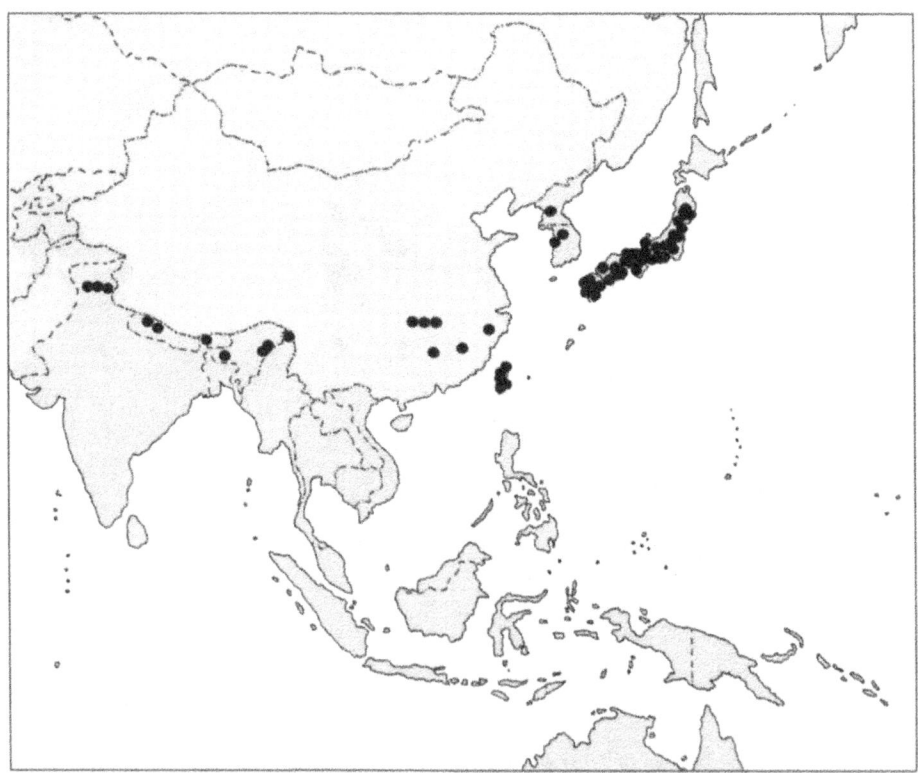

Figure 4.2. Distribution of Vigna angularis *var.* nipponensis.

with brown hairs (0.5 - 1.0 mm); rachis 1 - 6 cm long, covered with short white hairs (0.2 - 0.4 mm). *Primary bract* peltate, ovate, acute at the apex, 5.1 x 2.7 mm. *Secondary bract* (pedicel bract) basifixed, elliptic, acute at the apex, rounded at the base, 9.0 x 3.5 mm, concealing young flower bud completely, covered with white hairs (0.6 - 1.5 mm). *Bracteoles* basifixed, narrowly elliptic, acuminate at the apex, obtuse at the base, 6.8 x 1.5 mm, much longer than calyx, covered with long white hairs (0.6 - 1.2 mm). *Pedicels* ascending, 2 - 3 mm long when in flower, 3- 5 mm long when in fruit.

Flowers golden yellow. *Calyx* campanulate, 4.2 mm long, tube 3.0 mm long. *Standard* asymmetrical, transversely broadly elliptic, 12.8 mm long, 16.9 mm wide, emarginate at the apex, prominent appendage inside. *Right wing* half concealing upper portion of the keel petals, lamina obliquely elliptic, 10.0 mm long, 6.7 mm wide, claws 2.0 mm long, auricle 1.4 mm long. *Left wing* spreading horizontally, supported by a pocket on left hand keel petal, lamina obliquely elliptic, 9.7 mm long, 7.1 mm wide, claws 2.6 mm long, auricle 1.9 mm long. *Keel-petals* spirally incurved to the left through 260°, 16.3 mm long, left hand petal with a long horn-like pocket outside, the pocket 4.3 mm long. *Ovary* 4.9 mm long, ovules 8 - 12; style 18.7 mm long, filiform; style beak 0.9 mm long.

Fruits pendulous, linear, cylindrical, 5 - 8 cm long, 0.3 - 0.5 cm wide, glabrous, blackish when mature, 5 - 10 seeded. *Seed* oblong, grayish brown and variegated with black when mature, surface smooth, 3.7 x 2.9 x 2.6 mm; hilum linear, not protruding, 2.1 mm long, 0.5 mm wide; aril not developed. Germination hypogeal. *First and second leaves* simple, cordate, 1.7 cm long, 1.4 cm wide, long petiolated.

V. angularis var. *angularis*: cultivated form

Geographic distribution: This is a traditional crop of East Asia and northeastern Vietnam. It seems also to be an ancient crop in Nepal and Bhutan. Recently, it has been introduced as a crop to North America and Australia.

Origin:

It has been considered that azuki bean was domesticated in China. However, we now consider Japan to be more probable place of domestication because of the abundance of the wild form and frequent occurrence of an intermediate form between wild and cultivated azuki which is called weedy azuki in Japan. Recently, AFLP analysis has revealed that azuki bean in Nepal and Bhutan are genetically quite distinct from that in East Asia, suggesting the independent domestication in these areas (Zong *et al.*, 2002 in process).

Vernacular names:

Bhutan: semchen roka

Chinese: chi xiao dou, hung dou

English: adzuki bean, azuki bean

French: haricot adzuki

German: adzukibohne

Japanese: azuki, shouzu

Korean: pat

Figure 4.3. Flowering and fruiting shoots of Vigna angularis *var.* nipponensis.
Drawn from JP87905 (Japan) by Kyoko Motoyoshi.

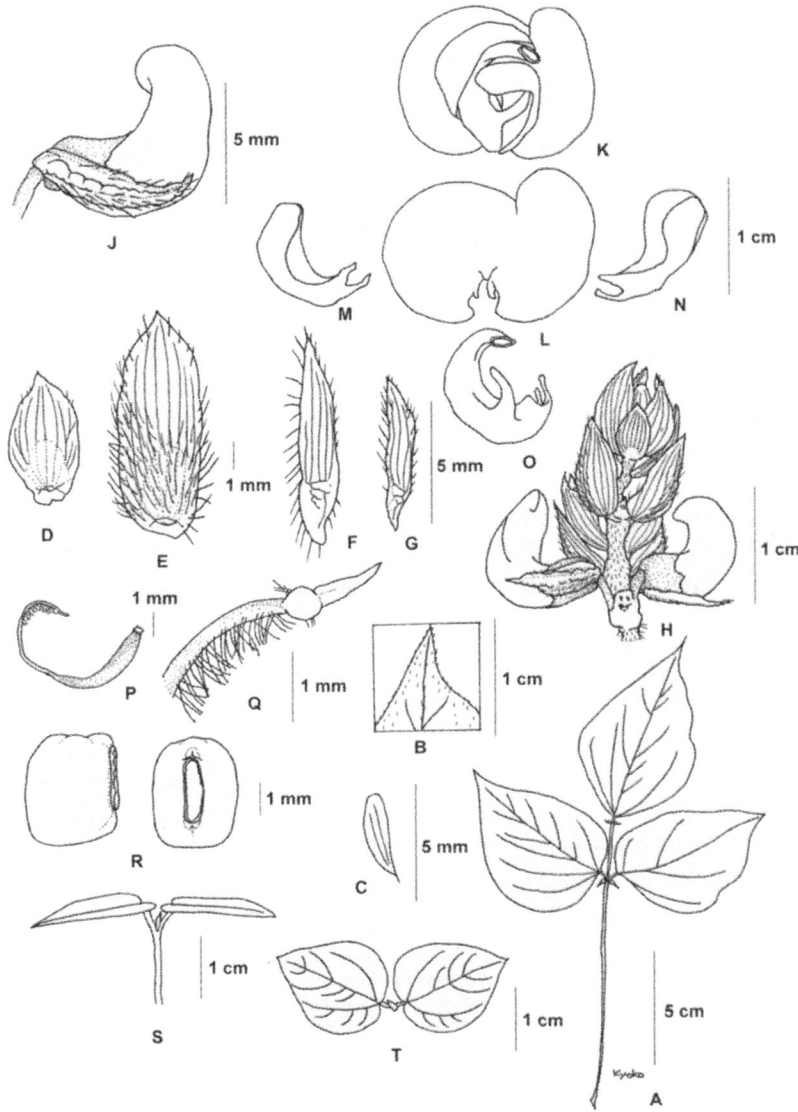

Figure 4.4. Plant parts of V. angularis *var.* nipponensis.
A: leaf; B: leaf apex; C: stipel; D: primary bract; E: secondary bract; F: stipule; G: stipule; H: inflorescence; J: flower bud and bracteole; K: flower; L: standard; M: right wing; N: left wing; O: keel; P: style; Q: stigma and style beak; R: seed; S & T: seedling shoot. Drawn from JP107861 (Japan), except F from JP87905 (Japan) by Kyoko Motoyoshi.

Nepali: simi, mash, rato mash
Spanish: frijol adzuki, judia adzuki
Vietnamese: dau do
Uses: After soybean this is the most important grain legume in East Asia. It is often cooked with glutinous rice at times of celebration in Japan. It is also used in the preparation of sweet bean jam and sweet bean soup.
Ethnobotanical notes: Azuki beans are used in Chinese medicine for kidney trouble, boils, abscesses, certain tumors, threatened miscarriage, difficult labor, retained placenta and non-secretion of milk. Leaves are used to lower fever. Sprouts used to treat threatened abortion caused by injury (Sacks, 1977).

V. angularis: weedy form. Weedy azuki shows a diverse range of morphology, sometimes has characteristics similar to the wild form, sometimes similar to cultivated, and sometimes shows characteristics intermediate between wild and cultivated azuki.
Ecology: Weedy azuki often grows in abandoned fields, and also grows in or at the edge of streams/drainage channels.
Vernacular names:
Japan: kusa-azuki, masara, yama-azuki, inmame, nouraku-azuki, no-azuki, norakko, nora-azuki, no-sasa, kuro-azuki, kitsune-azuki, taito-azuki, ishimame, karasu no azuki. The words nora, nouraku, no, norakko means "of cultivated areas" in Japanese.
Uses: In Japan weedy azuki used to be consumed as sweet beans known as 'an' or 'anko' but not used in the preparation of red rice. In some areas, the seeds of weed forms are used to make beanbags.

V. angularis var. *nipponensis* (Ohwi) Ohwi & Ohashi, J. Jap. Bot. 44:30 (1969): wild form
Geographical distribution: Japan, the Korean peninsula, China including Taiwan, Himalayan region of India as far west as the border with Pakistan (but not recorded in Pakistan), Nepal, Bhutan and northern-most part of Myanmar.
Altitude: Ranges from sea level in Japan to 2700 m in Nepal, mean 276 m (143 records) This variety shows clear differences in altitude with the latitude.
India (25-31°N) 1447-2000 m (3 records) mean 1682 m
Nepal (26°N) 1450-1975 m (3 records) mean 1691 m
Bhutan (27°N) 890-2485 m (8 records) mean 1970 m (Murata *et al.*, 1995)
Taiwan (23°N) 1000-1650 m (6 records) mean 1442 m
Korea (36°N) 200-235 m (2 records) mean 217 m
Japan (32-38°N) 0-700 m (129 records) mean 143 m
Phenology: Japan – September to October; India – August to September.
Ecology: Wild azuki is found in many disturbed habitats such as roadside and railway embankments, recently abandoned paddy fields, edge of drainage ditches. Population size is variable from 10 to more than 100 plants. It is found in mantel and sleeve communities (Yamaguchi, 1992).
Associated vegetation: *Achyranthes bidentata* Blume, *Ambrosia trifida* L.,

Amphicarpaea bracteata (L.) C.F.Reed, *Artemisia princeps* Pampan., *Bidens bitemata* Merr. et Sherff, *Cayratia japonica* (Thunb.) Gagnep., *Commelina communis* Engelm. ex Kunth., *Erigeron canadensis* Brot., *Glycine soja* Sieb. & Zucc., *Helianthus tuberosus* L., *Humulus japonicus* Sieb. et Zucc., *Kalimeris yomena* Kitamura, *Miscanthus sinensis* Anderss., *Persicaria longiseta* (De Bruyn) Kitagawa, *Pleioblastus simoni* Nakai, *Pueraria lobata* (Willd.) Ohwi, *Reynoutria japonica* Houtt., *Rubus hirsutus* Hayata, *Setaria viridis* Beauv., *Solidago altissima* Ait., *Sonchus oleraceus* L., *Xanthium occidentale* Bertol. (modified from Yamaguchi, 1992).

Vernacular names:

Japanese: yabu tsuru azuki

2.1.2. V. dalzelliana *(O. Kuntze) Verdcourt, Kew Bull. 24: 558 (1970)*

Chromosome number: 2n=2x=22

Taxonomic affinities and diagnosis: This species appears to be related to *V. exilis, V. tenuicaulis, V. hirtella* and *V. minima.* Useful characters to identify *V. dalzelliana* are

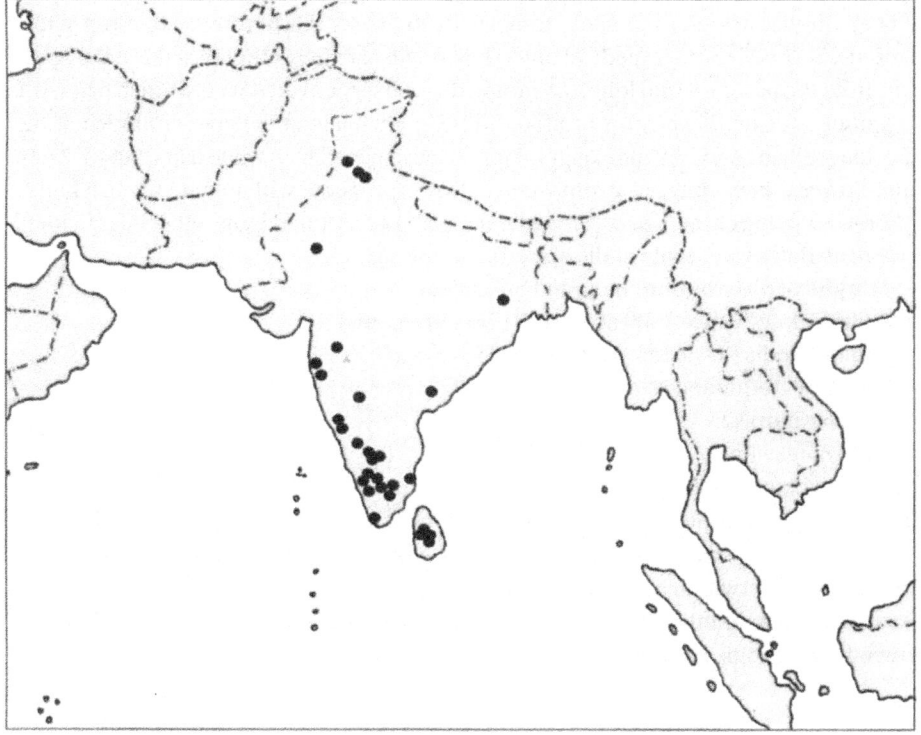

Figure 4.5. Distribution of Vigna dalzelliana.

flat style-beak; glabrous mature pod; smooth seed with linear hilum having developed aril. Diagnostic characters that distinguish it from *V. exilis* are elliptic seed shape; flat style-beak. Differs from *V. hirtella* and *V. tenuicaulis* by flat style-beak. Differs from *V. minima* by stipule morphology; less protruded hilum and less developed aril. Since living material has not been analyzed its relationship with other species remains to be clarified.

Description: Based on Tomooka *et al.* 2001S33 from Sri Lanka.

A twining herb. *Stems* slender, covered with thread-like white hairs (c. 0.5 mm long). *Stipules* peltate, narrowly elliptic, attenuate at the apex, acute at the base, 4.8 mm long, 1.1 mm wide, sparsely pubescent with white hairs around the edge, 0.2 - 0.3 mm long. *Leaf petioles* 3 - 5 cm long, sparsely covered with white hairs, 1.0 - 1.2 mm long. *Leaflets* chartaceous, very thin, entire, sparsely covered with white hairs on both surfaces, 0.4 - 0.8 mm long; *terminal leaflets* ovate, acuminate at the apex, rounded at the base, 5 - 6 cm long, c. 3 cm wide; *lateral leaflets* obliquely ovate, acuminate at the apex, rounded at the base, 4.0 - 5.0 cm long, 2.0 - 3.0 cm wide; stipels linear, acute at the apex, 1.8 mm long, 0.4 mm wide.

Inflorescence axillary; peduncles sparsely covered with white hairs. *Primary bract* peltate, elliptic, acute at the apex, truncate at the base, 0.8 mm long, 0.5 mm wide, glabrous. *Secondary bract* (pedicel bract) basifixed, narrowly elliptic, acute at the apex, truncate at the base, 1.8 mm long, 0.5 mm wide, sparsely covered with whitish hairs, 0.1 - 0.2 mm, outside. *Bracteoles* basifixed, linear, attenuate at the apex, rounded at the base, longer than calyx, 3.9 mm long, 1.4 mm wide, covered with white hairs, 0.6 - 1.2 mm long. *Pedicels* ascending, 1 - 2 mm when in flower, covered with minute whitish hairs.

Flowers pale yellow. *Calyx* campanulate, glabrous, 3.1 mm long, tube 3.0 mm long. *Style beak* flat, ovate, acute at the apex, 0.5 mm long.

Geographical distribution: India and Sri Lanka.

Altitude: Below 50 m to 2500 m, mean 917m (30 records).

Phenology: India - October to December.

(notes below from Babu *et al.*, 1985)

Vernacular names:

Gujarati: mugavaine.

Maranti: magauel, ranmug.

Ecology: Forms a component of the ground flora of monsoon forest, particularly of Sal forests. Grows in shady wet highland slopes where this species can root from stem nodes (field observation by N.T., in Sri Lanka).

Uses: Livestock often graze on the foliage. Reported a good soil binder and can be grown for soil conservation.

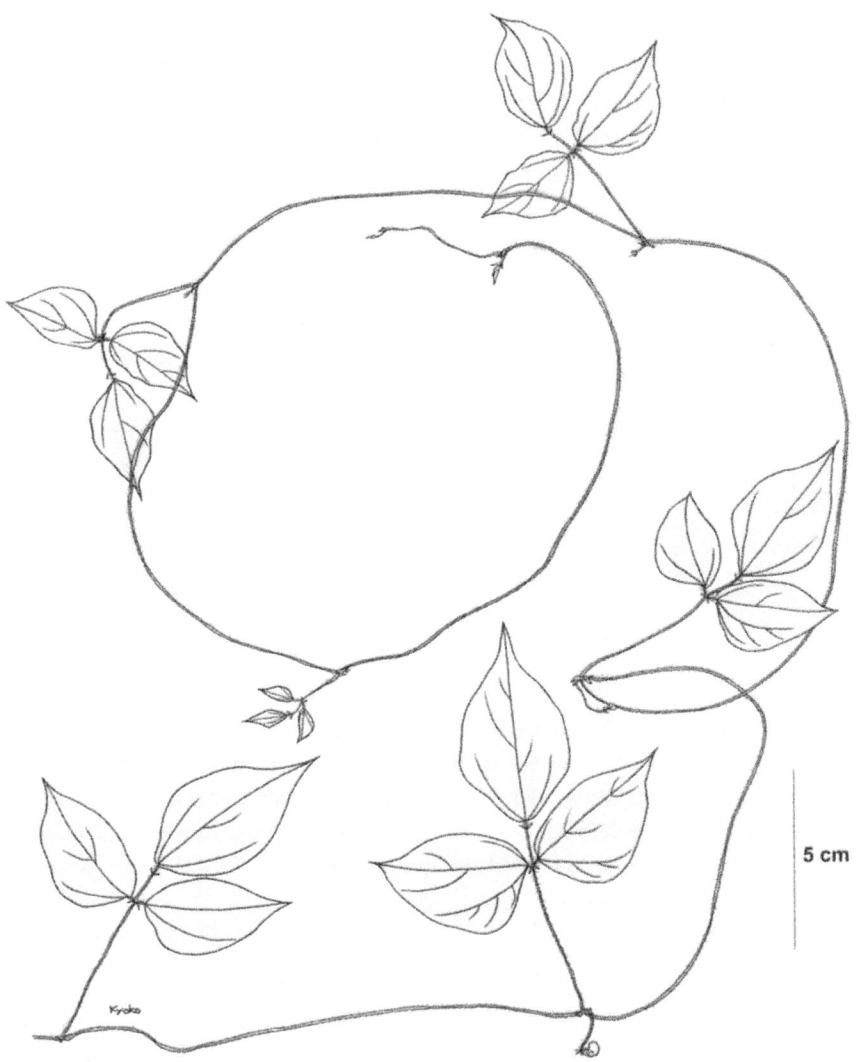

5 cm

Figure 4.6. Flowering shoot of Vigna dalzelliana.
Drawn from Tomooka et al. 2001S33 (Sri Lanka) (K) by Kyoko Motoyoshi.

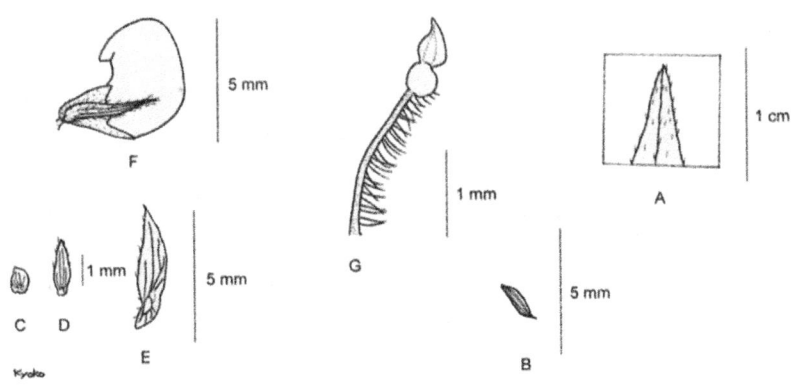

Figure 4.7. Plant parts of Vigna dalzelliana.
A: leaf apex; B: stipel; C: primary bract; D: secondary bract; E: stipule; F: flower bud and
bracteole; G: stigma and style beak. A, B & E drawn from Tomooka et al. 2001SL33 (Sri
Lanka)(K), C, D, F & G from Tomooka et al. 2001SL31B (Sri Lanka) by Kyoko Motoyoshi.

2.1.3. V. exilis *Tateishi & Maxted, Kew Bull. 57(3):625-633 (2002)*
Chromosome number: 2n=2x=22
Taxonomic affinities and diagnosis: *V. exilis* is most closely related to *V. hirtella, V.
dalzelliana* and *V. umbellata* within section *Angulares*. Useful characters to identify *V.
exilis* are nearly glabrous stem; glabrous mature pod; linear seed with protruding hilum
having well developed aril. Diagnostic character that distinguishes it from *V. dalzelliana,
V. hirtella* and *V. umbellata* is linear seed morphology.
Description: Based on JP205884 from Thailand
A twining herb. *Stems* slender, glabrous or sometimes sparsely pubescent with long
spreading or retrorse whitish hairs (0.3 - 0.4 mm long) when young, later glabrous.
Stipules, 5.5 x 2.4 mm, narrowly ovate or oblong, acute or acuminate at the apex, 5-9
nerved, surface glabrous, sparsely covered with minute whitish hairs outside (0.1 - 0.2
mm). *Leaf petioles* 3 - 5 cm long, sparsely covered with spreading whitish hairs (0.3 - 0.4
mm long). *Leaflets* very sparsely covered with spreading whitish fine hairs (0.3 - 0.9 mm
long) on both surfaces; *terminal leaflet* ovate to broadly ovate, 5 - 6 x 3 - 4 cm, entire or
faintly 3-lobed, acuminate at the apex, rounded at the base; *lateral leaflet* somewhat
asymmetric, entire or faintly 2-lobed, 5 - 6 x 3 - 4 cm; stipels linear, attenuate at the apex,

2.0 x 0.5 mm.

Inflorescence axillary; 2 - 6 flowered; peduncles short and very slender, 1 - 2 cm long, sparsely covered with whitish hairs (0.2 - 0.3 mm long); rachis 1-2 mm long. *Primary bract* peltate, but attached to the rachis near the lower end, triangular-ovate, acute at the apex, 1.1 x 0.7 mm, glabrous, caducous. *Secondary bract* (pedicel bract) nearly basifixed, elliptic, acute at the apex, truncate at the base, 1.7 x 0.7 mm, covered with white hairs (0.1 - 0.2 mm) outside, caducous. *Pedicels* ascending, slender, 2 - 2.5 mm long in flower, 2 - 3 mm long in fruit, glabrous. *Bracteoles* narrowly ovate, attenuate or acute at the apex, 3 - 3.5 x 0.8 - 1 mm, longer than calyx, glabrous.

Flowers pale yellow. *Calyx* campanulate, 2 - 2.8 mm long, glabrous outside; tube 1.5 - 2 mm long. *Standard* asymmetrical, obliquely and broadly elliptic, 6 - 8 x 10 - 11.5 mm, emarginate at the apex, with an appendage at the centre inside. *Right wing* concealing the upper portion of the keel-petals, claws 1.3 - 2 mm long; lamina obliquely obovate, 6 - 7 x 3 - 5.5 mm, auricle 0.4 - 0.8 mm long. *Left wing* spreading, claws 1.5 - 2 mm long; lamina obliquely oblong, 6.0 - 7.5 x 3.0 - 5.0 mm, auricle 1.0 - 1.2 mm long. *Keel-petals*

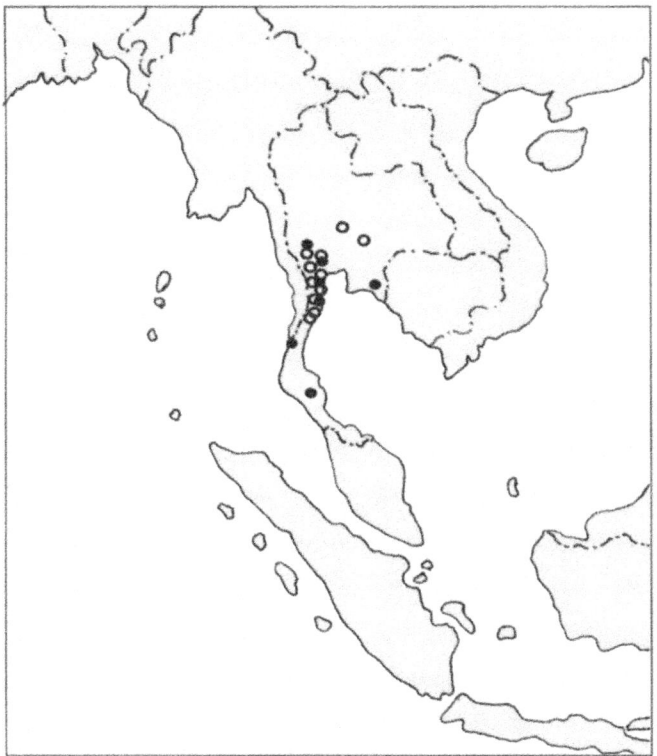

Figure 4.8. Distribution of Vigna exilis *based on herbarium specimens (●) and direct collection (○) by author (N.T.).*

Figure 4.9. Flowering and fruiting shoot of Vigna exilis.
Drawn from JP207983 (Thailand) by Kyoko Motoyoshi.

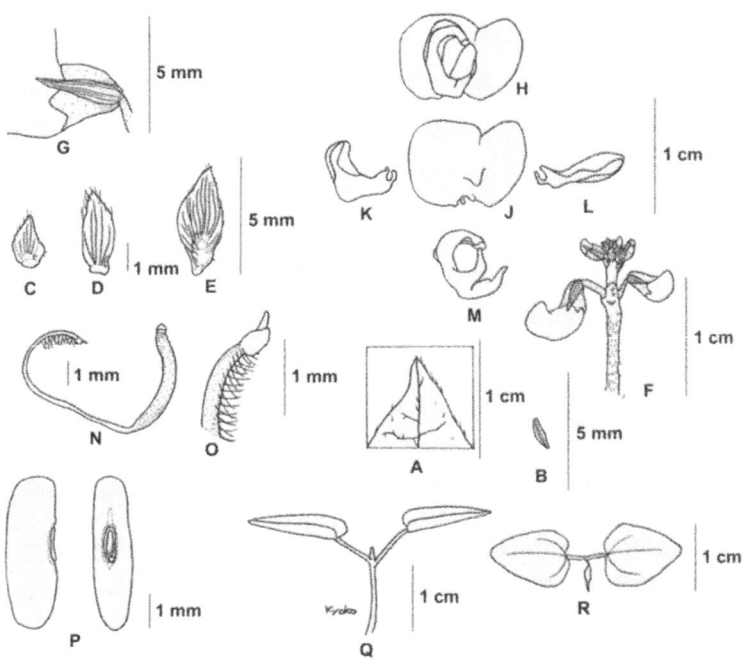

Figure 4.10. Plant parts of Vigna exilis.
A: leaf apex; B: stipel; C: primary bract; D: secondary bract; E: stipule; F: inflorescence;
G:lower part of flower bud and bracteole; H: flower; J: standard; K: right wing; L: left
wing; M: keel; N: style; O: stigma and style beak; P: seed; Q & R: seedling shoot. Drawn
from JP205884 (Thailand), except A & E from JP207983 (Thailand) by Kyoko Motoyoshi.

spirally incurved, 9.0 - 10.0 mm long, pocket on the left hand petal 2.0 - 3.0 mm long.
Ovary 4.0 - 5.0 mm long, 6 - 7 ovuled; style 11.0 - 13.5 mm long, filiform, beaked beyond
the stigma, the beak 0.3 - 0.4 mm long.

    Fruits spreading, linear, cylindrical, 2 - 4 x 0.2 - 0.3 cm, nearly glabrous, brownish
yellow when mature, 7 - 8 seeded. *Seeds* linear, 5.0 x 1.6 x 1.3 mm, smooth, pale brown
with blackish mottle when mature; hilum oblong, 1.1 mm long, 0.5 mm wide, aril developed;
Germination hypogeal. *First and second foliage leaves* simple, cordate, 1.2 x 0.8 cm,
with petioles.

Geographical distribution: Thailand.

Altitude: Range 20 - 400 m, mean 135 m (21 records).
Phenology: Thailand – October to November.
Ecology: Grows on limestone rocks in open evergreen forests of oak and laurels, where this species grows in cracks in the rocks in open or slightly shaded habitats of low disturbance. Population size reflects on the area of open limestone habitat.
Vernacular names:
Thai: tua jan, thua pee (Tomooka *et al.*, 2000e).
Uses: Relieves pain (Tomooka *et al.*, 2000e).

2.1.4. V. hirtella *Ridley, J. Fed. Mal. States Mus. 10:132 (1920)*
Chromosome number: 2n=2x=22
Taxonomic affinities and diagnosis: *V. hirtella* belongs to a group of closely related species that includes *V. angularis, V. tenuicaulis, V. umbellata* and *V. nepalensis* within section *Angulares*. Useful characters to identify *V. hirtella* are hairy stem; small secondary (pedicel) bract; glabrous mature pod; oblong seed with slightly protruding hilum with more or less developed aril. Diagnostic characters that distinguish it from *V. angularis, V. nepalensis* and *V. tenuicaulis* are small secondary bract which is nearly the same size as primary bract; protruding hilum with aril. Differs from *V. umbellata* by oblong seed with less protruding hilum having less developed aril.
Description: Based on JP108851 from Malaysia
A twining herb. *Stems* rather thick, densely clothed with long retrorse or spreading whitish brown hairs (1.1 - 1.5 mm). *Stipules* peltate, elliptic, acute at the apex, 6.2 mm long, 2.2 mm wide, sparsely ciliate with short white hairs outside (0.5 mm long). *Leaf petioles* 7 - 12 cm, covered with short (0.5 - 0.7 mm) white hairs on the ridge. *Leaflets* rather densely covered with whitish hairs on both surfaces (0.5 - 1.0 mm long); *terminal leaflets* ovate, acute or acuminate at the apex, obtuse at the base, 7.5 - 10.0 x 5.5 - 6.8 cm; *lateral leaflets* obliquely ovate, acute at the apex, rounded at the base, 6.2 - 8.6 x 5.0 - 6.2 cm; stipels elliptic acuminate at the apex, 2.8 x 1.0 mm.

Inflorescence axillary, 8 - 12 flowered; peduncles long, 20 - 25 cm long, covered with short (0.3 - 0.5 mm) white retrorse hairs; rachis 1.1 cm long, nearly glabrous. *Primary bract* small, rhombic, acute at the apex, truncate at the base, 2.2 x 1.4 mm, glabrous. *Secondary bract* nearly same size as primary bract, ovate, acute at the apex, rounded at the base, 2.9 x 1.5 mm, glabrous. *Bracteoles* subulate, as long as calyx, 3.0 x 1.0 mm. *Pedicels* ascending, 2.7m long in flower, 4.2 - 5.3 mm long in pod.

Flowers golden yellow. *Calyx* campanulate, glabrous, 4 mm long, tube 3.2 mm long. *Standard* asymmetrical, obliquely broadly elliptic, 15.1 x 19.4 mm, emarginate at the apex, prominent appendage inside. *Right wing* half concealing upper part of the keel-petals, lamina obliquely obovate, 12.6 x 9.7 mm, claws 3.3 mm long, auricle 2.0 mm long. *Left wing* spreading horizontally, supported by a pocket on the left-hand keel-petal, lamina obliquely obovate, 12.4 x 10.1 mm, claws 3.2 mm long, auricle 2.1 mm long. *Keel-*

petals spirally incurved to the left through 270°, 19.7 mm long, the left hand petal with a long horn-like pocket outside, the pocket 4.4 mm long. *Ovary* 7.1 mm long, nearly glabrous, ovules 15 - 19; style 20.3 mm long, filiform; style beak linear, 1.2 mm long.

   Fruits pendulous, linear, cylindrical, glabrous, blackish brown when mature, 15 - 19 seeded. *Seed* oblong, brown densely mottled with black when mature, surface smooth, 3.3 x 2.4 x 1.8 mm; hilum linear, 1.8 x 0.6 mm, aril more or less developed. Germination hypogeal. *First and second leaves* simple, cordate, acute at the apex, 0.9 x 0.6 cm, with petiole.

Geographic distribution: Northern India, Bangladesh, south China, Southeast Asia - Myanmar, Thailand, Laos, Vietnam and Malaysia.

Altitude: Range 25 m (Thailand) to 1700 m (China), mean 874 m (24 records)

Phenology: Thailand – October to November.

Ecology: Grows in wet highland sites in northern Thailand. Sometimes sympatric with *V. minima* and *V. tenuicaulis*.

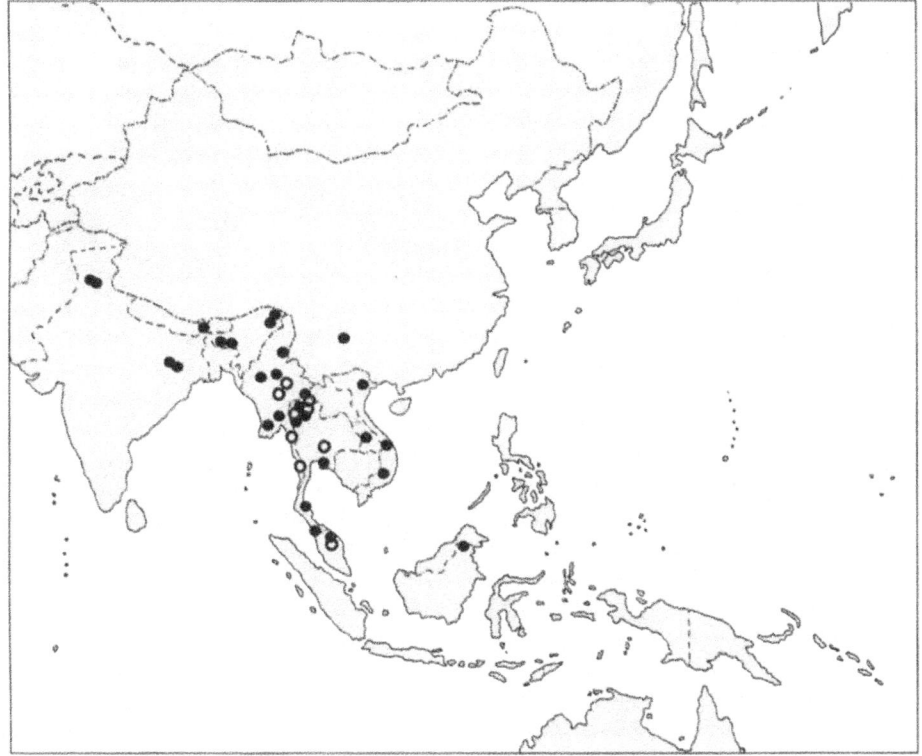

Figure 4.11. Distribution of Vigna hirtella *based on herbarium specimens (●) and direct collection (○) by author (N.T.).*

Fogure 4.12. Flowering and fruiting shoot of Vigna hirtella.
Drawn from JP108515 (Thailand) by Kyoko Motoyoshi.

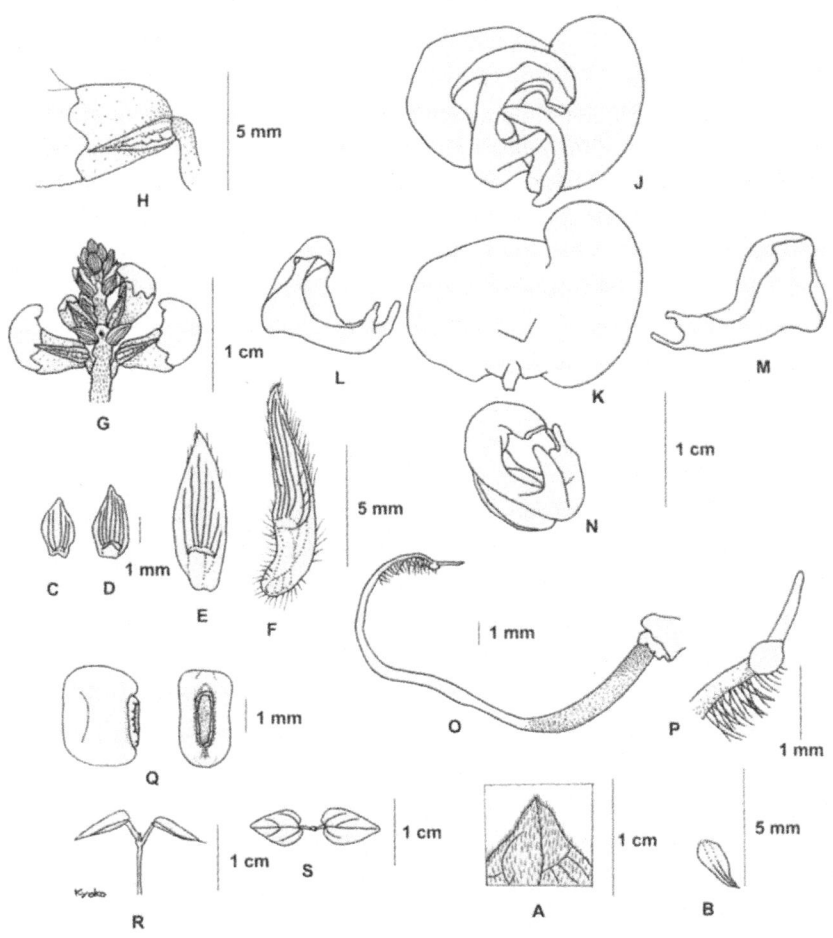

Figure 4.13. Plant parts of Vigna hirtella.
A: leaf apex; B: stipel; C: primary bract; D: secondary bract; E & F: stipule; G: inflorescence; H: lower part of flower bud and bracteole; J: flower; K: standard; L: right wing; M: left wing; N: keel; O:style; P: stigma and style beak; Q: seed; R & S: seedling shoot. Drawn from JP108851 (Malaysia), except A & F from Tomooka et al. 96120503 (Thailand) by Kyoko Motoyoshi.

2.1.5. V. minima *(Roxb.) Ohwi & Ohashi, J. Jap. Bot. 44: 30 (1969)*

Chromosome number: 2n=2x=22

Taxonomic affinities and diagnosis: *V. minima* is closely related to *V. nakashimae* and *V. riukiuensis* of East Asia within section *Angulares*. In Southeast Asia, it is similar and sometimes confused with *V. hirtella.* Useful characters to identify *V. minima* are small stipule; small primary and secondary bracts, small bracteole; glabrous spreading pod; elliptic seed with much protruding hilum having well developed aril. Diagnostic characters which distinguish it from *V. hirtella* are smaller bracteole and much protruding hilum with well developed aril. Differs from *V. nakashimae* by larger golden yellow flower. Differs from *V. riukiuensis* by larger leaflet with acuminate apex; longer pod.

Description: Based on JP205886 from Thailand

A twining herb. *Stems* slender, sparsely covered with pale brown hairs, c. 0.5 mm long.

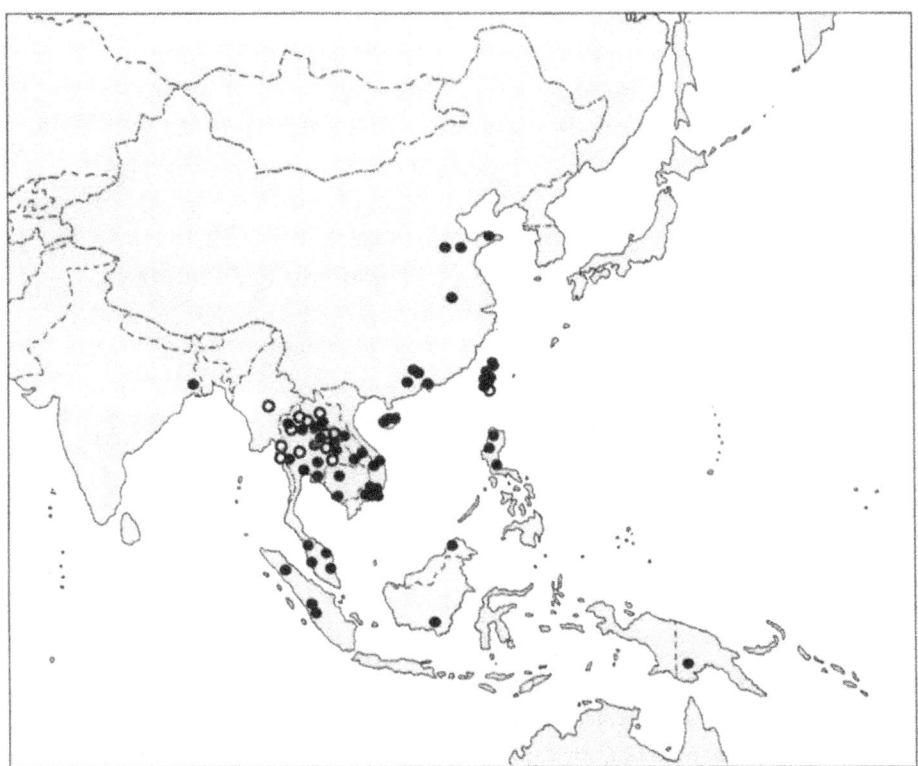

Figure 4.14. Distribution of Vigna minima *based on herbarium specimens (●) and direct collection (○) by author (N.T.).*

Stipules peltate, small, elliptic, acute at the apex, rounded or truncate at the base, 4.1 x 1.7 mm, sparsely ciliate with short hairs outside. *Leaf petioles* 2.5 - 6.7 cm long, covered with short (0.4 - 0.5 mm) whitish brown hairs on the ridges. *Leaflets* nearly glabrous, ciliate with short (0.2 - 0.3 mm) white hairs on leaf vein of both sides; *terminal leaflets* ovate or narrowly ovate, acuminate at the apex, obtuse at the base, 7 - 9 x 3 - 5 cm; *lateral leaflets* obliquely ovate, acuminate at the apex, rounded at the base, 6 - 7 x 3 - 5 cm; stipels narrowly elliptic, acuminate at the apex, 1.4 x 0.6 mm.

Inflorescence axillary, 12 - 16 flowered; peduncles 5 - 10 cm long, sparsely covered with short (0.3 - 0.4 mm) white retrorse hairs; rachis 1.5 - 4.6 cm long, covered with short (0.2 - 0.3 mm) white appressed hairs. *Primary bract* small, 1.6 x 1.1 mm, broadly ovate, acute at the apex, rounded at the base, glabrous. *Secondary bract* narrowly ovate, acute at the apex, rounded at the base, 2.5 x 0.8 mm, ciliate with short (0.1 mm long) white hairs outside. *Bracteoles* broadly subulate, shorter than calyx, 2.6 x 1.0 mm, glabrous. *Pedicels* ascending, 2 - 3mm long, sparsely pubescent with short whitish hairs.

Flowers golden yellow. *Calyx* campanulate, 4.2 mm long, glabrous, tube 3.0 mm long. *Standard* asymmetrical, obliquely elliptic, 11.3 x 17.7 mm, emarginate at the apex, with a prominent appendage inside. *Right wing* half concealing the upper portion of the keel-petal, lamina broadly obovate, 11.5 x 17.7 mm, claws 1.8 mm long, auricle 1.6 mm long. *Left wing* spreading, supported by a pocket on the left-hand keel-petal, lamina broadly obovate, 13.1 x 11.6 mm, claws 2.3 mm long, auricle 1.5 mm long. *Keel-petals* incurved to the left through 270°, 18.5 mm long, pocket 6.6 mm long. *Ovary* 5.2 mm long, 10 ovuled; style c. 20 mm long, filiform; style-beak linear or flat, c. 0.7 mm long.

Fruit spreading, linear, cylindrical, 6 - 7 x 0.3 - 0.4 cm, glabrous, blackish brown when mature, c. 10 seeded. *Seed* elliptic, pale brown densely mottled with black spots, surface smooth, 4.0 x 2.5 x 1.9 mm; hilum narrowly elliptic, prominently protruding, aril well developed. Germination hypogeal. *First and second leaves* simple, narrowly elliptic, acute at the apex, cordate at the base, 1.8 x 0.6 cm, with petiole.

Geographic distribution: Southeast Asia – Cambodia, Indonesia, Laos, Myanmar, Philippines, Thailand, Vietnam; South Asia – northern India; East Asia – China including Taiwan; Papua New Guinea.

Altitude: Range 25 m (Thailand) to 2000 m (China), mean 547 m (51 records)

Phenology: Cambodia – October to November; China – September to October; Indonesia – April to June; Laos – October to November; Malaysia – January to February; Philippines – October to November; Thailand – October to November; Vietnam – October to December.

Ecology: Grows at rather shady place in dry deciduous forest, twining to the trees in northeast Thailand (Tomooka *et al.*, 2000e)

Pests and diseases: Wilted pods are sometimes observed (may be caused by stink bug) (Tomooka *et al.*, 2000e)

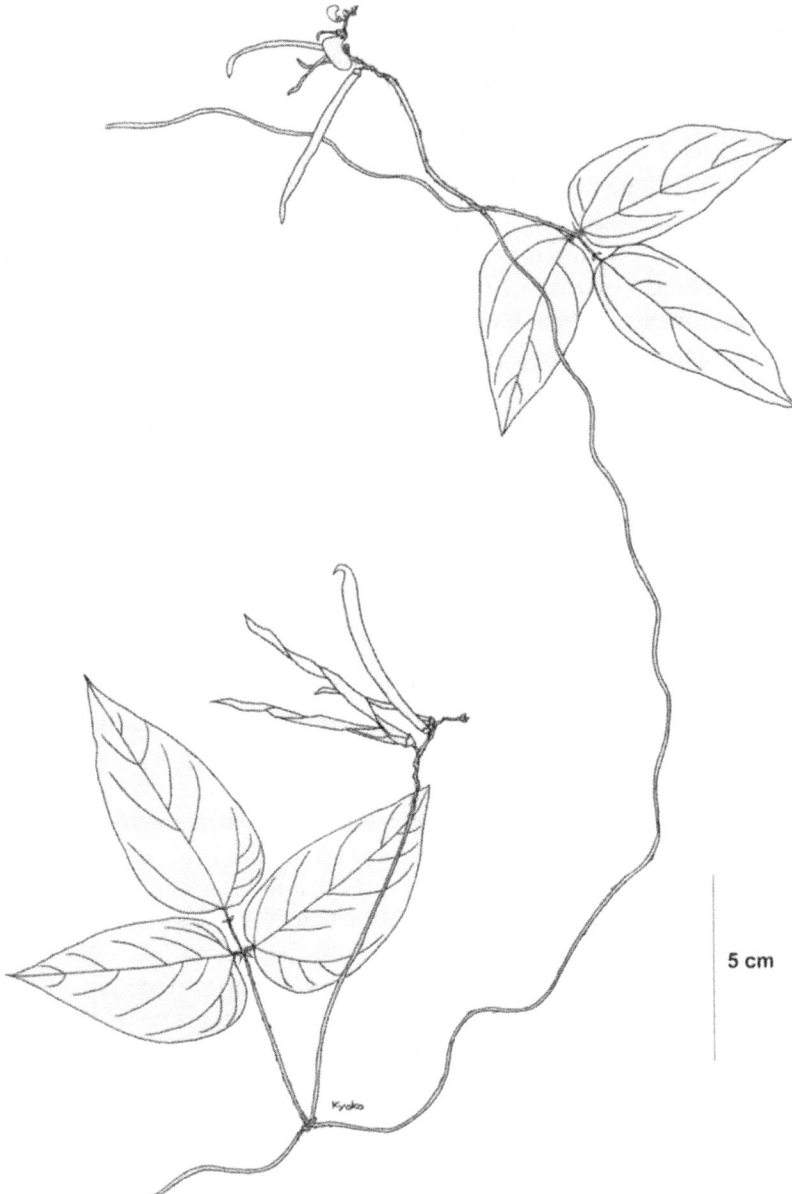

Figure 4.15. Flowering and fruiting shoot of Vigna minima.
Drawn from JP205886 (Thailand) by Kyoko Motoyoshi.

5 cm

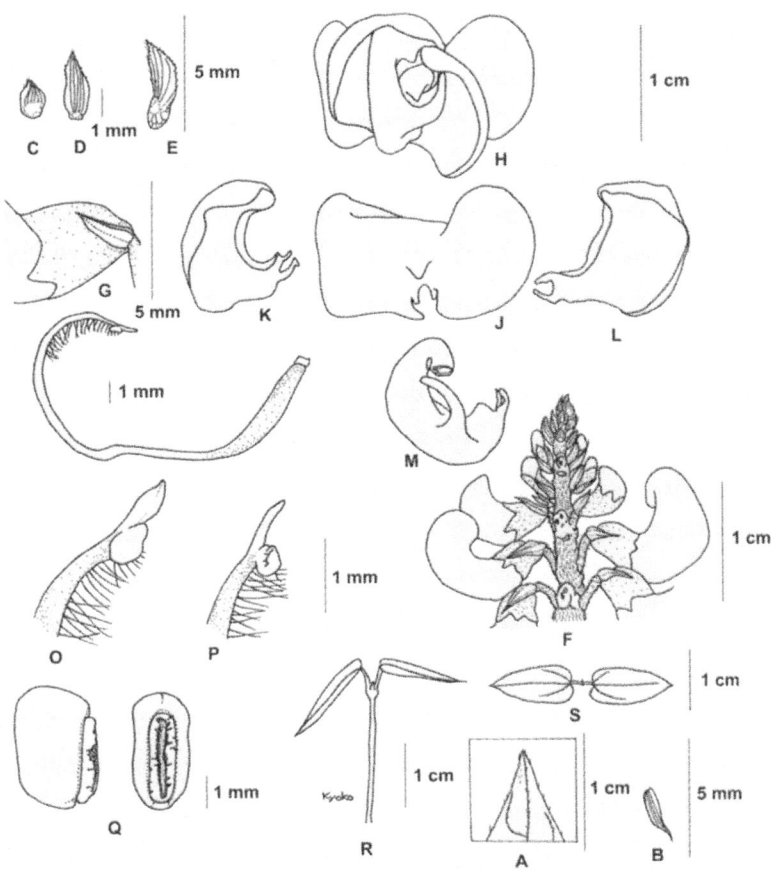

Figure 4.16. Plant parts of Vigna minima.

A: leaf apex; B: stipel; C: primary bract; D: secondary bract; E: stipule; F: inflorescence;
G: lower part of flower bud and bracteole; H: flower; J: standard; K: right wing; L: left
wing; M: keel; N:style; O & P: stigma and style beak; Q: seed; R & S: seedling shoot.
Drawn from JP205886 (Thailand), except A & B from JP210670 (Thailand), E from
JP205891 (Thailand), P from JP107869 (Thailand) by Kyoko Motoyoshi.

2.1.6. V. nakashimae *(Ohwi) Ohwi & Ohashi, J. Jap. Bot. 44: 30 (1969)*

Chromosome number: 2n=2x=22

Taxonomic affinities and diagnosis: *V. nakashimae* is most closely related to *V. riukiuensis* and *V. minima* within section *Angulares*. Useful characters to identify *V. nakashimae* are small stipule, small primary and secondary bracts; small bracteole; pale yellow flower; glabrous pod; smooth seed with much protruding hilum having well developed aril. Diagnostic character that distinguishes it from *V. riukiuensis* and *V. minima* is smaller pale yellow flower.

Description: Based on JP107879 from Japan

An annual twining herb. *Stems* slender, very sparsely covered with short (0.5 - 0.7 mm long) white hairs. *Stipules* peltate, rather rhombic, acute at the apex, obtuse at the base, small (4.6 x 2.2 mm), sparsely ciliate with short (c. 0.5 mm) white hairs outside. *Leaf petioles* 2.0 - 7.5 cm long, very rarely covered with short (c. 0.6 mm) pale brown hairs.

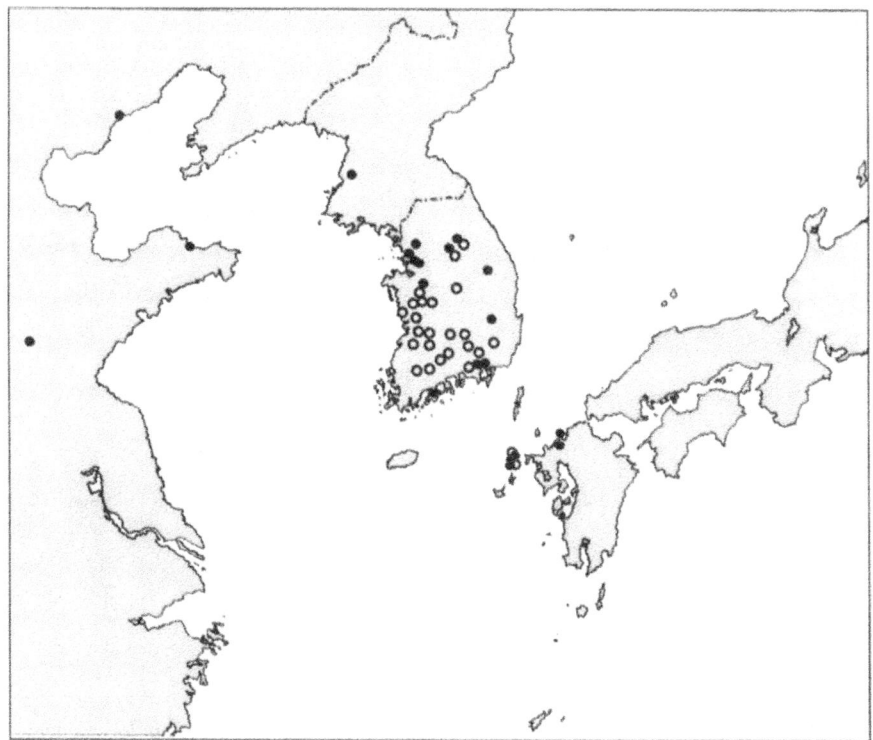

Figure 4.17. Distribution of Vigna nakashimae *based on herbarium specimens (●) and direct collection (○) made by author (N.T.) in Japan and by staff of RDA, Korea in Korea.*

Leaflets membranous, sparsely covered with pale brown hairs (0.5 - 0.7 mm long) on upper surface, ciliate with white hairs (c. 0.5 mm long) on veins of lower surface; *terminal leaflets* narrowly ovate, acute at the apex, obtuse at the base, 2.0 - 6.5 x 1.2 - 3.7 cm; *lateral leaflets* obliquely ovate, acuminate at the apex, rounded at the base, 1.5 -5.0 x 1.2 - 3.1 cm; stipels narrowly elliptic, acuminate at the apex, 2.1 x 0.8 mm.

Inflorescence axillary, c. 10 flowered; peduncles 1.3 cm long, almost glabrous; rachis c. 1 cm long, covered with short (0.2 - 0.3 mm) white hairs. *Primary bract* narrowly ovate, attenuate at the apex, rounded at the base, c. 1.5 x 0.6 mm, glabrous. *Secondary bract* narrowly triangular, attenuate at the apex, truncate at the base, c. 2.6 x 0.8 mm, covered with whitish hairs, c. 0.2 - 0.4 mm long. *Bracteoles* narrowly elliptic, shorter than calyx, 3.0 x 0.8 mm, covered with white hairs, 0.2 - 0.4 mm long. *Pedicels* ascending, c. 1.5 mm in flower, c. 2.2 mm in fruit, sparsely covered with white hairs, 0.2 - 0.3 mm long.

Flowers pale yellow. *Calyx* campanulate, c. 3.8 mm long, tube c. 2.3 mm long. *Standard* asymmetrical, obliquely elliptic, c. 9.4 x 11.7 mm, emarginate at the apex, with an appendage inside. *Right wing* concealing the upper portion of the keel-petals, lamina obliquely elliptic, c. 6.3 x 5.2 mm, claws 1.8 mm long, auricle 1.6 mm long. *Left wing* spreading, lamina obliquely ovate, 9.9 x 5.5 mm, claws c. 1.8 mm long, auricle c. 1.4 mm long. *Keel-petals* spirally incurved to the left, c. 13.2 mm long, the left-hand petal with a horn-like keel-pocket, the keel-pocket 3.8 mm long. *Ovary* c. 3.8 mm long, 7 ovules; style 12.9 mm long, style-beak linear, c. 1.0 mm long.

Fruits spreading, linear, 5 - 7 cm long, 0.4 - 0.5 cm wide, glabrous, blackish brown when mature, 5 - 8 seeded. *Seed* elliptic, pale brown densely mottled with black when mature, 4.4 x 2.9 x 2.6 mm, hilum linear and protruding, 2.8 x 0.8 mm, aril well developed. Germination hypogeal. *First and second leaves* narrowly cordate, 1.4 x 1.0 cm with petioles.

Geographic distribution: Korean peninsula, northeast China and eastern Kyushu and nearby islands, Japan.

Altitude: Range 5 - 100m, average 53 m (2 records)

Phenology: Korea - September to October; Japan – September to October.

Ecology: This species grows in habitats similar to *V. angularis* with which it is sometimes sympatric (Korea).

Vernacular names:

Japan: Hime tsuru azuki.

Pests and diseases: Resistant to brown stem rot (Kaga *et al.*, 2000a)

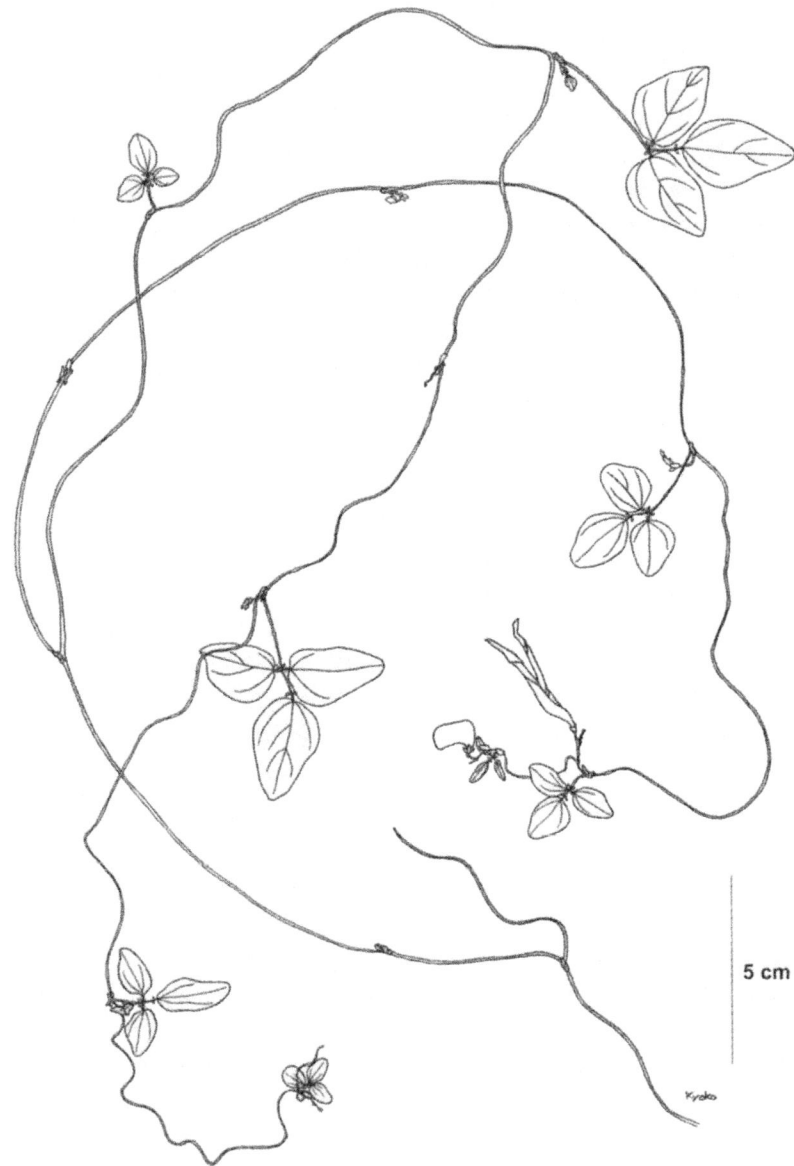

Figure 4.18. Flowering and fruiting shoot of Vigna nakashimae.
Drawn from JP110356 (Japan) by Kyoko Motoyoshi.

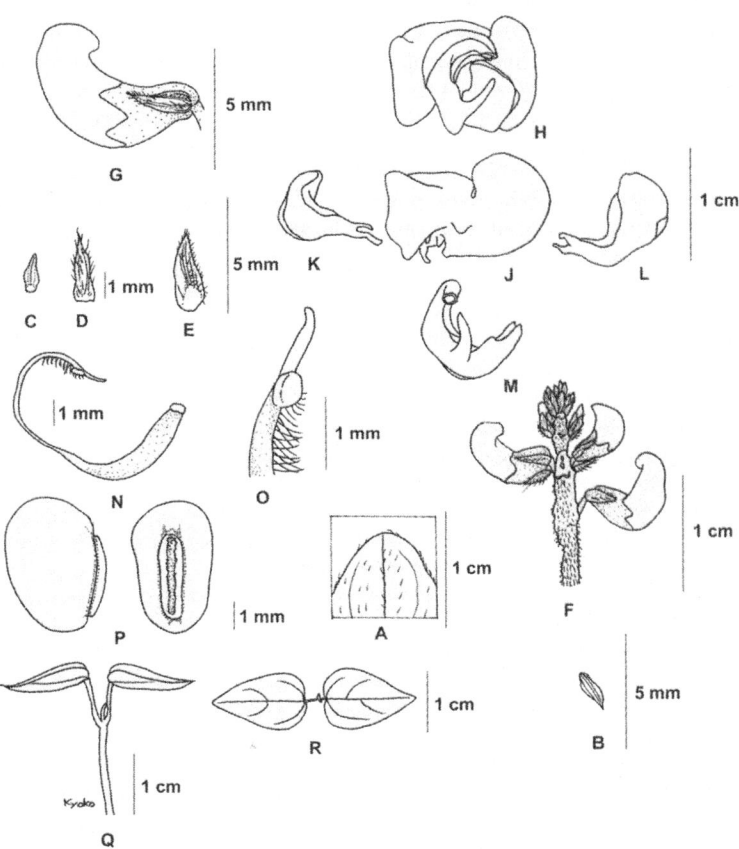

Figure 4.19. Plant parts of Vigna nakashimae.
A: leaf apex; B: stipel; C: primary bract; D: secondary bract; E: stipule; F: inflorescence;
G: flower bud and bracteole; H: flower; J: standard; K: right wing; L: left wing; M: keel;
N:style; O : stigma and style beak; P: seed; Q & R: seedling shoot. Drawn from JP107879
(Japan), except A & B from JP110356 (Japan) by Kyoko Motoyoshi.

2.1.7. V. nepalensis *Tateishi & Maxted, Kew Bull. 57(3):625-633 (2002)*

Chromosome number: 2n=2x=22

Taxonomic affinities and diagnosis: *V. nepalensis* belongs to a group of closely related species that includes *V. hirtella*, *V. tenuicaulis* and *V. angularis* within section *Angulares*. Useful characters to identify *V. nepalensis* are large glabrous secondary bract; glabrous pod; smooth seed with non-protruding linear hilum without aril development. Diagnostic characters that distinguish it from *V. hirtella* are non protruding hilum without aril development. Differs from *V. angularis* and *V. tenuicaulis* by glabrous shorter bracteole as long as calyx.

Description: Based on JP107881 from Nepal

A twining herb. *Stems* slender, sparsely covered with white hairs (0.6 - 0.8 mm long). *Stipules* peltate, narrowly elliptic, attenuate at the apex, rounded at the base, 8.1 x 2.7 mm, ciliate with spreading whitish hairs outside, hairs 0.6 - 0.7 mm long. *Leaf petioles* 3.8 - 13.2 cm long, ciliate with pale brown hairs (0.7 - 0.9 mm long) on the ridge. *Leaflets*

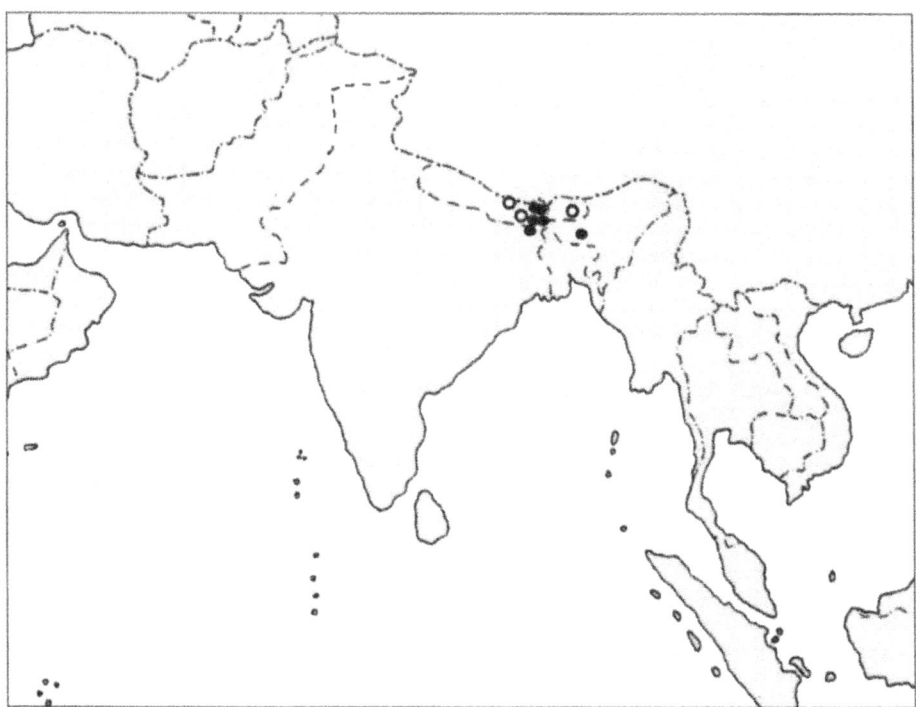

Figure 4.20. Distribution of Vigna nepalensis *based on herbarium specimens (●) and direct collection (○).*

sparsely covered with white hairs (0.5 - 0.6 mm long) on upper surface, ciliate with white hairs (0.5 - 0.7 mm long) on veins of lower surface; *terminal leaflets* ovate, acuminate at the apex, obtuse to rounded at the base, 8.2 - 11.9 x 4.9 - 8.7 cm; *lateral leaflets* obliquely ovate, acuminate at the apex, rounded at the base, 6.6 - 11.0 x 4.0 - 7.9 cm; stipels ovate, acuminate at the apex, rounded at the base, 2.5 x 1.0 mm.

Inflorescence axillary, 10 - 20 flowered; peduncles 4 - 5 cm long, very sparsely covered with pale brown hairs, 0.6 - 0.8 mm long; rachis c. 1 cm long, glabrous. *Primary bract* orbiculate to ovate, 4.5 x 4.1 mm, glabrous. *Secondary bract* (pedicel bract) ovate, 7.3 x 4.7 mm, glabrous. *Bracteoles* narrowly triangular, as long as calyx, 4.4 x 1.5 mm, glabrous. *Pedicels* ascending, 2.9 mm long in flower, 2.5 - 4.3 mm in fruit, glabrous.

Flowers golden yellow. *Calyx* campanulate, 4.4 mm long, glabrous, tube 3.6 mm long. *Standard* asymmetrical, obliquely elliptic, 12.7 x 19.9 mm, emarginate at the apex, with a prominent appendage at the centre of the inner surface. *Right wing* obovate, concealing the upper portion of keel-petals, lamina 10.6 x 10.9 mm, claws 1.9 mm long, auricle 1.8 mm long. *Left wing* obliquely elliptic, lamina 13.3 x 9.7 mm, claws 2.4 mm long, auricle 2.3 mm long. *Keel-petals* incurved to the left, 18.2 mm long, the left petal with a long pocket, the keel-pocket 5.2 mm long. *Ovary* 5.0 mm long; style 16.8 mm long, style-beak linear, 0.9 mm long.

Fruits pendulous, 7.9 - 8.6 cm long, 0.3 - 0.5 cm wide, glabrous, blackish when mature, 12 - 17 seeded. *Seed* oblong, pale brown densely mottled with black when mature, surface smooth, 3.5 x 2.2 x 1.8 mm, hilum linear, 1.6 x 0.4 mm, not protruding, aril not developed. Germination hypogeal. *First and second leaves* cordate, 1.2 x 1.1 cm with petioles.

Geographic distribution: Eastern Nepal and adjacent parts of India and Bhutan.

Altitude: Range 350 - 1650 m, mean 913 m (10 records) (Tateishi, 1983; Murata *et al.*, 1995).

Phenology: India – September to October.

Associated vegetation: *Cucumis sativus, Momordica charantia, Eupatorium, Artemisia, Bidens, Cyperus, Castanopis,* thickets and margin of *Shorea robusta* forest (Tateishi, 1983).

Habitat: Grows in disturbed habitats beside paths, around rice fields, forest margins or rocky areas near rivers.

Vernacular name: Nepali: Mas

5 cm

Figure 4.21. Flowering and fruiting shoot of Vigna nepalensis.
Drawnn from JP107881 (Nepal) by Kyoko Motoyoshi.

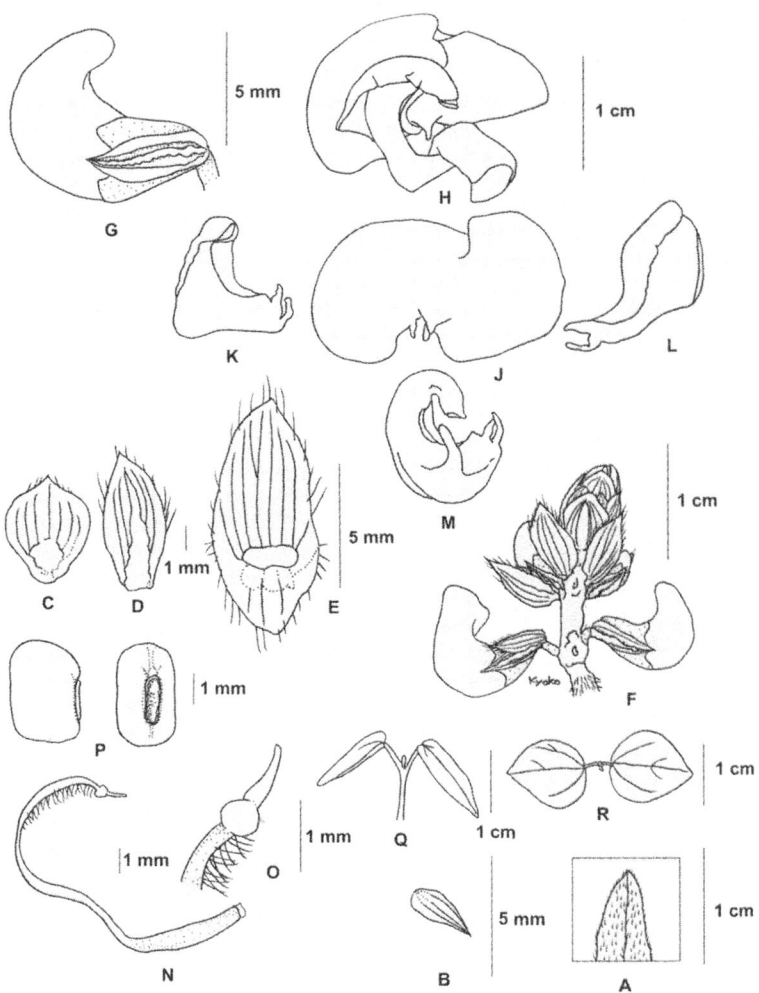

Figure 4.22. Plant parts of Vigna nepalensis.
A: leaf apex; B: stipel; C: primary bract; D: secondary bract; E: stipule; F: inflorescence;
G: flower bud and bracteole; H: flower; J: standard; K: right wing; L: left wing; M: keel;
N:style; O : stigma and style beak; P: seed; Q & R: seedling shoot. Drawn from JP107881
(Nepal) by Kyoko Motoyoshi.

2.1.8. V. reflexo-pilosa *Hayata, J. Coll. Sci. Imp. Univ. Tokyo 30: 82 (1911)*

Chromosome number: 2n=4x=44

Taxonomic affinities and diagnosis: This species is a tetraploid species within section *Angulares*. One genome donor to this species is *V. trinervia,* the other genome donor is most likely *V. hirtella* or closely related species. This species has been confused with *V. radiata* var. *sublobata.* Useful characters to identify *V. reflexo-pilosa* are pod covered with short brown hairs (cultivated form, var. *glabra,* has glabrous pod); smooth oblong truncate seed with non-protruding hilum without aril development. Diagnostic characters which distinguished it from *V. radiata* var. *sublobata* are much larger golden yellow flower; smooth seed surface. Differs from *V. trinervia* by smooth seed surface.

Description: Based on wild form var. *reflexo-pilosa*, JP108815 from Japan

Wild form - A twining herb. *Stems* densely covered with long (1.6 - 1.9 mm) retrorse white hairs. *Stipules* peltate, oblong, 11.3 x 5.6 mm, acute at the apex, obtuse at the base, covered with long whitish hairs, 0.9 - 1.4 mm long. *Leaf petioles* with petiole 3.9 - 18.4 cm long, covered with long whitish hairs, 0.9 - 1.1 mm long. *Leaflets* densely covered with

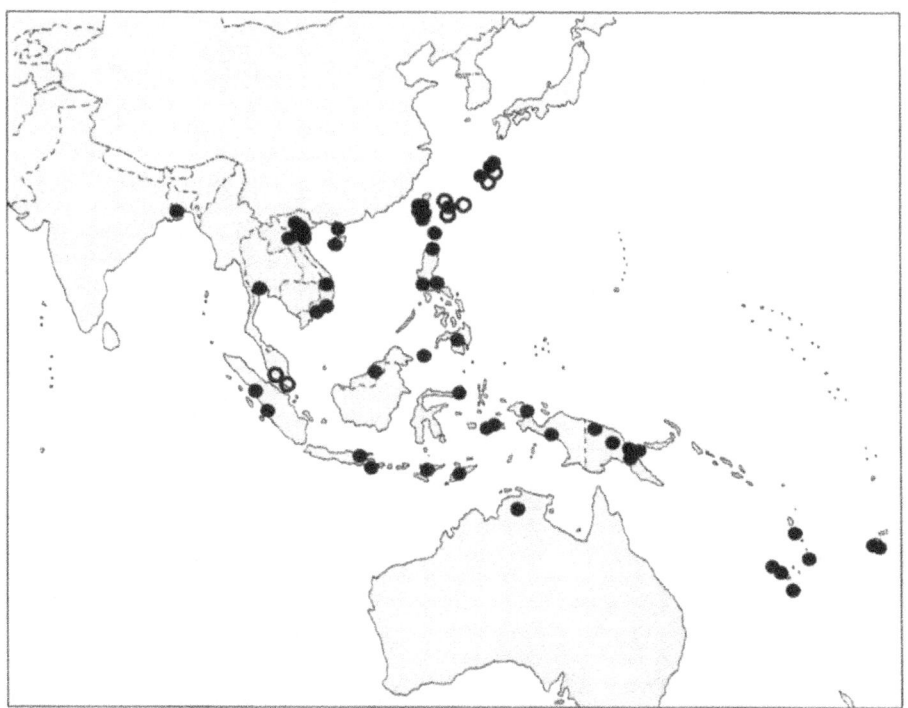

Figure 4.23. Distribution of Vigna reflexo-pilosa *var.* reflexo-pilosa *based on herbarium specimens (●) and direct germplasm collection (○) by authors (N.T., D.A.V.).*

white hairs on upper surface, 0.4 - 0.7 mm long, ciliate with white hairs (0.6 - 0.7 mm long) on veins and sparsely covered with white hairs (0.4 - 0.5 mm long) on lower surface; *terminal leaflets* ovate to narrowly ovate, sometimes faintly 3 lobed, acuminate at the apex, obtuse at the base, 6.3 - 11.1 x 3.9 - 9.6 cm; *lateral leaflets* obliquely ovate, acuminate at the apex, rounded at the base, 5.2 - 10.6 x 3.0 - 8.9 cm; stipels narrowly ovate, acuminate at the apex, 3.5 x 1.7 mm long.

Inflorescence axillary, 10 - 20 flowered; peduncles 14 cm long, densely covered with appressed retrorse white hairs (0.7 - 0.8 mm long); rachis 1.2 - 3.8 cm long, covered with short (0.3 mm) white hairs. *Primary bract* ovate to widely ovate, acute at the apex, rounded at the base, 3.5 x 3.4 mm, glabrous. *Secondary bract* (pedicel bract) elliptic, acute at the apex, obtuse at the base, 4.1 x 2.8 mm, ciliate with white hairs (0.2 - 0.4 mm long) on the margin. *Bracteoles* ovate, acute at the apex, obtuse at the base, as long as calyx, 4.3 x 2.4 mm, ciliate with short whitish hairs (0.2 - 0.4 mm) on the margin. *Pedicels* ascending, 1.3 - 1.5 mm long in flower, 2.7 - 4.4 mm long in fruit.

Flowers golden yellow. *Calyx* campanulate, 4.4 mm long, tube 3.5 mm long, glabrous.

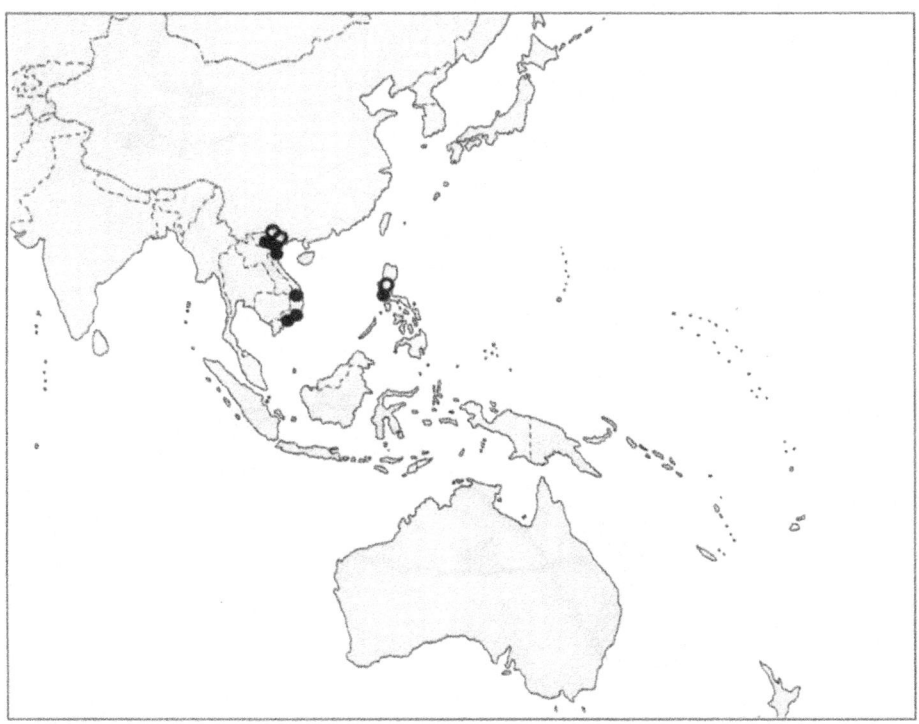

Figure 4.24. Distribution of Vigna reflexo-pilosa *var.* glabra *based on herbarium specimens (●) and direct germplasm collection (○).*

Figure 4.25. Flowering and fruiting shoot of Vigna reflexo-pilosa *var.* reflexo-pilosa.
Drawn from JP108815 (Japan) by Kyoko Motoyoshi.

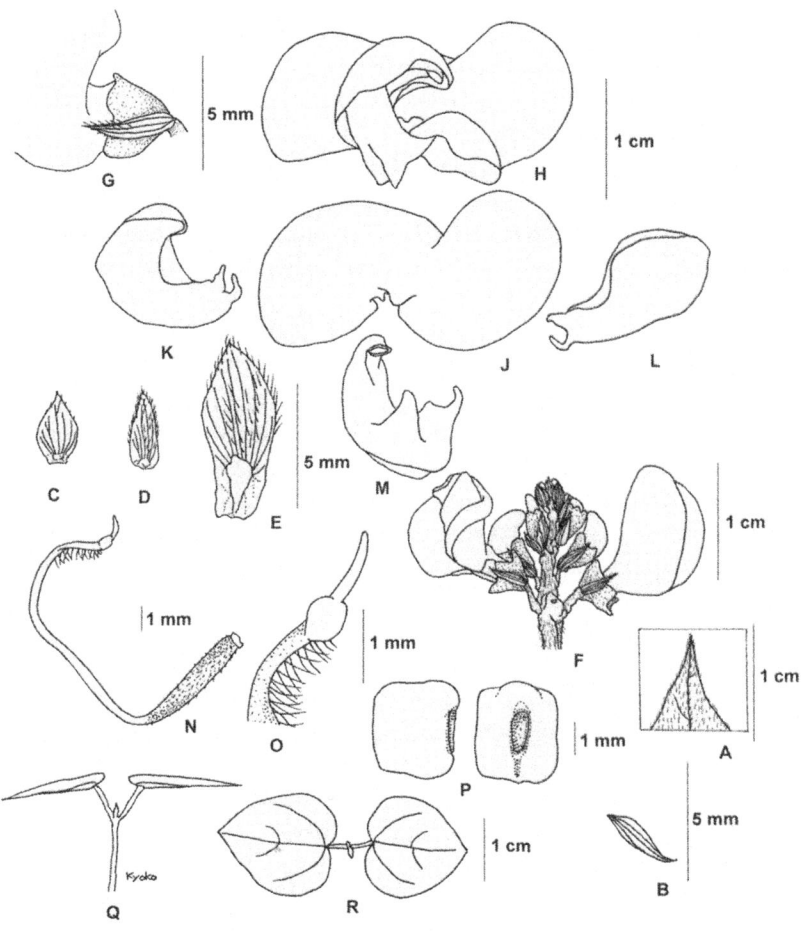

Figure 4.26. Plant parts of Vigna reflexo-pilosa *var.* reflexo-pilosa.
A: leaf apex; B: stipel; C: primary bract; D: secondary bract; E: stipule; F: inflorescence; G: lower part of flower bud and bracteole; H: flower; J: standard; K: right wing; L: left wing; M: keel; N:style; O : stigma and style beak; P: seed; Q & R: seedling shoot. Drawn from JP108815 (Japan), except F from CED01011 (Japan) by Kyoko Motoyoshi.

Standard asymmetrical, obliquely elliptic, 9.9 x 14.0 mm, emarginate at the apex, with an appendage at the centre inside. *Right wing* half concealing the upper portion of the keel-petals, lamina obliquely obovate, 14.8 x 8.1 mm, claws 2.6 mm long, auricle 1.5 mm long. *Left wing* spreading, lamina obovate, 13.4 x 8.2 mm, claws 2.4 mm long, auricle 1.4 mm long. *Keel-petals* spirally incurved to the left, 18.1 mm long, left keel with a horn-like pocket, the keel pocket 4.1 mm long. *Ovary* 5.3 mm long, ovules 12 - 15; style 16.8 mm long, style-beak linear, 0.9 mm long.

Fruits spreading, 5 - 6 cm long, blackish and pubescent with short (0.1 - 0.2 mm) brown hairs when mature, 10 - 14 seeded. *Seed* oblong, brown densely mottled with black when mature, surface smooth, 3.1 x 2.2 x 2.1 mm, hilum narrowly elliptic, 1.6 x 0.5 mm, not protruding, aril not developed. Germination hypogeal. *First and second leaves* petiolate, cordate, 1.7 x 1.6 cm.

Distribution: var. *reflexo-pilosa*: wild form - widely distributed across the islands of the western Pacific islands from New Caledonia, Tonga and Vanuatu in the southeast to Okinawa, Japan in the north. East Asia - southern China and Taiwan; Southeast Asia - Cambodia, Indonesia, Laos, Thailand and Vietnam; South Asia - Bangladesh. It is also reported from Papua New Guinea and northern Australia.

var. *glabra*: cultivated form – confirmed reports from Vietnam and the Philippines.

Altitude: Wild form: Range 3 - 1700 m, average 401 m (55 records). Altitude highest in the tropics. Average altitude in Japan 61 m (12 records). Average altitude in Taiwan 341 m (29 records).

var. *reflexo-pilosa*: wild form

Phenology: Japan (in Okinawa) – January to February; Indonesia – March to June; Malaysia – January to February; Papua New Guinea – July to August; Philippines – July to August.

Ecology: Grows in disturbed habitat such as abandoned field and roadside (Japan: Tomooka *et al.*, 2000d, Malaysia: Tomooka *et al.*, 1993)

Vernacular names:

Japanese: oo-yabu-tsuru-azuki

var. *glabra* (Maréchal, Mascherpa & Stainier) N. Tomooka & Maxted, Kew Bull. 57(3):613-624 (2002a): cultivated form

This cultivated variety has thick glabrous erect stem.

This species was formerly considered as a glabrous variety of mungbean, *V. radiata* var. *glabra* (Roxb.) Verdc. (Verdcourt, 1970). Maréchal *et al.* (1978) treated this as a distinct species and named *V. glabrescens* Maréchal, Mascherpa & Stainier. Tomooka et al. (2002a) proposed this taxon as a cultivated variety of *V. reflexo-pilosa* and named it *V. reflexo-pilosa* var. *glabra*.

Vernacular names:

Vietnamese: dau moi

Mauritius: lentille de Créole

Uses: In West Bengal *V. reflexo-pilosa* var. *glabra* has been used as a forage crop and is cultivated under the name "Lentille de Créole" in Mauritius (Baudoin and Maréchal, 1988). In Vietnam, it is used in the same way as mung bean (Kobayashi *et al.*, 1994). At AVRDC (Taiwan) attempts have been made to use this species as an insect resistance gene source for the mungbean breeding program (Fernandez and Shanmugasundar, 1988). Reported to be cultivated in Mlingano, Tanzania (Verdcourt, 1970). Babu *et al.* (1985) report that it is often cultivated in West Bengal and Orissa in India.

2.1.9. V. riukiuensis *(Ohwi) Ohwi & Ohashi, J. Jap. Bot. 44: 31 (1969)*

Chromosome number: 2n=2x=22

Taxonomic affinities and diagnosis: *V. riukiuensis* is most closely related to *V. nakashimae* and *V. minima* within section *Angulares*. Useful characters to identify *V. riukiuensis* are small stipule, small primary and secondary bracts; small bracteole; small glossy leaflets with obtuse or rounded apex; glabrous pod; elliptic smooth seed with slightly protruding hilum having developed aril. Diagnostic character that distinguishes it from *V. nakashimae* is larger golden yellow flower. Differs from *V. minima* by smaller glossy leaflets; shorter pod.

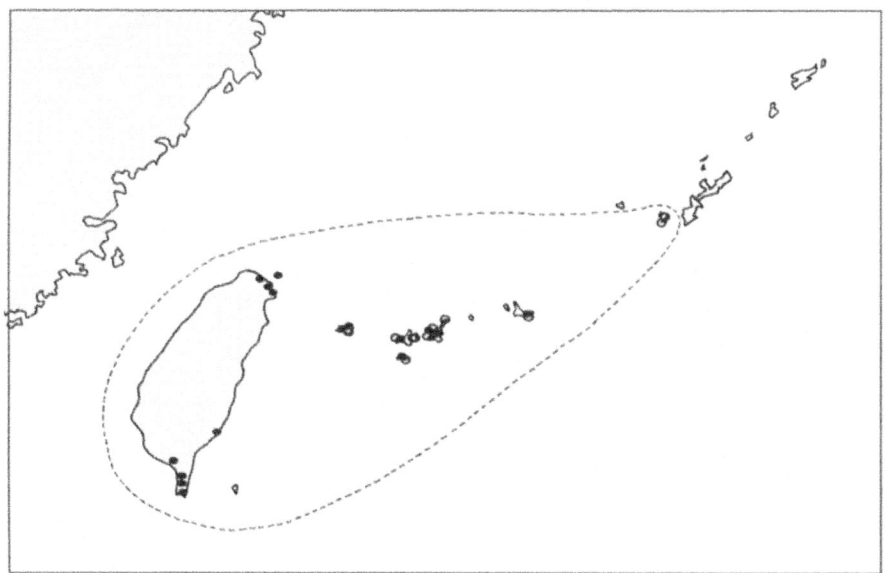

Figure 4.27. Distribution of Vigna riukiuensis *based on herbarium specimens (●) and direct germplasm collection (○) by authors (N.T., D.A.V.).*

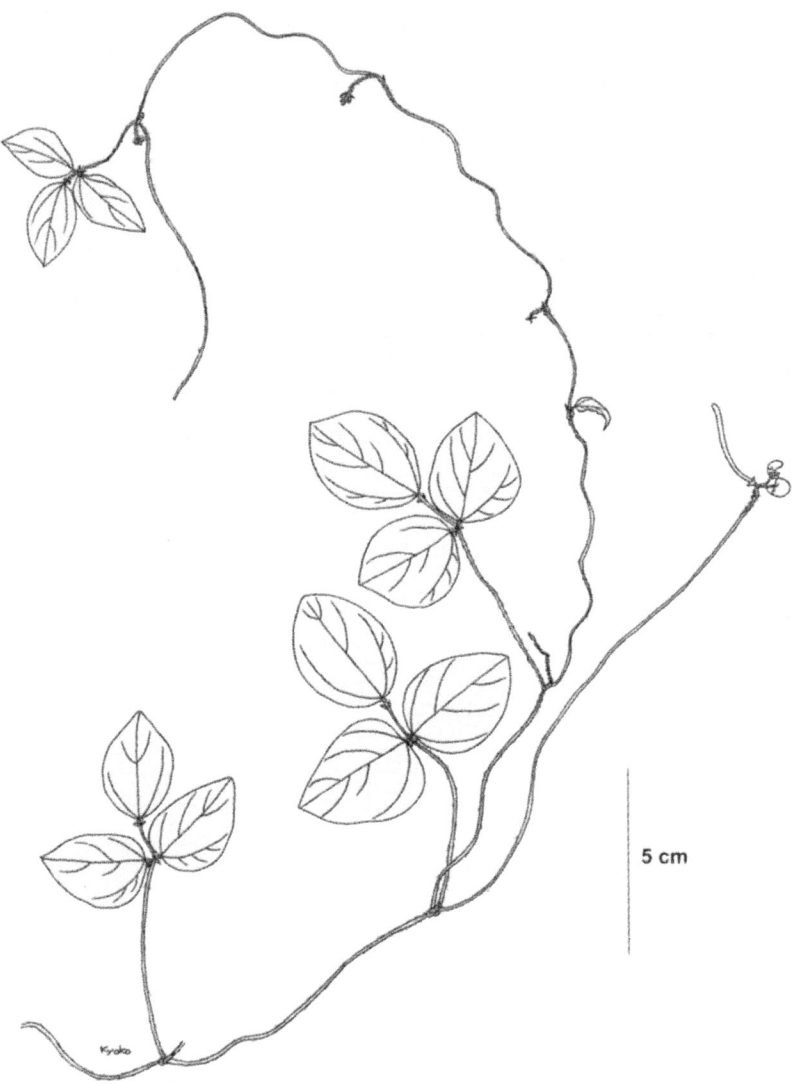

Figure 4.28. Flowering and fruiting shoot of Vigna riukiuensis.
Drawn from JP108810 (Japan) by Kyoko Motoyoshi.

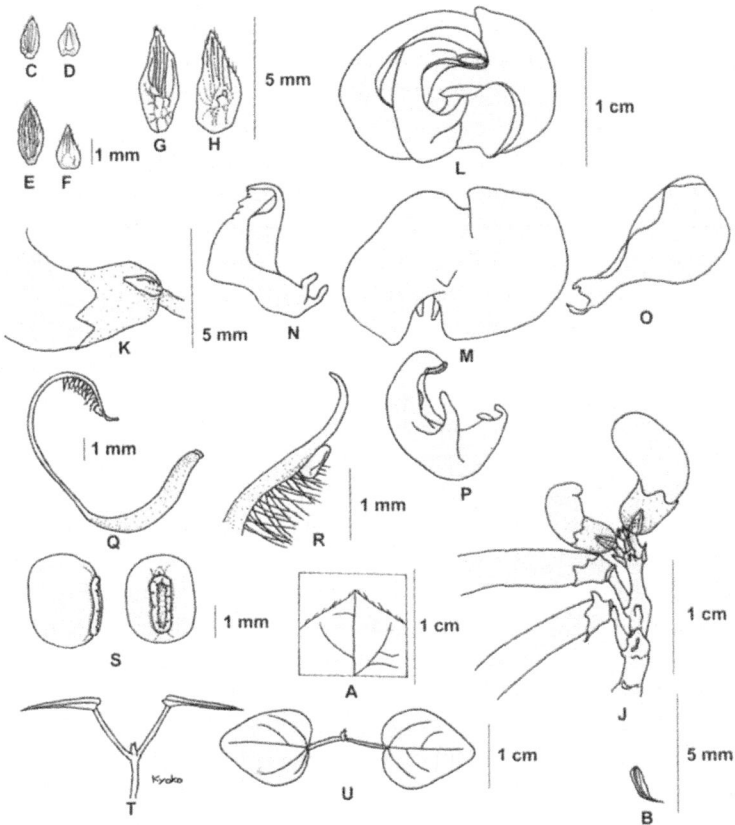

Figure 4.29. Plant parts of Vigna riukiuensis.
A: leaf apex; B: stipel; C & D: primary bract; E & F: secondary bract; G & H: stipule; J: inflorescence; K: lower part of flower bud and bracteole; L: flower; M: standard; N: right wing; O: left wing; P: keel; Q:style; R : stigma and style beak; S: seed; T & U: seedling shoot. Drawn from JP108810 (Japan), except B & H from JP201477 (Japan), A, C, E & G from JP108804 (Japan) by Kyoko Motoyoshi.

Description: Based on JP108810 from Japan

A trailing or twining herb. *Stems* sparsely covered with pale brown hairs (0.5 - 0.7 mm long). *Stipules* peltate, ovate, acute or acuminate at the apex, obtuse or rounded at the base, small, 4.3 x 1.7 mm, ciliate with pale brown hairs (0.3 - 0.8 mm) on the margin. *Leaf petioles* 4.8 - 5.4 cm long, sparsely covered with pale brown hairs (0.9 mm long). *Leaflets* glossy, upper surface glabrous, ciliate with short (0.3 mm) pale brown hairs on the margin, sparsely covered with pale brown hairs (0.4 - 0.5 mm) on the lower surface; *terminal leaflets* ovate, acute or obtuse at the apex, obtuse at the base, 3.0 - 3.5 x 1.9 - 2.5 cm; *lateral leaflets* obliquely ovate, acute at the apex, rounded at the base, 2.6 - 3.3 x 1.7 - 2.6 cm; stipels widely ovate, acuminate at the apex, 2.1 x 0.8 mm.

Inflorescence axillary, 4 - 10 flowered; peduncles 14.8 cm long, very sparsely covered with white hairs (0.4 - 0.5 mm long); rachis 8.5 - 9.1 mm long, glabrous. *Primary bracts* ovate, acute at the apex, rounded at the base, 1.1 x 0.7 mm, glabrous. *Secondary bracts* (pedicel bract) broadly ovate, acuminate at the apex, rounded at the base, 1.7 x 1.0 mm, glabrous. *Bracteoles* triangular-ovate, acute at the apex, obtuse or rounded at the base, small, shorter than calyx, 1.7 x 0.6 mm, glabrous. *Pedicels* ascending, 1.8 mm long in flower, 2.8 - 3.3 mm long in fruit, glabrous.

Flowers golden yellow. *Calyx* campanulate, 3.7 mm long, tube 2.9 mm long, glabrous. *Standard* asymmetrical, obliquely elliptic, 12.6 x 18.0 mm, emarginate at the apex, with a prominent appendage at the centre inside. *Right wing* half concealing the upper portion of the keel-petals, lamina obliquely obovate, 11.0 x 9.7 mm, claws 2.0 mm long, auricle 1.7 mm long. *Left wing* spreading forward, supported by a pocket on left keel-petal, lamina obliquely obovate, 13.1 x 7.7 mm, claws 2.4 mm long, auricle 0.7 mm long. *Keel-petals* spirally incurved to the left, 19.0 mm long, left keel-petal with a long (5.7 mm long) horn-like pocket. *Ovary* 5.1 mm long, glabrous, ovules 5 - 7; style filiform, 19.3 mm long, long-beaked beyond stigma, the style-beak linear, 1.1 mm long.

Fruit spreading, linear, 3.0 - 4.5 cm long, 0.3 - 0.5 cm wide, glabrous, blackish brown when mature, 5 - 7 seeded. *Seed* elliptic, pale brown mottled with black when mature, surface smooth, 2.4 x 1.8 x 2.2 mm; hilum narrowly elliptic, slightly protruding, 1.8 x 0.6 mm, aril developed. Germination hypogeal. *First and second leaves* petiolate, simple, cordate, 1.3 x 1.1 cm.

Geographic distribution: Restricted to the islands of southern Okinawa prefecture, Japan and Taiwan, China.

Altitude: Range from sea level to 150 m, average 42 m (18 records)

Phenology: Japan (Okinawa) – all year round.

Ecology:

Habitat: Growing on seaside cliffs or roadside, open sunny places. Sometimes sympatric with *V. reflexo-pilosa*.

Vernacular names:

Japanese: hina azuki.

2.1.10. V. tenuicaulis *N. Tomooka & Maxted, Kew Bull. 57(3):613-624 (2002a)*
Chromosome number: 2n=2x=22
Taxonomic affinities and diagnosis: *V. tenuicaulis* belongs to a group of closely related species that includes *V. nepalensis* and *V. angularis.* Useful characters to identify *V. tenuicaulis* are small flower and seed; glabrous pod; smooth oblong seed with non-protruding short linear hilum without aril development. Diagnostic characters that distinguish it from *V. nepalensis* are hairy bracteole longer than calyx; smaller plant parts. Differs from *V. angularis* by smaller plant parts, especially smaller secondary bract.
Description: Based on JP108552 from Thailand
A twining herb. *Stems* slender, sparsely pubescent with long (1 - 1.4 mm) rather retrorse white or yellowish hairs. *Stipules* peltate, prolonged below the point of insertion, narrowly elliptic to narrowly ovate, acute at the apex, 4 - 6 x 1.2 - 1.6 mm, 4 - 5 nerved,

Figure 4.30. Distribution of Vigna tenuicaulis *based on
direct collection (O) by author (N.T.).*

Figure 4.31. Flowering and fruiting shoot of Vigna tenuicaulis.
Drawn from JP108552 (Thailand) by Kyoko Motoyoshi.

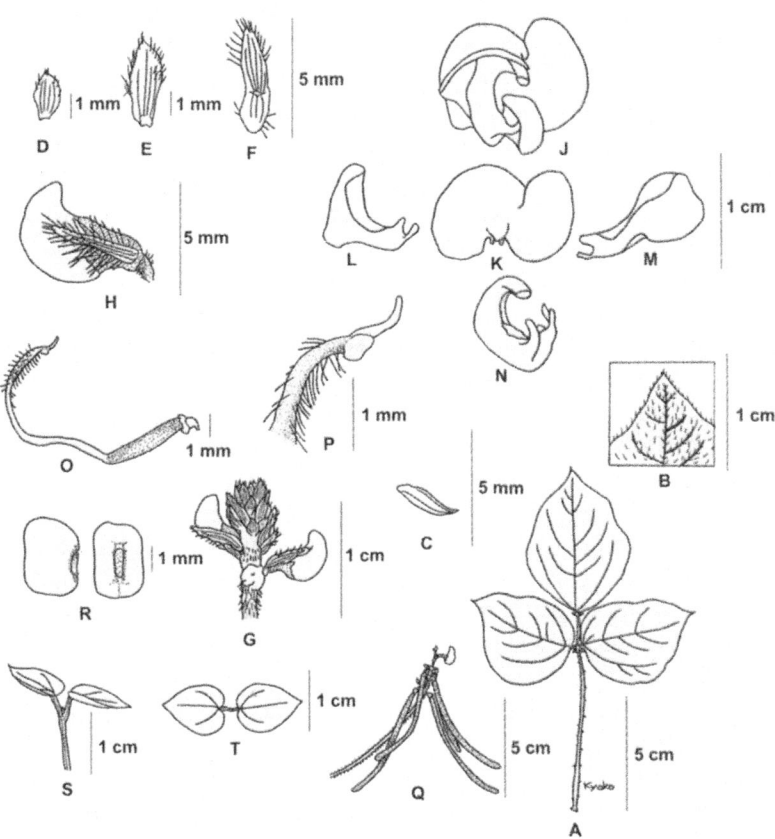

Figure 4.32. Plant parts of Vigna tenuicaulis.
A: leaf; B: leaf apex; C: stipel; D: primary bract; E: secondary bract; F: stipule; G: inflorescence; H: flower bud and bracteole; J: flower; K: standard; L: right wing; M: left wing; N: keel; O:style; P : stigma and style beak; Q: pods; R: seed; S & T: seedling shoot. Drawn from JP108552 (Thailand) by Kyoko Motoyoshi.

sparsely spreading pale brownish hairy (0.5 - 1.3 mm long) outside. *Leaf petioles* 1.5 - 10 cm long, sparsely to rather densely covered with retrorse to spreading long pale brownish hairs (0.5 - 1.7 mm long), rachis 0.5 - 1 cm long, pulvinus 1.7 - 2 mm long, spreading short whitish to pale brownish hairy (0.2 - 0.5 mm long). *Leaflets* membranous, sparsely to rather densely covered with long (0.5 - 1.5 mm) spreading whitish to yellowish hairs on both surfaces; *terminal leaflets* narrowly ovate to ovate, entire, acuminate at the apex, obtuse or rounded at the base, 3.5 - 5 x 1.7 - 3.2 cm; *lateral leaflets* obliquely ovate to obliquely narrowly ovate, acuminate at the apex, rounded at the base, 2.5 - 4.3 x 1.7 - 2.9 cm; stipels narrowly elliptic, acuminate at the apex, 1.5 - 2.8 mm long.

Inflorescence axillary, 7 - 9 flowered; peduncles 3 - 12 cm long, densely covered with retrorse or spreading long brownish hairs (0.8 - 1.5 mm long); rachis 6 - 12 mm long, sparsely to rather densely clothed with whitish or pale brownish spreading to retrorse hairs (0.5 - 1 mm long). *Primary bract* peltate, ovate to narrowly ovate, obtuse or acute at the apex, 1.6 - 1.9 x 0.7 - 1 mm, surface glabrous, sparsely ciliate with whitish or pale brownish hairs (0.2 - 0.6 mm long) outside, caducous. *Secondary bract* (pedicel bract) basifixed, oblong or somewhat cymbiform, obtuse at the apex, 3.5 - 4.3 x 1.2 - 1.8 mm, surface nearly glabrous, ciliate with long whitish or pale brownish long hairs (0.4 - 0.8 mm long) outside, caducous. *Bracteoles* basifixed, subulate or narrowly elliptic, acuminate at the apex, 3.8 - 4.6 mm long, 0.5 - 0.9 mm wide, longer than calyx, ciliate with long spreading whitish hairs (0.5 - 1.8 mm long) especially on the margin. *Pedicels* ascending, 1 - 3 mm long in flower, 3 - 4 mm long in fruit, glabrous or sparsely short whitish hairy (0.2 - 0.3 mm long).

Flowers golden yellow, 12 - 13 mm in diameter. *Calyx* campanulate, 2.5 - 3 mm long, glabrous, tube 1.5 - 2.1 mm long. *Standard* asymmetrical, obliquely broadly elliptic, 9.2 - 9.8 x 12.3 - 12.9 mm, emarginate at the apex, with an appendage (c. 1 mm long) at the centre inside. *Right wing* half concealing the upper portion of the keel-petals, claws 1.6 - 2 mm long, lamina obliquely elliptic, 6.8 - 8.7 x 6.7 - 7.1 mm, auricle 1.2 - 1.4 mm long. *Left wing* spreading horizontally, supported by a pocket on left keel-petal, claws 1.7 - 3.2 mm long, lamina obliquely elliptic, 7 - 9.1 x 6.1 - 9 mm, auricle 1 - 1.2 mm long. *Keel-petals* spirally incurved through 260°, the left-hand petal with a long horn-like pocket outside, the keel-pocket 3.3 - 3.9 mm long. *Ovary* 3.6 - 6.5 mm long, glabrous, ovules 9 - 15; style 12.2 - 17.3 mm long, filiform, long-beaked beyond stigma, the style-beak linear, 0.7 - 0.8 mm long.

Fruits spreading to rather pendulous, linear, cylindrical, 3.5 - 6.3 x 0.3 cm, glabrous, dark brown when mature, 9 - 15 seeded. *Seed* oblong, grayish brown variegated with black when mature, smooth, 2.4 - 2.8 x 1.8 - 2.1 x 1.6 - 1.7 mm, hilum linear, c. 1 x 0.4 mm, not protruding, aril not developed. Germination hypogeal. *First and second leaves* long-petiolate, simple, cordate, acute at the apex, 0.8 - 1 mm long, 0.7 - 0.8 mm wide.

Geographical distribution: Myanmar, Thailand

Altitude: Range 560 - 1365 m (average 833 m 4 records) in Thailand

Phenology: Thailand – October to November.

Ecology:
Associated vegetation: Grasses. Found growing together with *V. minima* and *V. hirtella*.
Habitat: Usually growing in full sun or light shade. Usually in highly disturbed habitats
such as roadside verges that may be subject to repeated cutting.
Habit: Twining plant. It also can grow trailing under heavy weeding condition. Stems
can root from nodes.

2.1.11. V. trinervia *(Heyne ex Wall.) Tateishi & Maxted, Kew Bull.57(3):625-633*
(2002)

Chromosome number: 2n=2x=22

Taxonomic affinities and diagnosis: *V. trinervia* is considered as one of the genome
donors to the tetraploid *V. reflexo-pilosa*. While placed within the section *Angulares*
this species is on the boundary between this section and section *Ceratotropis* having
characteristics of both sections. This species has been confused with *V. radiata* var.
sublobata. Useful characters to identify *V. trinervia* are long peduncle covered with
brown hispid hairs; spreading pods covered with brown hispid hairs; rough (with
mesh-like reticulation) rectangular seed with non-protruding hilum without aril
development. Diagnostic characters which distinguish it from *V. reflexo-pilosa* are longer

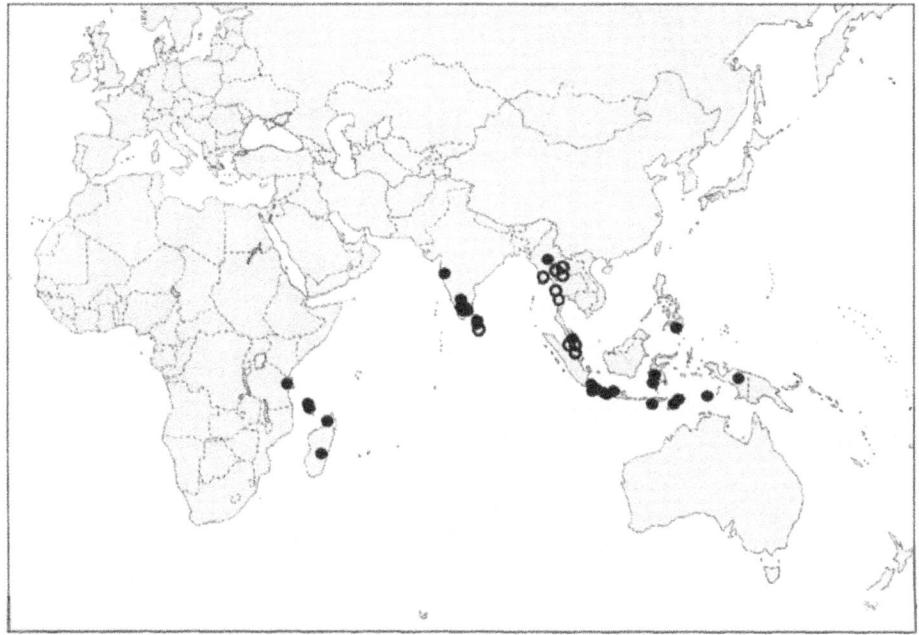

Figure 4.33 Distribution of Vigna trinervia var. trinervia *based on herbarium specimens*
(●) and direct collection (○) by author (N.T.).

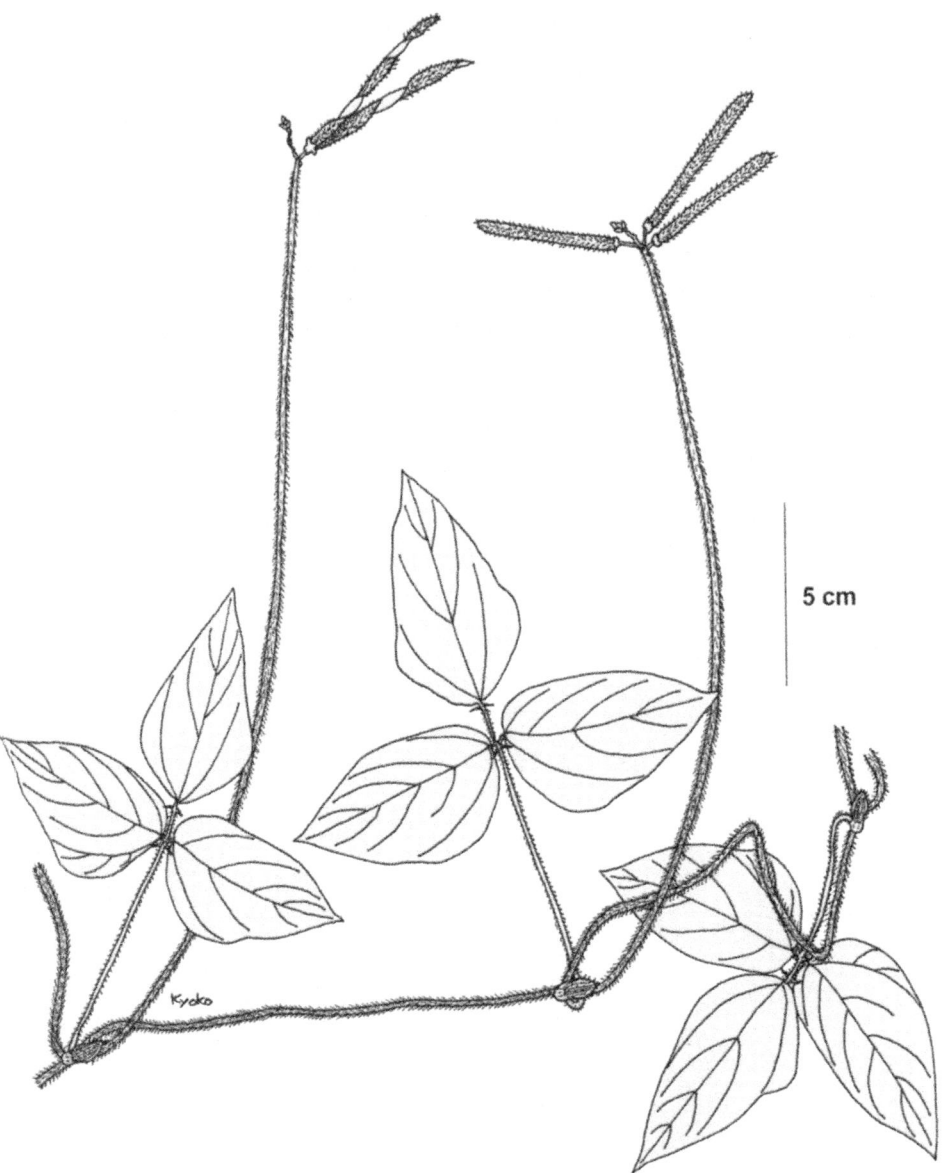

5 cm

Figure 4.34. Flowering and fruiting shoot of Vigna trinervia *var.* trinervia.
Drawn from JP207980 (Sri Lanka) by Kyoko Motoyoshi.

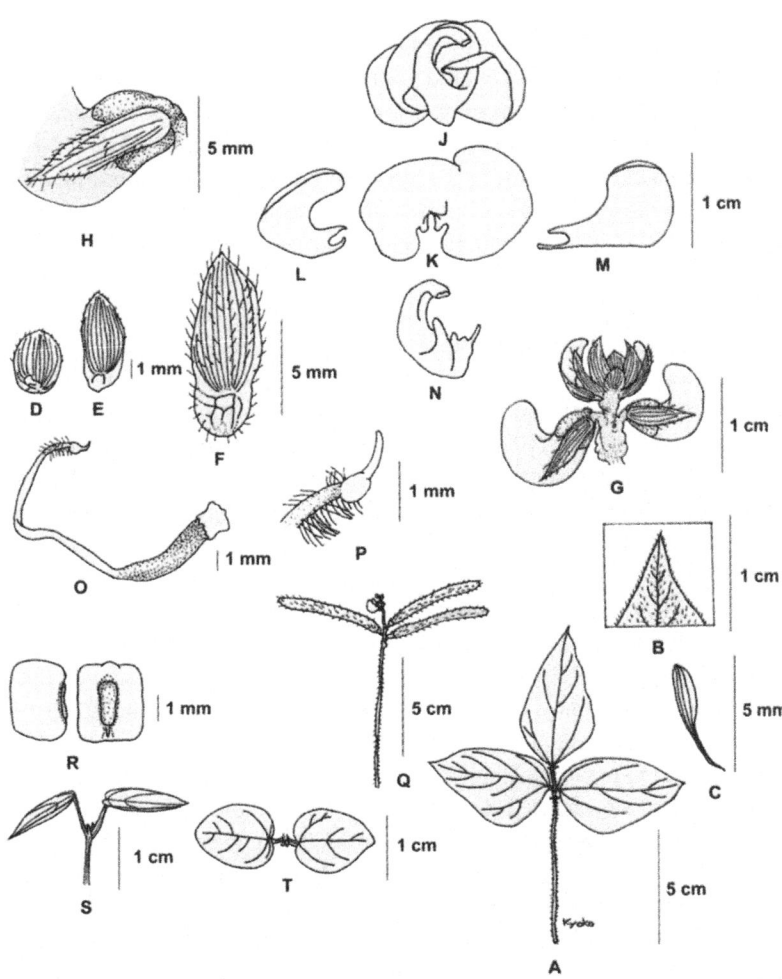

Figure 4.35. Plant parts of Vigna trinervia *var.* trinervia.
A: leaf; B: leaf apex; C: stipel; D: primary bract; E: secondary bract; F: stipule; G: inflorescence; H: lower part of flower bud and bracteole; J: flower; K: standard; L: right wing; M: left wing; N: keel; O: style; P : stigma and style beak; Q: pods; R: seed; S & T: seedling shoot. Drawn from JP108840 (Malaysia), except Q from 2001SL30D (Sri Lanka) by Kyoko Motoyoshi.

bracteole compared with calyx; rough seed surface. Differs from *V. radiata* var. *sublobata* by longer bracteole compared with calyx; golden yellow flower; hypogeal germination; cordate first and second leaves with petiole.

var. *trinervia*

Description: Based on JP108840 from Malaysia

A twining herb. *Stems* densely covered with long (1.7 - 2.1 mm) retrorse brown hairs. *Stipules* long elliptic, acute at the apex, rounded at the base, 9.8 x 3.6 mm, ciliate with pale brown hairs on the margin, 0.4 - 0.6 mm long. *Leaf petioles* 3.8 - 11.8 cm long, densely covered with very long (1.3 - 2.5 mm) brown retrorse hairs. *Leaflets* rather densely covered with short (0.4 - 0.5 mm) white hairs on upper surface, sparsely covered with white hairs on lower surface (0.3 - 0.8 mm long); *terminal leaflets* narrowly ovate or rather rhomboid, acute at the apex, acute or obtuse at the base, 6.1 - 7.5 x 3.4 - 7.0 cm; *lateral leaflets* obliquely ovate, acute at the apex, obtuse at the base, 5.4 - 7.1 x 3.2 - 5.4 cm; stipels long, narrowly elliptic, attenuate at the apex, 4.4 x 0.8 mm.

Inflorescence axillary, 4 - 14 flowered; peduncles very long, 22.0 - 26.7 cm long, sparsely covered with very short (0.2 - 0.3 mm) brown hispid retrorse hairs; rachis 8.1 mm long, covered with short (0.1 - 0.2 mm) brown hairs. *Primary bract* round or rather ovate, 3.1 x 2.3 mm, glabrous on the surface. *Secondary bract* (pedicel bract) elliptic, acute at the apex, obtuse at the base, 4.8 x 2.1 mm, glabrous on the surface. *Bracteoles* elliptic, attenuate at the apex, truncate at the base, longer than calyx, 7.8 x 1.7 mm, ciliate with long (0.6 - 1.1 mm) pale brown hairs especially on the margin. *Pedicels* ascending, 1.3 mm long in flower, 2.1 - 6.5 mm long in fruit.

Flowers golden yellow. *Calyx* campanulate, 4.8 mm long, tube 3.4 mm long, glabrous. *Standard* asymmetrical, obliquely elliptic, 11.0 x 17.5 mm, emarginate at the apex, with a prominent appendage centre inside. *Right wing* half concealing the upper portion of the keel-petals, lamina obliquely obovate, 9.2 x 10.2 mm, claws 1.5 mm long, auricle 0.7 mm long. *Left wing* spreading, supported by a keel pocket, lamina obovate, 11.7 x 8.6 mm, claws 2.3 mm long, auricle 1.3 mm long. Keel-petals spirally incurved to the left, 17.5 mm long, left keel petal with a long horn-like pocket, the pocket 4.2 mm long. *Ovary* 6.7 mm long; style 19.2 mm long, prolonged beyond stigma to make style-beak, the style-beak linear, 0.8 mm long.

Fruit spreading or rather upwards, linear, 5 - 6 cm long, 0.3 - 0.5 cm wide, covered densely with long (0.5 - 0.7 mm) hispid brown hairs, blackish when mature, 10 - 14 seeded. *Seed* rectangular, surface rough, 3.2 x 2.1 x 2.6 mm; hilum linear, 1.8 x 0.6 mm, not protruding, aril not developed. Germination hypogeal. *First and second leaves* simple, petiolate, leaflets cordate, 1.2 x 1.0 cm.

Geographic distribution: Tanzania, the Comoros islands and Madagascar, southern India, Sri Lanka, Myanmar, Thailand, the southern Philippines, Indonesia as far east as Irian Jaya.

Altitude: Range 35 - 1630 m; mean 872 m (22 records). This species appears to grow at higher elevations in Sri Lanka. In Sri Lanka (6-7°N) the average elevation of populations is 1188 m (11 records) whereas in Thailand (12-20°N) the average elevation is 388 m (5 records).

Phenology: India – September to November; Indonesia – March to May; Malaysia – January to February; Sri Lanka – January to February; Thailand - October to November.

Ecology:

Soil: Red clay or sandy.

Associated vegetation: Grasses.

Habitat: Usually in light or medium shade. Disturbance intermediate. In Sri Lanka this species has been found on steep rocky slopes and often very wet habitats of highland. In Thailand and Malaysia, it grows in lowland wet habitat.

Pests and diseases: In both Sri Lanka and Malaysia stinkbug is a reported pest resulting in shriveled seeds. Pods often exhibit insect bites. Flower buds observed with eggs of *Lepidoptera* species. In Malaysia flowers may be covered with ants that seem to deter stink bugs. Plants observed with mottle possibly virus infection.

Vernacular names:

Thai: tua pee

Uses: Cover crop for young rubber plantations – this species has been used as a cover crop for a long time, according to Thai farmers. The seeds of *V. trinervia* are expensive 50-60 Baht/kg (about US$1:20 in 2000) in the market in Thailand. Suppresses weed growth (Tomooka *et al.*, 2000e)

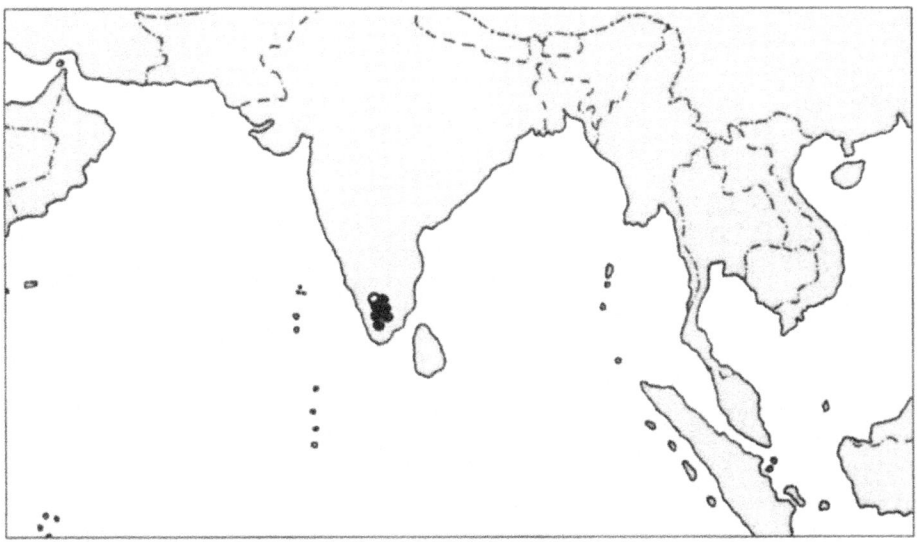

Figure 4.36. Distribution of Vigna trinervia *var.* bourneae *based on herbarium specimens (●) and direct collection by Dr. P. Saravanakumar, University of Perideniya, Sri Lanka (O).*

Figure 4.37. Flowering and fruiting shoot of Vigna trinervia *var.* bourneae.
Drawn from JP207981 (India) by Kyoko Motoyoshi.

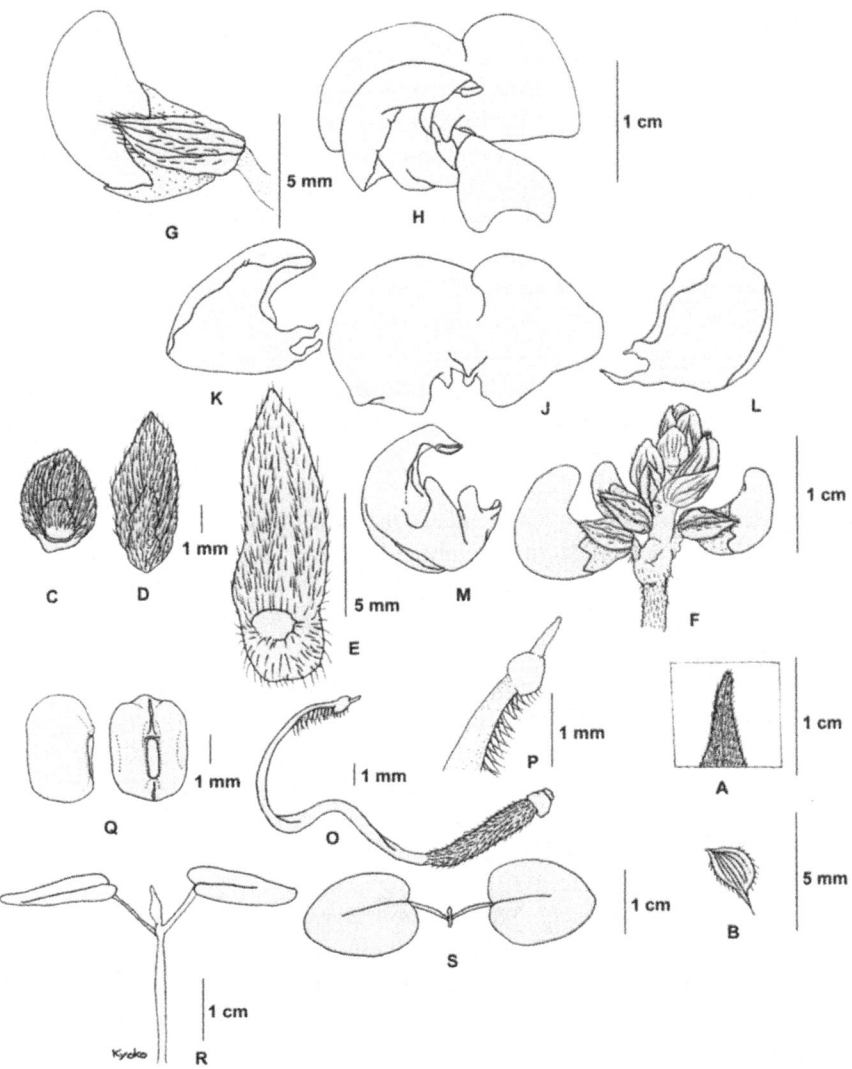

Figure 4.38. Plant parts of Vigna trinervia *var.* bourneae.
A: leaf apex; B: stipel; C: primary bract; D: secondary bract; E: stipule; F: inflorescence;
G: flower bud and bracteole; H: flower; J: standard; K: right wing; L: left wing; M: keel;
O:style; P: stigma and style beak; Q: seed; R & S: seedling shoot. Drawn from JP207981
(India) by Kyoko Motoyoshi.

var. *bourneae* (Gamble) Tateishi & Maxted, Kew Bull. 57(3):625-633 (2002)

Description: Based on JP207981 from India

Overall plants morphology is similar to *V. trinervia* var. *trinervia*. Only differentiating characters are described here. *Stems* heavily covered with rather shorter (0.7 - 1.0 mm) brown villose hairs. *Stipule* surface heavily covered with white villose hairs, 0.5 mm long. *Leaf petioles* heavily covered with white retrorse hairs, 0.5 - 0.8 mm long. *Leaflets*, upper surface densely covered with rather short white hairs (0.2 - 0.4 mm long), lower surface densely covered with longer white hairs, 0.5 - 0.6 mm long. *Peduncles* covered with longer white and brown hairs, 0.5 - 0.8 mm long. *Flower-rachis* glabrous. *Primary bract* larger and densely pubescent, 4.2 x 3.1 mm. *Secondary bract* larger and more hairy, 6.6 x 2.7 mm. *Bracteole* wider, 6.6 x 2.7 mm. *Flowers* golden yellow and all flower parts and petals larger. *Calyx*, 4.8 mm long, tube 3.4 mm long. *Standard* 13.0 x 21.0 mm. *Right wing* lamina 12 x 14 mm, claws 3 mm long, auricle 3 mm long. *Left wing* lamina 12 x 12 mm, claws 4.4 mm long, auricle 1.7 mm long. *Keel-petals* 21 mm long, the keel pocket 5.5 mm long. *Style* longer, 25 mm long, style beak 0.7 mm long. *Seed* size, 3.5 x 2.2 x 2.6 mm, hilum smaller, 1.4 x 0.4 mm.

Geographic distribution: Southern India in the western Ghats.

Phenology: India – September to December.

2.1.12. V. umbellata *(Thunb.) Ohwi & Ohashi, J. Jap. Bot. 44: 31 (1969)*

Chromosome number: 2n=2x=22

Taxonomic affinities and diagnosis: *V. umbellata* is closely related to *V. minima* and *V. hirtella* in the section *Angulares*. Useful characters to identify *V. umbellata* are very fine pale brown hairs on stem; glabrous mature pod, smooth elliptic seed with protruding hilum having well developed aril; lanceolate first and second leaves. Diagnostic characters that distinguish it from *V. hirtella* are more protruding hilum and well developed aril. Differs from *V. minima* by having longer stipule and bracteole.

Description: Based on wild form JP109675 from Thailand

A twining herb. *Stems* sparsely to densely covered with retrorse pale brown very fine hairs, 1.0 - 1.1 mm long. *Stipules* peltate, narrowly elliptic to rather falcate, acuminate at the apex, rounded at the base, covered with fine hairs on the surface and margin. *Leaf petioles* 4.3 - 9.2 cm long, very sparsely covered with pale brown hairs, 0.6 mm long. *Leaflets* sparsely to densely covered with pale brown fine hairs on both surfaces, 0.2 - 0.8 mm long; *terminal leaflets* narrowly ovate to ovate, acuminate at the apex, obtuse at the base, sometimes faintly 3 lobed, 6.2 - 7.7 x 3.7 - 4.3 cm; *lateral leaflets* obliquely ovate, acuminate at the apex, rounded at the base, sometimes faintly 3 lobed, 5.4 - 6.4 x 3.7 - 4.2 cm; stipels linear, attenuate at the apex, 2.7 x 0.6 mm.

Inflorescence axillary, 10 - 30 flowered; peduncles 13.7 - 17.6 cm long, sparsely covered with short (0.3 mm) white hairs; rachis 30 mm long, covered with short (0.1 - 0.2 mm) white hairs. *Primary bract* narrowly elliptic, attenuate at the apex, truncate at the base, 3.0 x 1.0 mm, glabrous. *Secondary bract* (pedicel bract) narrowly elliptic, attenuate at the apex, rounded at the base, 3.8 x 0.7 mm, ciliate with short (0.1 - 0.2 mm) white hairs on the margin. *Bracteoles* subulate, attenuate at the apex, rounded or truncate at the base, as long as or a little longer than calyx, 6.2 x 0.7 mm, ciliate with short (0.1 - 0.2 mm) white hairs. *Pedicels* ascending, 3.6 mm long in flower, 3.9 - 7.7 mm long in fruit.

Flowers golden yellow. *Calyx* campanulate, 4.4 mm long, tube 3.0 mm long. *Standard* asymmetrical, obliquely elliptic, 11.9 x 17.7 mm, emarginate at the apex, with a prominent appendage at the centre inside. *Right wing* concealing the upper portion of keel-petals, lamina obliquely obovate, 12.7 x 10.4 mm, claws 2.2 mm long, auricle 1.6 mm long. *Left wing* spreading forward, supported by a keel-pocket on the left keel-petal, lamina obovate, 12.4 x 8.0 mm, claws 2.4 mm long, auricle 1.7 mm long. *Keel-petals* spirally incurved to the left, 19.0 mm long, left keel with a horn-like pocket, the keel-pocket 5.5 mm long. *Ovary* 4.9 mm long; style 16.8 mm long, long beaked beyond stigma, the style-beak linear, 0.7 mm long.

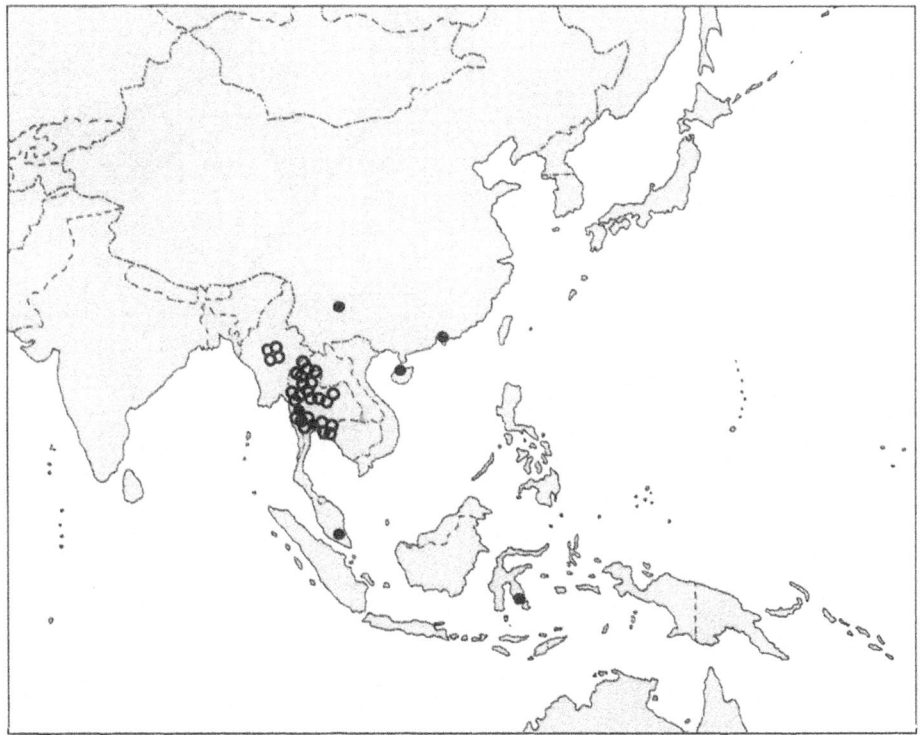

Figure 4.39. Distribution of wild Vigna umbellata *based on herbarium specimens (●) and direct collection (○) by author (N.T.).*

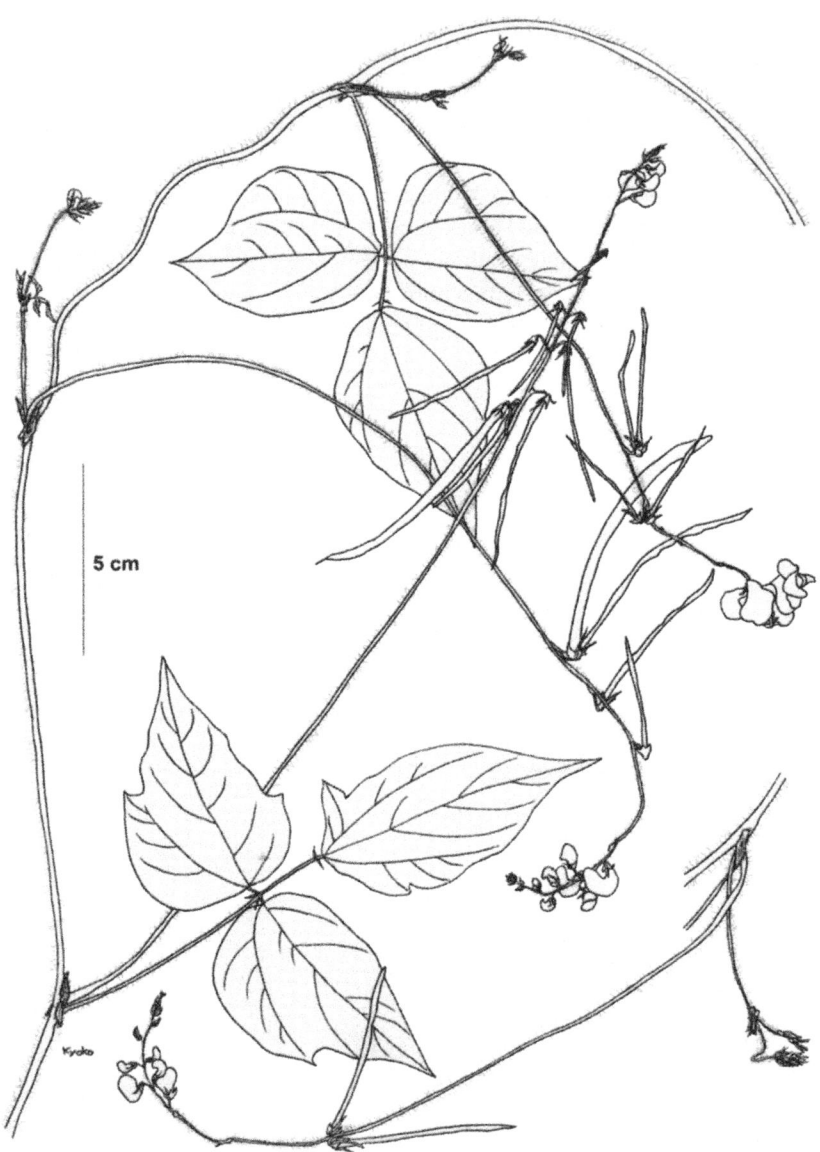

Figure 4.40 Flowering and fruiting shoot of Vigna umbellata *(wild).*
Drawn from JP207982 (Thailand) by Kyoko Motoyoshi.

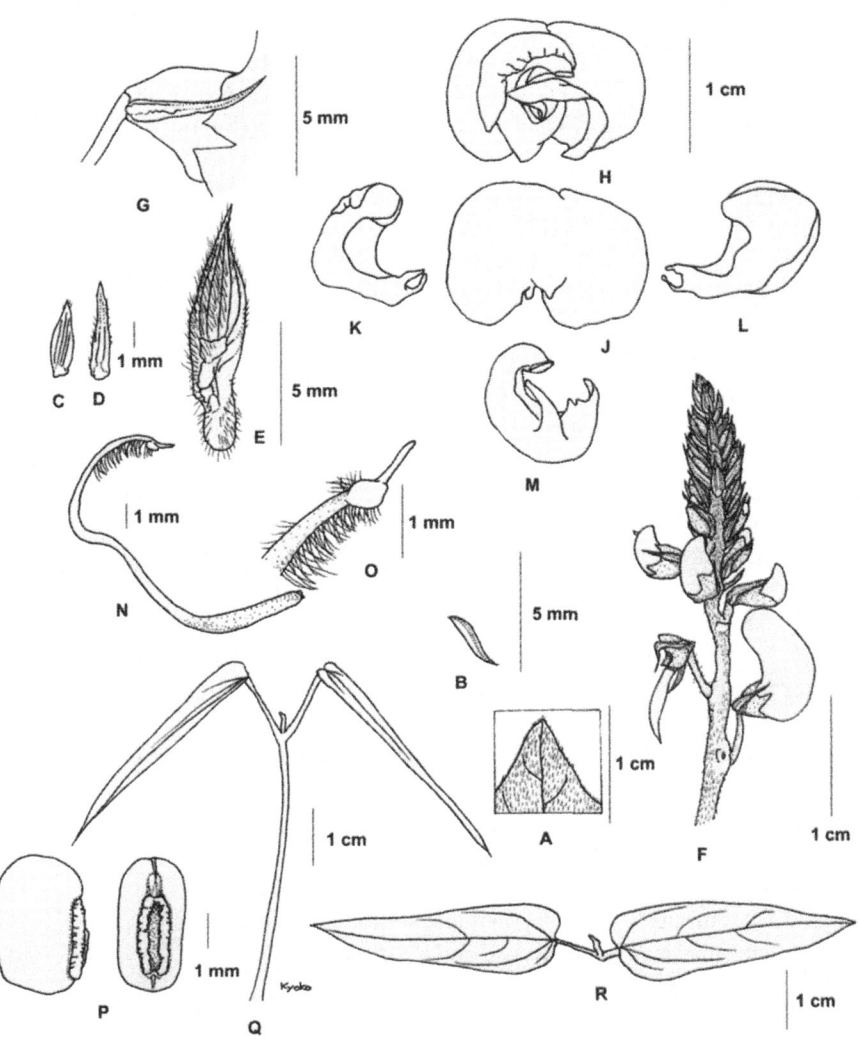

Figure 4.41. Plant parts of Vigna umbellata.

A: leaf apex; B: stipel; C: primary bract; D: secondary bract; E: stipule; F: inflorescence; G: lower part of flower bud and bracteole; H: flower; J: standard; K: right wing; L: left wing; M: keel; N:style; O : stigma and style beak; P: seed; Q & R: seedling shoot. Drawn from JP109675 (Thailand), except A & E from JP207982 (Thailand) by Kyoko Motoyoshi.

Fruits pendulous, linear, 4.6 - 7.1 x 0.2 - 0.4 cm, blackish brown when mature, glabrous, 6 - 10 seeded. *Seed* elliptic, 4.4 x 2.6 x 2.2 mm, surface smooth; hilum elliptic, 2.5 x 1.0 mm, protruding, aril well developed. Germination hypogeal. *First and second leaves* petiolate, leaflet lanceolate, 3.8 x 1.2 cm.

Geographic distribution: Cultigen is widely grown across tropical Asia as far north as southern Japan. It is also reported to be cultivated in Fiji, Mauritius and East Africa. Wild form is found in Myanmar, Thailand, Malaysia, Indonesia (Sulawesi) and southeast China.

Altitude: Cultigen is grown between 150 m and 1850 m, mean 550 m (8 records). Wild form is grown between 25 m and 1680 m, mean 438 m (65 records)

Uses: Cultigen: In Chiang Rai market, northern Thailand, immature pods are briefly boiled with salt and sold. Apparently, this is a traditional way of eating rice bean among the hill tribes of the region. Dried seeds are boiled and eaten with rice. Seeds are also used to make sweets. (Takeya and Tomooka, 1997).

The dried pulse is used for human consumption in India, Myanmar, Malaysia, China, Fiji, Mauritius and the Philippines. The beans are usually boiled and used as a vegetable. The whole plant is used as a fodder. It has been tried as a green manure and cover crop (Purseglove, 1974).

Wild form: Suppresses weed growth; cows eat foliage (Tomooka *et al.*, 2000e). Leaf boiled and used as uretic medicine, also said to be good for liver in Myanmar (Tomooka, based on interview).

Cultigen

Phenology: Japan – September to October; Thailand, Myanmar – October to November.

Vernacular names:

Cambodian: sandaek angkat mieh, sandaek riech mieh

Chinese: mu-tsa (Shanghai), crab-eye or lazy-man pea (Soochow), and climbing mountain bean (Yachow)

Cuba: little devil or mambi bean

English: rice bean

French: haricot riz

Hindi: ghurush, gurounsh, gurush, pau maia, shiltong, sita-mas, sutari

India: sutri, sita-mas, pau maia, gurush and gurounsh

Indonesian: kacang uci

Japanese: tsuru azuki, menaga (long eye) on Tsushima, sasage or kani no me (crab eye) on Goto island, Nagasaki prefecture, and Yamaguchi prefecture, baka azuki (fools azuki) in Kagoshima prefecture, kage azuki (shade azuki) in Tottori prefecture

Laotian: thoua land tek (in Luang Prabang), (khua 'mak) 'thoua pa (Xieng Khouang)

Malaysia: katjang sepalit

Myanmar: pe-yin

Nepali: masyang

Philippines: anilei, anipai, kapilan, pagsei

Thai: thua daeng (red bean), thua ae, ma pee

Vietnamese: dau gao (Phu Khanh)

Wild form

Phenology: Myanmar, Thailand – October to November.

Ecology: Grows in disturbed habitat such as road side, beside paddy field and near the stream. Sometimes makes very large population.

Vernacular names:

Indochinese ethnic groups: bo chua a tan (Thuan Hai)

Laotian: 'thoua 'sadet pa (Louang Prabang); 'thoua 'phi (Vientiane)

Thai: tua pee, yaa (weed)

Pests and diseases: *Cercospora* leaf spot observed. MAFF Japan accession CED99-T34 at collection site had no nodules but was heavily infested with nematodes. (Tomooka *et al.*, 2000e)

2.2. Section Ceratotropis *N. Tomooka & Maxted, Kew Bull. 57(3):613-624 (2002a)*

2.2.1. V. grandiflora *(Prain) Tateishi & Maxted, Kew Bull.* 57(3):625-633 (2002)

Chromosome number: 2n=2x=22

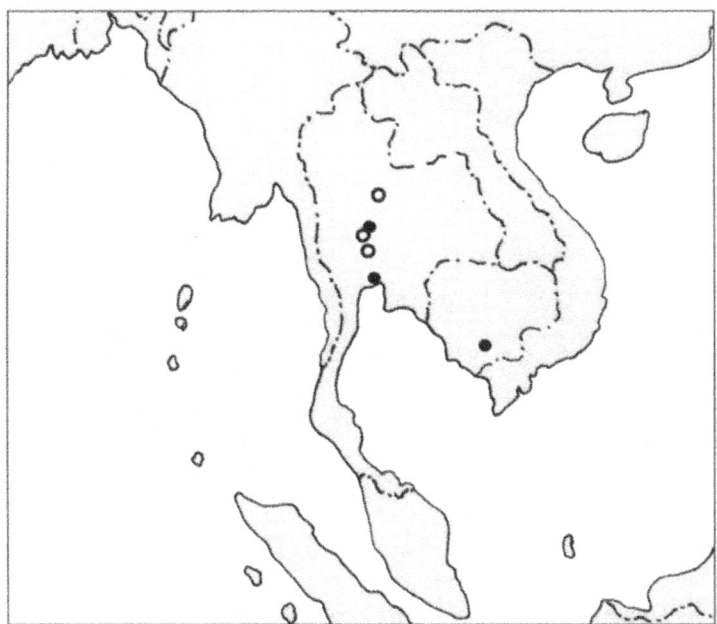

Figure 4.42. Distribution of Vigna grandiflora *based on herbarium specimens (●) and direct collection (○) by author (N.T.) .*

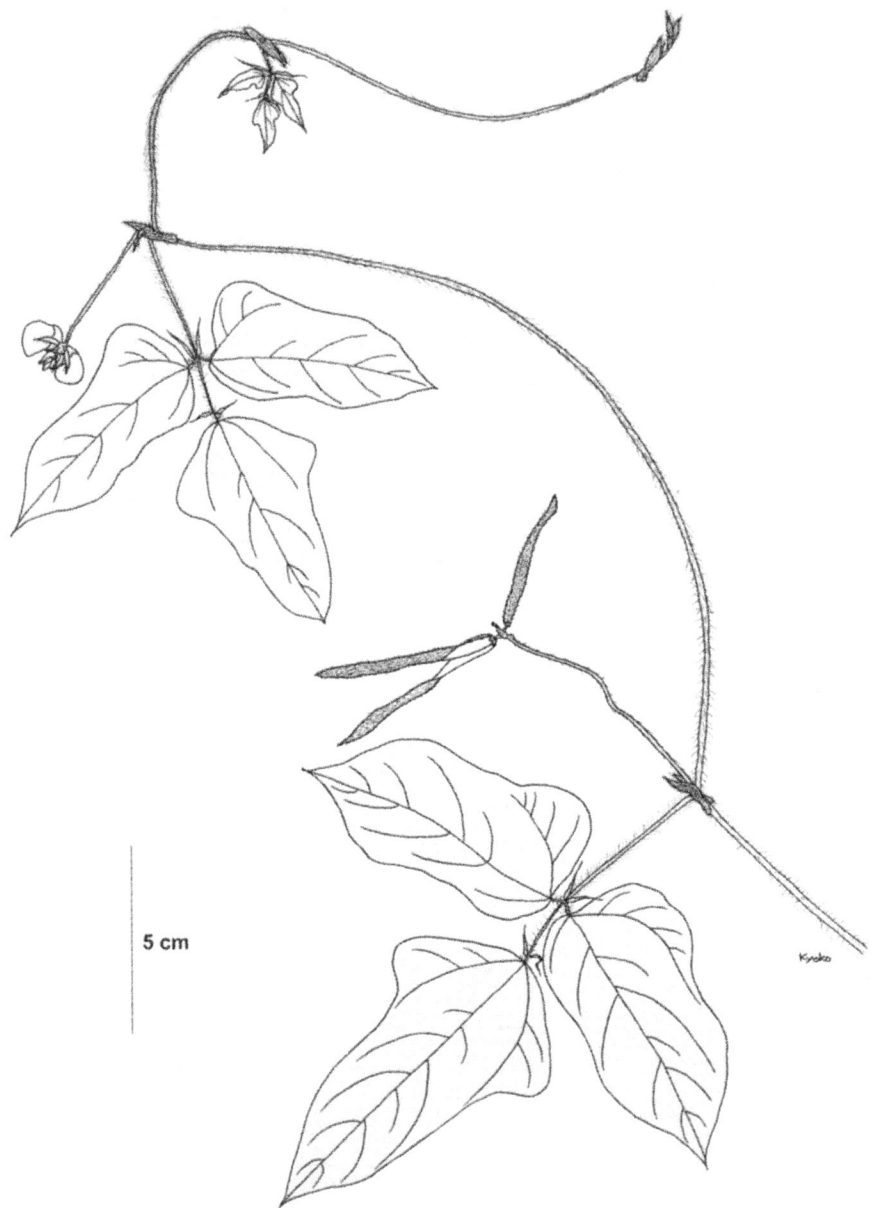

5 cm

Figure 4.43. Flowering and fruiting shoot of Vigna grandiflora.
Drawn from JP207984 (Thailand) by Kyoko Motoyoshi.

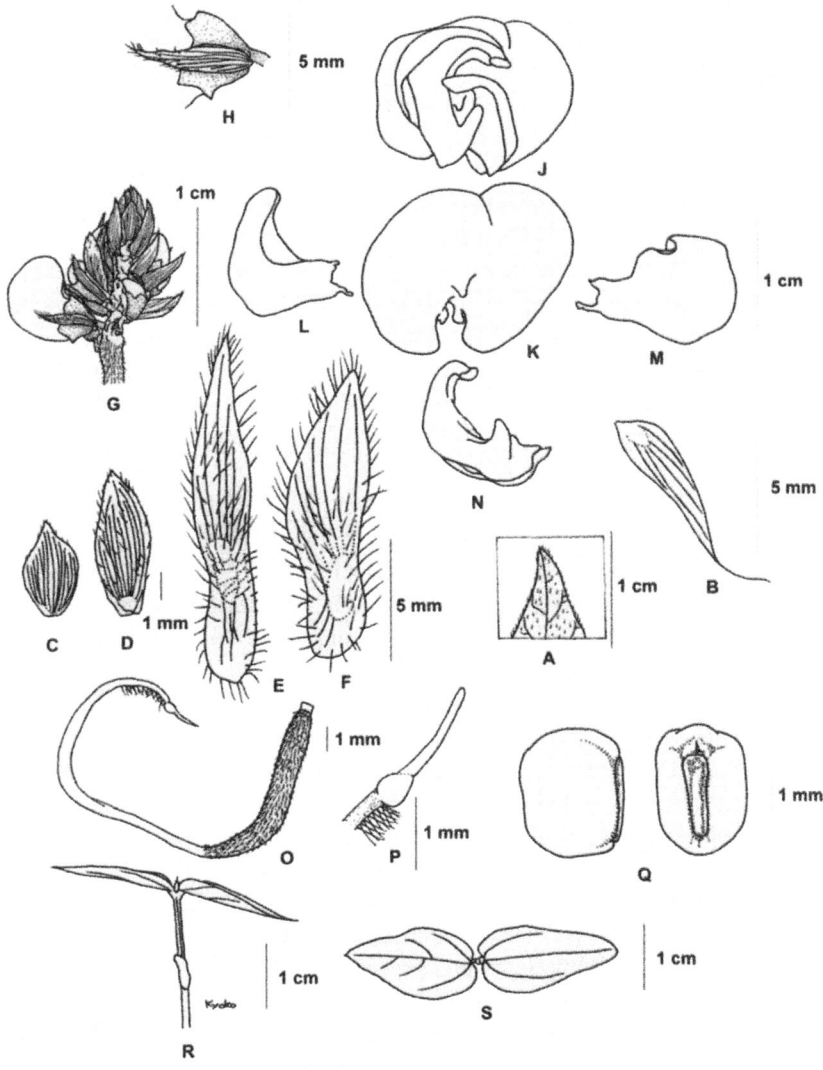

Figure 4.44. Plant parts of Vigna grandiflora.
A: leaf apex; B: stipel; C: primary bract; D: secondary bract; E & F: stipule; G: inflorescence; H: lower part of flower bud and bracteole; J: flower; K: standard; L: right wing; M: left wing; N: keel; O:style; P: stigma and style beak; Q: seed; R & S: seedling shoot. Drawn from JP107862 (Thailand), except A, C, D, E, G & H from JP207984 (Thailand) by Kyoko Motoyoshi.

Taxonomic affinities and diagnosis: According to molecular data, *V. grandiflora* is most closely related to *V. mungo* within section *Ceratotropis*. However, this species has been confused with *V. radiata* var. *sublobata*. Useful characters to identify *V. grandiflora* are pale creamy yellow flower; rather ascending mature pod covered with brown hispid hairs; rough seed with slightly protruding hilum having slightly developed aril. Diagnostic characters which distinguish it from *V. radiata* var. *sublobata* are narrowly elliptic stipule; longer style-beak; longer slightly protruding hilum. Differs from *V. mungo* var. *silvestris* by pale creamy yellow flower; less protruding hilum with less developed aril.

Description: Based on JP107862 from Thailand

A twining herb. *Stems* densely covered with long (0.7 - 1.8 mm) brown hispid retrorse hairs. *Stipules* narrowly elliptic, attenuate at the apex, rounded at the base, 11.8 x 3.9 mm, covered with rather long (0.8 - 1.0 mm) white hairs on outer surface and at the margin. *Leaf petioles* 1.7 - 4.2 cm long, very heavily covered with long (1.2 - 1.6 mm) brown hispid retrorse hairs. *Leaflets* membranous, covered with white hairs (0.3 - 0.8 mm) on both surfaces; *terminal leaflets* narrowly ovate, acuminate at the apex, obtuse or rounded at the base, shallowly 3 lobed, 7.5 - 9.5 x 4.1 - 5.8 cm; *lateral leaflets* obliquely narrowly ovate, acuminate at the apex, truncate at the base, very shallowly 3 lobed, 6.0 - 7.7 x 3.2 - 4.5 cm; stipels subulate, very long, 8.6 x 1.7 mm.

Inflorescence axillary, 4 - 8 flowered; peduncles 15 cm long, densely covered with long (1.0 - 1.2 mm) brown retrorse hispid hairs; rachis sparsely covered with short (0.3 - 0.4 mm) white hairs. *Primary bract* rhomboid or rather orbiculate, acute or obtuse at the apex, obtuse at the base, 4.9 x 3.6 mm, glabrous. *Secondary bract* (pedicel bract) elliptic, acute at the apex, truncate at the base, 6.2 x 3.4 mm, covered with long (0.8 mm) white hairs. *Bracteoles* narrowly elliptic or subulate, as long as or a little longer than calyx, 4.8 x 1.3 mm, glabrous or short hairs on the margin. *Pedicels* ascending, 1.7 - 2.0 mm long in flower, 2.6 mm long in fruit, glabrous.

Flowers pale creamy yellow. *Calyx* campanulate, 3.9 mm long, tube 2.5 mm long, glabrous. *Standard* asymmetrical, obliquely and broadly elliptic, 12.7 x 16.2 mm, emarginate at the apex, with an appendage at the centre inside. *Right wing* half concealing the upper portion of the keel-petals, lamina obliquely obovate, 10.3 x 8.8 mm, claws 1.7 mm long, auricle 1.0 mm long. *Left wing* spreading forward, supported by a keel-pocket, lamina broadly obovate, 11.0 x 8.4 mm, claws 2.0 mm long, auricle 1.2 mm long. *Keel-petals* spirally incurved to the left, 16.0 mm long, the left keel-petal with a horn-like pocket, the pocket 3.6 mm long. *Ovary* 5.6 mm long; style 20.5 mm long, long style-beak linear, 1.2 mm long.

Fruits spreading or rather ascending, linear, 4.7 - 5.1 x 0.3 - 0.4 cm, densely covered with brown hispid hairs (0.4 - 0.7 mm), 8 - 10 seeded, dark brown when mature. *Seed* elliptic, 4.2 x 3.1 x 2.8 mm, surface rough with a mesh-like reticulation; hilum linear, 2.8 x 0.7 mm, a little protruding; aril somewhat developed. Germination epigeal. *First and second leaves* without petiole, leaflets simple, narrowly elliptic, acute at the apex, rounded at the base, 2.1 x 1.0 cm.

Distribution: Thailand and Cambodia

Altitude: Range 25-50 m in Thailand, mean 35 m (7 records)

Phenology: Thailand - October to November

Ecology: Grows in disturbed habitats such as road sides or near the cultivated fields in rather dry hot lowland sites (Thailand).

Pests and diseases: *Cercospora* leaf spot, powdery mildew, ants visit extra floral nectaries (Tomooka *et al.*, 2000e)

2.2.2. V. mungo *(L.) Hepper, Kew Bull. 11: 128 (1956)*

Chromosome number: 2n=2x=22

Taxonomic affinities and diagnosis: *V. mungo* is most closely related to *V. grandiflora* within section *Ceratotropis*. Useful characters to identify *V. mungo* are golden yellow flower; ascending mature pod covered with long brown hairs; rough seed with much protruding ovate hilum having well developed aril. Diagnostic characters that distinguish it from *V. grandiflora* are golden yellow flower; protruding ovate hilum with well developed aril.

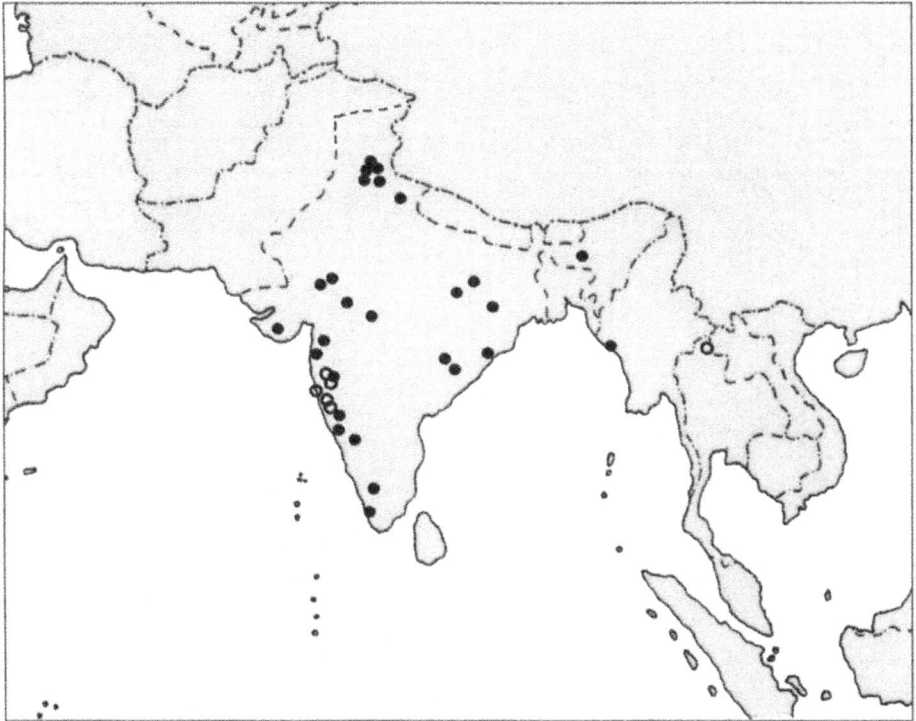

Figure 4.45. Distribution of Vigna mungo var. silvestris *based on herbarium specimens (●) and direct collection by author (N.T.) in Thailand and (by Dr. H. Kobayashi, Kyoto University, Japan) in India (○).*

Figure 4.46. Flowering and fruiting shoot of Vigna mungo *var.* silvestris.
Drawn from JP107874 (India) by Kyoko Motoyoshi.

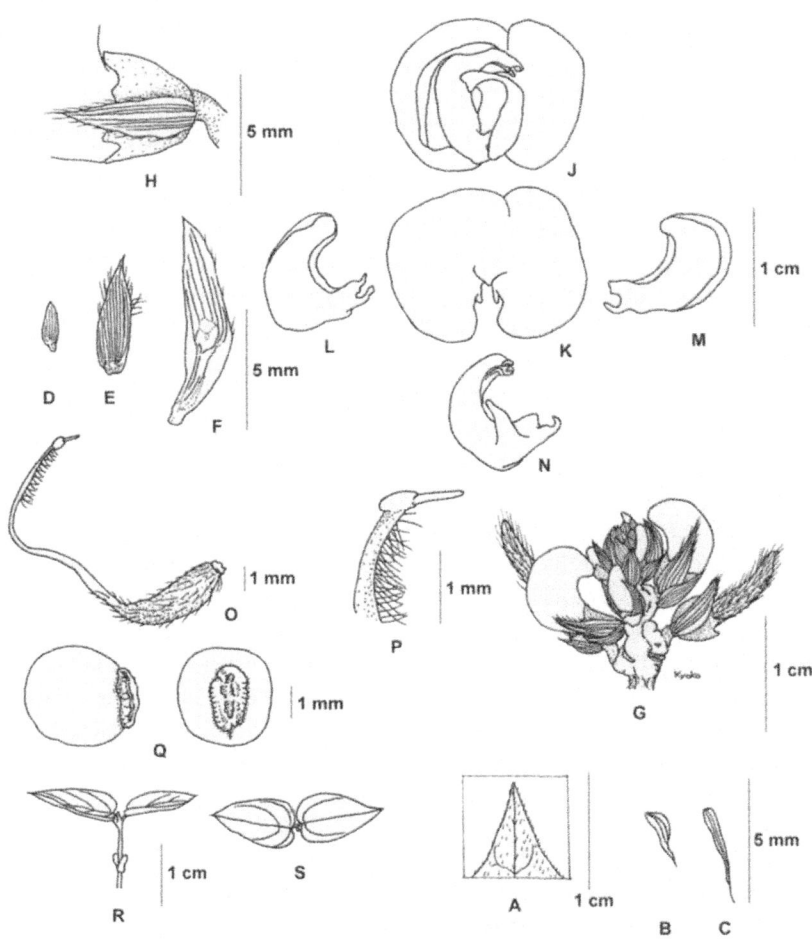

Figure 4.47. Plant parts of Vigna mungo *var.* silvestris.
A: leaf apex; B & C: stipel; D: primary bract; E: secondary bract; F: stipule; G: inflorescence;
H: lower part of flower bud and bracteole; J: flower; K: standard; L: right wing; M: left
wing; N: keel; O: style; P: stigma and style beak; Q: seed; R & S: seedling shoot. Drawn
from JP107874 (India), except A, C, D, E, & F from JP107873 (India) by Kyoko Motoyoshi.

Description: Based on wild form, var. *silvestris*, JP107874 from India

A twining herb. *Stems* slender, very sparsely covered with white hairs, 0.9 - 1.0 mm long. *Stipules* narrowly elliptic or rather falcate, acute at the apex, obtuse or rounded at the base, 8.8 x 3.9 mm, glabrous. *Leaf petioles* 1.6 - 6.0 cm long, very rarely ciliate with pale brown hairs (0.6 mm long). *Leaflets* covered with short (0.3 - 0.4mm) white hairs on both surfaces; *terminal leaflets* ovate or narrowly ovate, acuminate at the apex, acute or obtuse at the base, 3.8 - 8.2 x 3.3 - 5.7 cm; *lateral leaflets* obliquely ovate, acuminate at the apex, obtuse or rounded at the base, 3.4 - 6.4 x 2.4 - 5.4 cm; *stipels* narrowly elliptic, attenuate at the apex, 2.2 x 0.7 mm.

Inflorescence axillary, 4 - 8 flowered; peduncles 6.7 cm, very rarely covered with short (0.6 mm) white hairs; rachis 0.8 cm, glabrous. *Primary bract* elliptic, acute at the apex, obtuse at the base, 3.0 x 3.4 mm, glabrous. *Secondary bract* (pedicel bract) elliptic, acuminate at the apex, obtuse at the base, 4.5 x 3.4 mm, ciliate with 0.3 - 0.5 mm long white hairs on the margin. *Bracteoles* broadly subulate, longer than calyx, 5.2 x 2.4 mm, glabrous. *Pedicels* ascending, 1.5 mm long in flower, 1.5 - 2.0 mm long in fruit, glabrous.

Flowers golden yellow. *Calyx* campanulate, 3.9 mm long, tube 2.8 mm long. *Standard* asymmetrical, obliquely broadly elliptic, 11.5 x 14.5 mm, emarginate at the apex, an appendage at the centre inside. *Right wing* half concealing the upper portion of keel-petals, lamina obliquely obovate, 10.1 x 10.0 mm, claws 1.8 mm long, auricle 1.5 mm long. *Left wing* spreading, lamina obliquely elliptic, 11.5 x 6.3 mm, claws 1.9 mm long, auricle 0.9 mm long. *Keel-petal* spirally incurved to the left, 14.9 mm long, left petal with a horn-like pocket, the pocket 3.0 mm long. *Ovary* 4.1 mm long; style 14.9 mm long, prolonged beyond stigma to form style-beak, the beak linear, 0.7 mm long.

Fruits ascending, linear, 3.2 - 3.7 x 0.4 - 0.6 cm, covered densely with long (1.8 - 3.0 mm) white hairs when young, brown hairs on black surface when mature, 6 - 8 seeded. *Seed* round, 3.2 x 3.1 x 3.0 mm, surface rough; hilum ovate, 2.1 x 1.3 mm, much protruding, aril very well developed. Germination epigeal. *First and second leaves* without petioles, leaflets simple, narrowly long elliptic, 1.5 x 0.7 cm.

var. *mungo*: cultivated form

Origin: Black gram was domesticated in India from its wild relative *V. mungo* var. *silvestris* (Chandel *et al.*, 1984). Center of genetic diversity of the crop is India.

Geographic distribution: Traditionally cultivated in India and surrounding areas and recently spread to tropical areas worldwide where Indians have migrated.

Ecology: In India this species is grown from sea level to 2000m. Drought resistant. Growing where rainfall is not more than 900mm. Grows in clay or heavy black cotton soils in India.

It is one of the most highly prized pulses in India, particularly in the vegetarian diet of the high caste Hindus. Used in India for local consumption and is particularly important in Mysore (Purseglove, 1974).

Vernacular names:

Australia: komin (Rockhampton Aborigines), kadolo (Cleveland bay, Aborigines)

Bengali: mash, kalai, tikari
Chinese: lu tou
English: black gram , urd bean, black matpe
French: haricot mungo, amberique
German: urdbohne
Gujurati: adad, arard
Hindi: urd, urid
Japanese: ke-tsuru azuki
Kanda: uddu, udni bele
Malay: katjang hitam, katang hijai
Marati: maga, udid
Myanmar: mat-pe
Nepali: mash
Philippines: balatong, mungo
Portuguese: feijao-da-China
Spanish: frijol mungo
Tamil: ulundu, ulutham, paruppu.
Telagu: minumukulu, karuminimulu, manipa pappu
Vietnamese: dau ma, dau muong an
West Indies: wooly pyrol
Uses: Black gram is cooked as dhal soup (split dehusked bean soup) in South Asia and adjacent regions. Japan imports black gram for producing bean sprouts.

var. *silvestris* Lukoki, Maréchal & Otoul, Bull. Jard. Bot. Nat. Belg. 50: 390 (1980): wild form
Geographic distribution: India, Myanmar, Thailand.
Altitude: 70 - 3500 m, mean 850 m (20 records)
Phenology: India – September to October.
Ecology: Grows in disturbed habitat such as road side (Thailand).

2.2.3. V. radiata *(L.) Wilczek, Fl. Congo Belge 6: 386 (1954)*
Chromosome number: 2n=2x=22
Taxonomic affinities and diagnosis: This species is most closely related to *V. subramaniana*. This species has been confused with *V. grandiflora, V. trinervia* and *V. reflexo-pilosa*. Useful characters to identify *V. radiata* are ovate stipule; pale grayish yellow flower; spreading mature pods covered with brown hispid hairs; rough seed with non-protruding hilum without aril development. Diagnostic characters that distinguish it from *V. subramaniana* are larger flower and seed size; regular rectangular shaped seed coat reticulation. Differs from *V. trinervia* and *V. reflexo-pilosa* by grayish yellow flower; epigeal germination; sessile first and second leaves. Differs from *V. reflexo-pilosa* by rough seed coat.

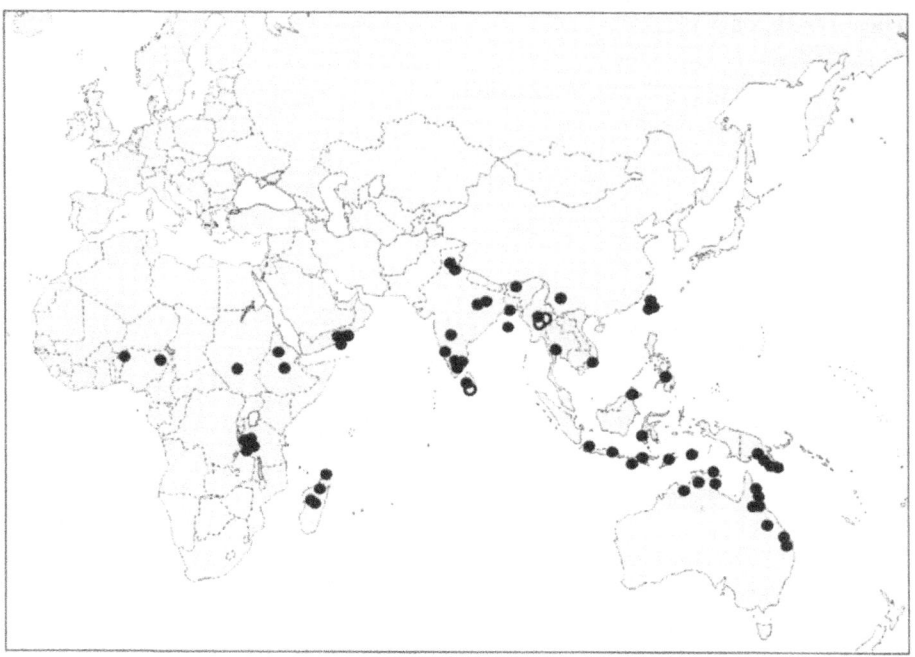

Figure 4.48. Distribution of Vigna radiata *var.* sublobata *based on herbarium specimens*
(●) and direct collection (○) by author (N.T.).

Description: Based on wild form, var. *sublobata*, JP107877 from Madagascar
A twining herb. *Stems* covered with long (1.2 - 1.9 mm) spreading brown hairs. *Stipules*
ovate, acute at the apex, rounded at the base, 12.2 x 6.4 mm, ciliate with long (1.0 - 1.5
mm) white hairs on the margin. *Leaf petioles* 6.4 - 15.3 cm long, covered with spreading
or rather retrorse whitish hairs, 0.4 - 1.6 mm long. *Leaflets* covered with shorter (0.2 - 03
mm) white hairs on upper surface and longer (0.7 - 0.9 mm) white hairs on veins of upper
surface, ciliate with longer (0.7 - 1.0 mm) white hairs on veins of lower surface; *terminal
leaflets* ovate, acute at the apex, obtuse at the base, sometimes shallowly to prominently
3 lobed, 3.4 - 9.1 x 2.3 - 8.2 cm; *lateral leaflets* obliquely ovate, acute at the apex, obtuse
at the base, 3.4 - 8.3 x 2.7 - 7.4 cm; stipels narrowly elliptic, 8.1 x 1.3 mm.

Inflorescence axillary, 4 - 10 flowered; peduncles 3.9 - 8.4 cm long, covered with
brown retrorse hairs, 0.6 - 1.0 mm long; rachis 0.7 - 1.1 cm long, covered with short (0.2
- 0.3 mm) white hairs. *Primary bract* orbiculate to ovate, acute or rounded at the apex,
truncate at the base, 2.5 x 1.9 mm, glabrous. *Secondary bract* narrowly ovate, acute at
the apex, truncate at the base, 3.6 x 1.4 m, glabrous. *Bracteoles* lanceolate, as long as
calyx, 4.8 x 1.2 mm, sparsely ciliate with white hairs mainly on the margin, 0.3 - 0.6 mm
long. *Pedicels* ascending, 1.3 mm long in flower, 1.5 mm long in fruit.

Flowers pale grayish yellow. *Calyx* campanulate, 4.1 mm long, tube 3.1 mm long.

Standard asymmetrical, somewhat obliquely elliptic, 9.9 x 14.0 mm, emarginate at the apex, with an appendage at the centre inside. *Right wing* obliquely broadly obovate, half concealing the keel-petal, lamina 8.3 x 7.0 mm, claws 2.3 mm long, auricle 1.7 mm long. *Left wing* spreading forward, obliquely obovate, lamina 7.7 x 8.4 mm, claws 2.0 mm long, auricle 1.5 mm long. *Keel-petals* spirally incurved to the left, upper portion of keel petals grayish, 14.7 mm long, with a horn-like pocket on the left keel-petal, the pocket 1.4 mm long. *Ovary* 5.4 mm long; style 17.1 mm long, prolonged beyond stigma to form style-beak, the beak linear, 0.5 mm long.

    Fruits spreading, linear, 4 - 5 x 0.3 - 0.4 cm, densely covered with hispid brown hairs (0.5 - 0.9 mm), blackish brown when mature, 10 - 13 seeded. *Seed* rather rectangular, 3.0 x 3.0 x 3.1 mm, surface rough, covered with mesh-like reticulation; hilum elliptic, 1.0 x 0.4 mm, not protruding, aril not developed. Germination epigeal. *First and second leaves* without petioles, leaflet simple, narrowly ovate, acute at the apex, somewhat cordate at the base, 2.4 x 1.4 cm.

var. *radiata*: cultivated form
Vernacular names:
Bengali: sonamung
Cambodian: sandaek bay, sandaek luong
Central Asia (Uzbekistan, Tajikistan, Kirgyzstan): mash
Chinese: lu dou, chih hsiao tou, tung tou
English: mungbean, green gram, golden gram
French: haricot mungo, haricot dore , amberique
German: mungbohne
Gujarati: mag
Hindi: dord, mung, moong, pessara
Indochinese ethnic groups: rota'-dong, rota'-tokuih (Jorai)
Indonesian: kacang hijau, arta ijo, kacang djong
Japanese: yaenari, bundou, fundou, ryokutou, ao mami (Okinawa), Kumami (Okinawa)
Kanda: hesaru
Korean: rogdu
Laotian: thoua khieo (Vientiane)
Malaysia: kacang hijau
Malayala: cerupayaru
Marati: mug
Myanmar: pe-di-sien
Nepali: mung
Pakistan: mash
Philippines: mongo, balatong
Russian: mas
Singhala (Sri Lanka): ulundu, mun, bu-me
Spanish: frijol mungo, judia mung

Figure 4.49. Flowering and fruiting shoot of Vigna radiata *var.* sublobata.
Drawn from JP107877 (Madagascar) by Kyoko Motoyoshi.

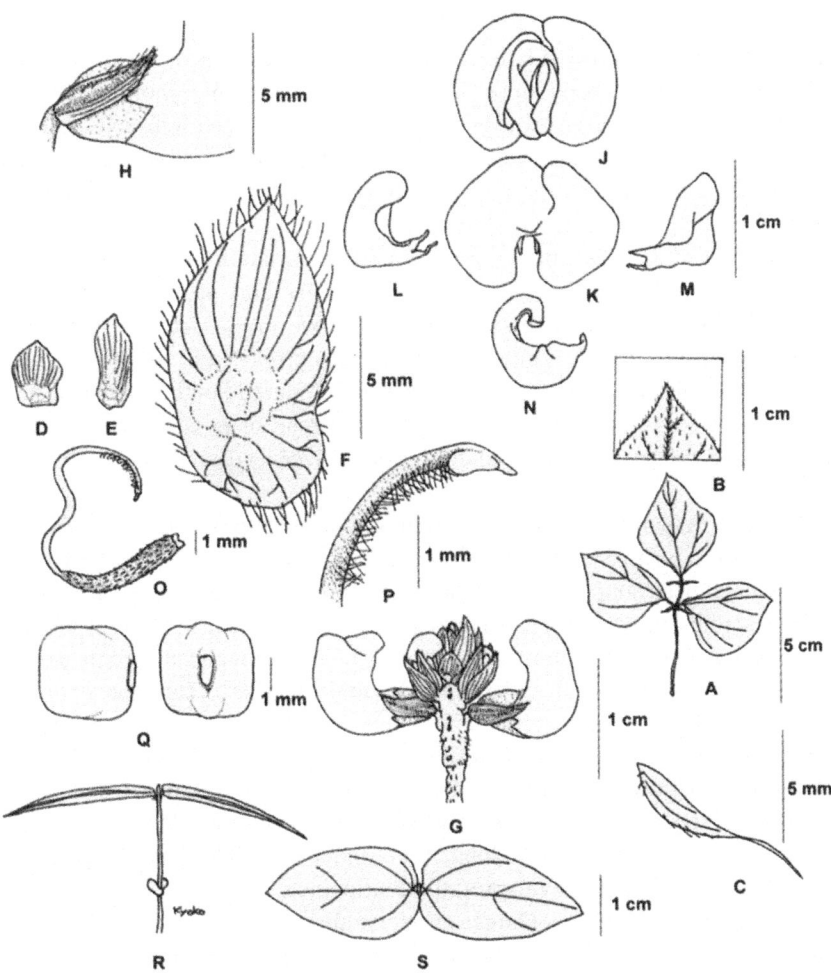

Figure 4.50. Plant parts of Vigna radiata *var.* sublobata.
A: leaf; B: leaf apex; C: stipel; D: primary bract; E: secondary bract; F: stipule; G:
inflorescence; H: lower part of flower bud and bracteole; J: flower; K: standard; L: right
wing; M: left wing; N: keel; O:style; P: stigma and style beak; Q: seed; R & S: seedling
shoot. Drawn from JP107877 (Madagascar) by Kyoko Motoyoshi.

Tamil (India): pachapayru
Tamil (Sri Lanka): uluntu, chiruppayaru
Tanzania: ndotodoto
Telagu: patchapesalu, pesalu
Thai: thua khieo (green), thua thong (golden)
Vietnamese: dau che, dau tam, dau xanh

Origin and dissemination: It is believed that mungbean was domesticated in India (Vavilov, 1926). This has been supported by studies based on morphology (Singh *et al.*, 1974), the existence of wild and weedy types (Chandel, 1984) and archaeological remains (Jain and Mehra, 1980) in India. However, the wild form of mungbean var. *sublobata* is widely distributed in Africa, Asia and Australia so domestication more than once cannot be ruled out.

Tomooka *et al.* (1992a), based on seed protein variation in mungbean landraces from Asia, identified regions of protein diversity and proposed two paths of dissemination. Protein type diversity is greatest in West Asia (Afghanistan-Iran-Iraq) rather than India. From this region, mungbean was proposed to have moved east by two routes one following a southern route through India to Southeast Asia. The other pathway followed a northern route along the silk road or possibly through India to China. Similar result has been obtained from enzyme diversity data (Dela Vina & Tomooka, 1994).

Uses: In South Asia mungbean is used to make dhal a soup of split seeds with spices. In Southeast and East Asia, mungbean is used to make various kinds of sweets, bean jam, sweetened bean soup, vermicelli and bean sprouts. In Japan, *V. radiata* is used to make bean sprouts (moyashi) or eaten mixed with steamed rice in the Nansei Archipelago southern Japan. On Tanegashima southern Japan, long mungbean sprouts are offered to the ancestors during the "Bon Festival" (Buddhist holiday) (Tomooka *et al.*, 1994).

var. *sublobata* (Roxb.) Verdcourt, Kew Bull. 24: 559 (1970): wild form

Geographic distribution: The most widely distributed of the wild species in the subgenus *Ceratotropis*; Africa – across equatorial Africa as far west as Benin, also Tanzania, Madagascar; Middle East – Oman; South Asia – India, Bangladesh, Sri Lanka; Southeast Asia –, Myanmar, Thailand, Vietnam, the Philippines, Indonesia; East Asia – China including Taiwan; Papua New Guinea and Australia.

Altitude: Range 20 m (Papua New Guinea) to 1650 m (India); mean 430 m (36 records)

Phenology: Australia – April to July; India – October to November; Indonesia – March to June; Madagascar – February to March; Myanmar – October to November; Oman and Sudan – September to October; Sri Lanka – February and March.

Ecology: A species of forest margins and grassland. Grows in disturbed habitat such as road side (Myanmar).

Vernacular names:
English: Jerusalem pea
Bengali: ghora mung

Australian aboriginal names: gabala, kadolo (kerdolo), komin, umida

Uses: Seeds are used as a pulse at times of food scarcity. Livestock graze on this plant that is a component of the Western Ghats grasslands in India. It may be used as a fodder, soil conservation cover crop (Babu *et al.*, 1985). Tuberous rooted perennials of var. *sublobata* have been used as a root crop by aborigines for a long time (Lawn and Cottrell, 1988).

2.2.4. V. subramaniana *(Babu ex Raizada) M. Sharma, J. Econ, Taxon, Bot. 6(3):736 (1985) et Geobis New Rep. 5(1): 54 (1986)*

Chromsome number: 2n=2x=22

Taxonomic affinities and diagnosis: This species is most closely related to *V. radiata* within section *Ceratotropis*. Useful characers to identify *V. subramaniana* are ovate stipule; pale grayish very small flower (c. 10 mm in diameter); hairy mature pod; rough seed with short hilum without aril development; epigeal germination. Diagnostic characters that distinguish it from *V. radiata* are smaller flower size; smaller seed size; irregular seed coat reticulation.

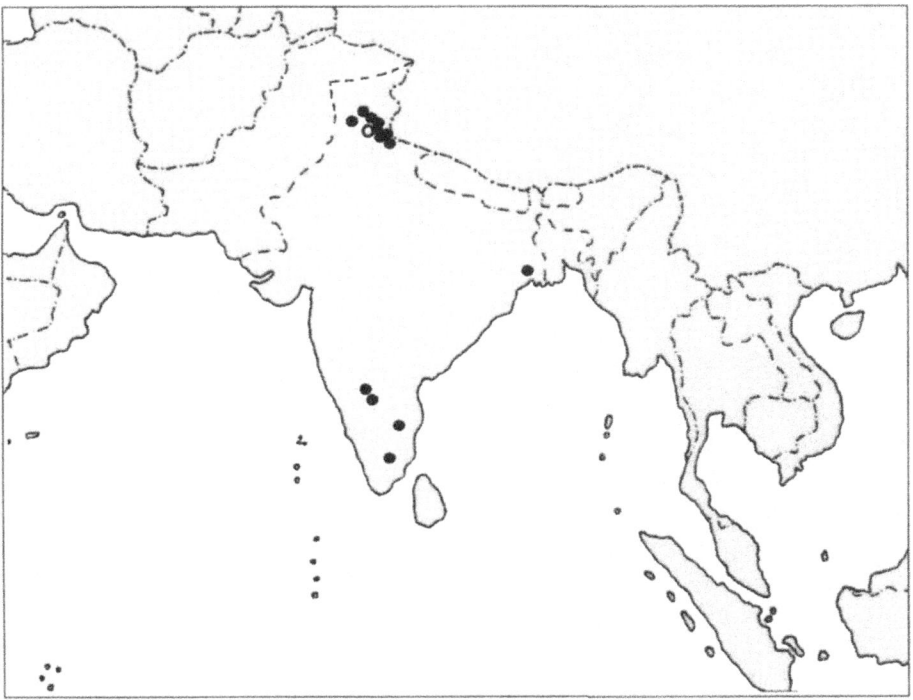

Figure 4.51. Distribution of Vigna subramaniana *based on herbarium specimens (●) and direct germplasm collection (○) by van der Maesen, ICRISAT.*

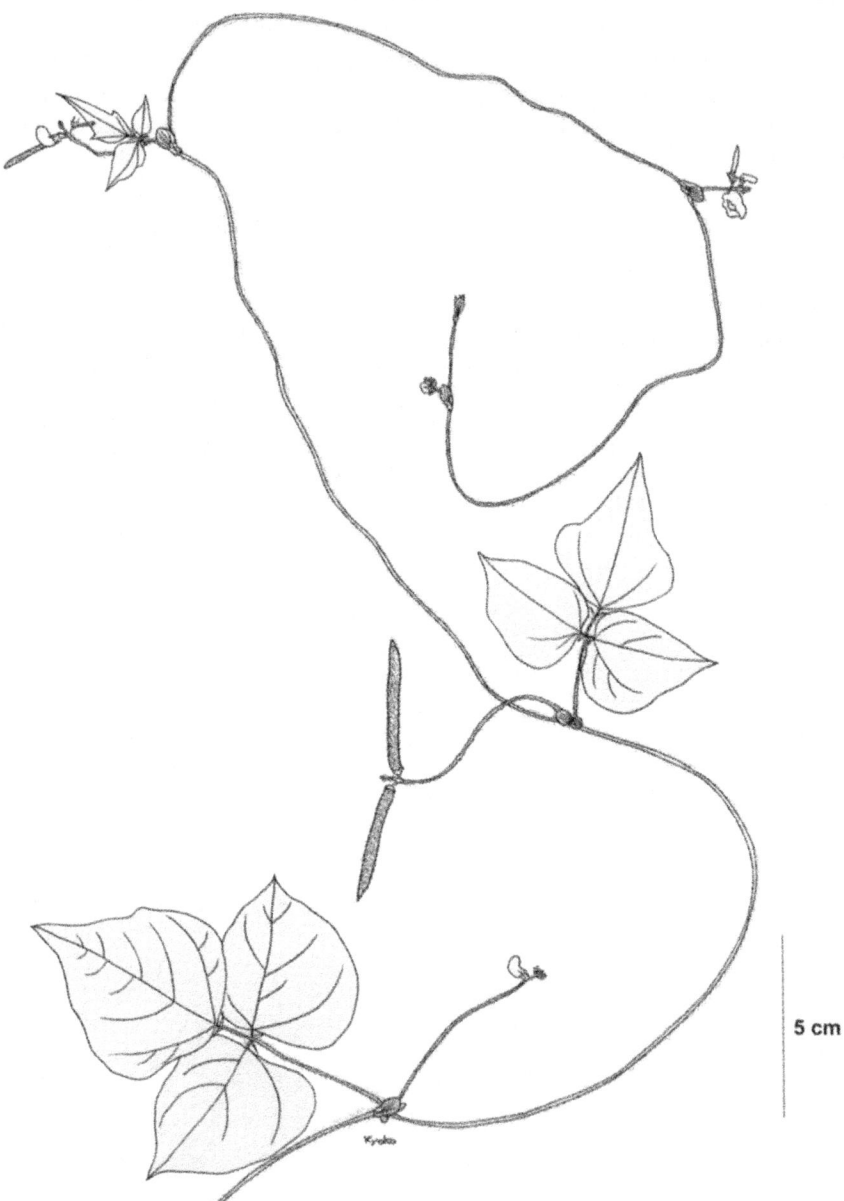

5 cm

Figure 4.52. Flowering and fruiting shoot of Vigna subramaniana.
Drawn from JP110836 (India) by Kyoko Motoyoshi.

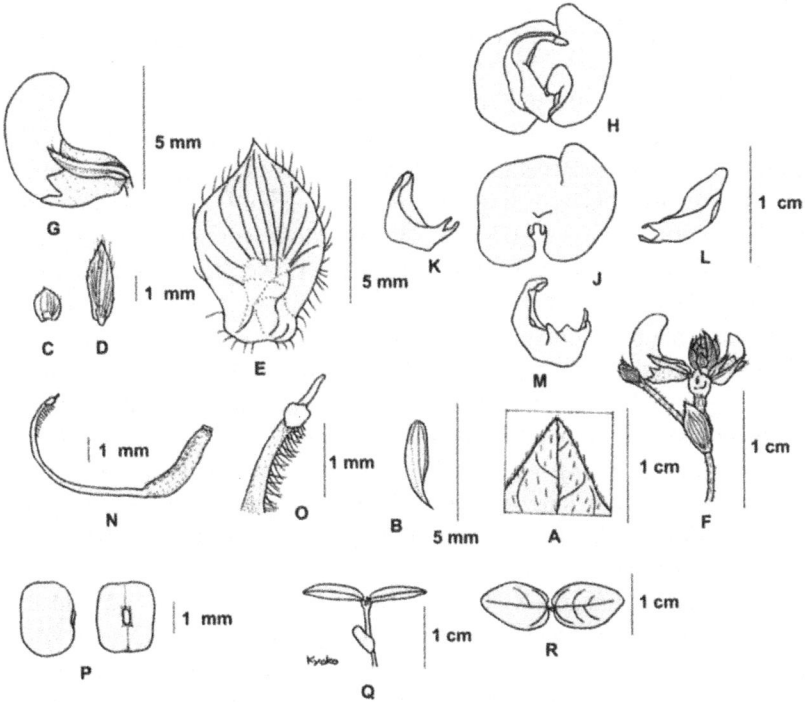

Figure 4.53. Plant parts of Vigna subramaniana.
A: leaf apex; B: stipel; C: primary bract; D: secondary bract; E: stipule; F: inflorescence;
G: flower bud and bracteole; H: flower; J: standard; K: right wing; L: left wing; M: keel;
N:style; O: stigma and style beak; P: seed; Q & R: seedling shoot. Drawn from JP110836
(India) by Kyoko Motoyoshi.

Description: Based on JP110836 from India
A twining herb. *Stems* slender, sparsely pubescent with long (1.1 -1.8 mm) retrorse pale brown hairs. *Stipules* peltate, broadly ovate, acuminate at the apex, rounded at the base, 7.9 x 4.7 mm, ciliate with long (0.6 - 1.1 mm) white hairs on the margin, nerves radiate from point of attachment to the stem. *Leaf petioles* 2.4 - 4.5 cm long, very sparsely covered with long (0.6 - 1.3 mm) pale brown hairs. *Leaflets* sparsely covered with white hairs (0.3 - 0.4 mm long) on upper surface, 0.6 - 0.7 mm long on lower surface); *terminal leaflets* broadly ovate, acuminate at the apex, obtuse at the base, sometimes faintly 3 lobed, 4.4 - 7.1 x 3.3 x 5.5 cm; *lateral leaflets* obliquely broadly ovate, acuminate

at the apex, truncate at the base, 3.4 - 5.9 x 2.8 - 4.9 cm; stipels lanceolate, acuminate at the apex, 3.5 x 0.8 mm.

Inflorescence axillary, 6 - 8 flowered; peduncles 5.4 cm long, densely covered with brown retrorse appressed hairs, 0.3 - 0.5 mm long; rachis 2.7 - 5.2 mm long, sparsely covered with short (0.1 - 0.2 mm) brown hairs. *Primary bract* ovate, acute at the apex, truncate at the base, 1.2 x 1.0 mm, ciliate with very short white hairs on the margin, 0.1 mm long. *Secondary bract* (pedicel bract) narrowly elliptic, acute at the apex, acute at the base, 3.3 x 1.0 mm, ciliate with white hairs on the margin, 0.5 - 0.6 mm long. *Bracteoles* subulate, as long as calyx, 3.6 x 0.8 mm, sparsely ciliate with short (0.2 mm) white hairs on the upper margin. *Pedicels* ascending, 0.2 mm in flower, 1.3 - 1.8 mm in fruit.

Flowers pale yellow. *Calyx* campanulate, 3.1 mm long, tube 2.1 mm long. *Standard* asymmetrical, obliquely broady elliptic, small appendage on the centre inside, emarginate at the apex, 9.4 x 11.1 mm. *Right wing* half concealing right side of keel-petals, lamina obliquely obovate, 6.2 x 4.8 mm, claws 1.6 mm long, auricle 1.1 mm long. *Left wing* spreading forward, lamina obliquely narrowly obovate, 6.3 x 4.8 mm, claws 1.9 mm long, auricle 1.3 mm long. *Keel-petals* spirally incurved to the left, upper portion of the keel brightly grayish, 12.3 mm long, left keel with a horn-like pocket, the pocket 1.8 mm long. *Ovary* 3.2 mm long; style 11.0 mm long, prolonged beyond stigma to form style beak, the beak linear, 0.5 mm long.

Fruits spreading, linear, 2.8 - 4.0 cm long, 0.2 - 0.3 cm wide, 8 - 12 seeded, densely covered with brown bristle hairs, 0.4 - 0.7 mm long, brown when mature. *Seed* oblong or rather rectangular, 2.3 x 1.8 x 2.0 mm, surface rough, with a mesh-like reticulation, dark brown when mature, hilum short elliptic, 0.5 x 0.3 mm, not protruding, aril not developed. Germination epigeal. *First and second leaves* without petioles, leaflets elliptic, acute at the apex, rather cordate at the base, 1.1 x 0.7 mm.

Geographic distribution: India. Most collections have been from the Himalayan foothills of northwest India.

Altitude: Range from 330 m to 1250 m, mean 807 m (4 records)

Phenology: India – September to December.

2.3. Section Aconitifoliae *N. Tomooka & Maxted, Kew Bull. 57(3):613-624 (2002a)*

2.3.1. V. aconitifolia *(Jacq.) Maréchal, Bull. Jard. Bot. Natl. Belg. 39: 160 (1969)*
Chromosome number: 2n=2x=22
Taxonomic affinities and diagnosis: *V. aconitifolia* is most closely related to *V. aridicola* within section *Aconitifoliae*. Based on habit (wild prostrate versus cultigen erect or decumbent ascending) and pod shattering (wild shattering and cultigen non-shattering) it seems likely that the wild and cultivated forms of *V. aconitfolia* warrant recognition of varietal status as with other crops and their wild progenitors in the subgenus. However, we have seen insufficient material of the wild form to make a judgment. Useful characters to identify *V. aconitifolia* are long fine spreading brown hairs on stem; subulate or lanceolate stipule; deeply 5 lobed terminal leaflet; deeply 4 lobed lateral leaflet; long subulate stipel; bright yellow small flower; smooth seed with non-protruding short linear hilum; epigeal germination with first and second petiolate leaves. Diagnostic characters which distinguish it from *V. aridicola* are 5 lobed terminal leaflet; bright yellow flower; glabrous pod; smooth seed; petiolate first and second leaves.

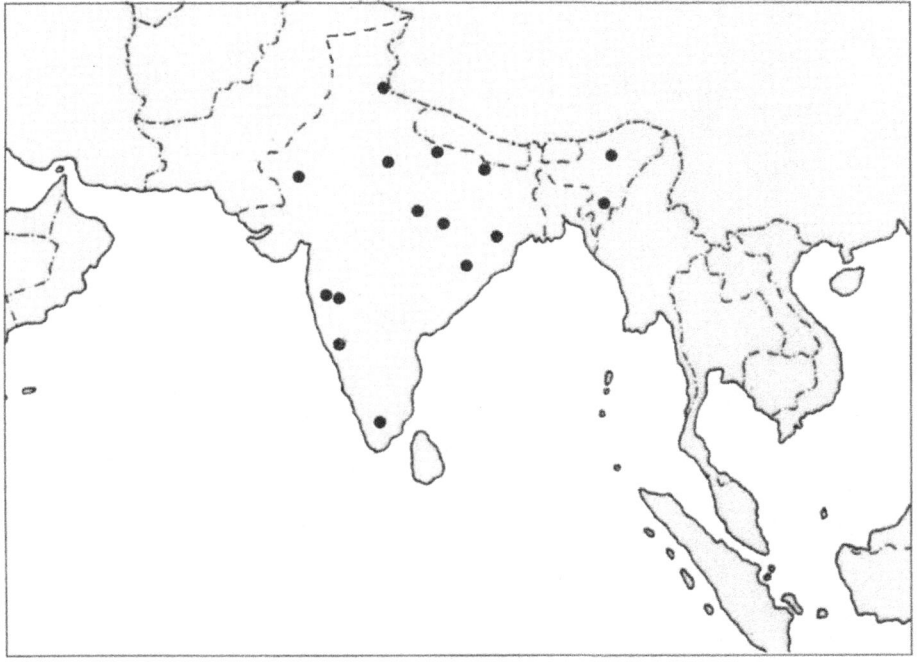

Figure 4.54. Distribution of wild form of Vigna aconitifolia.
Information from Arora and Nayar (1984).

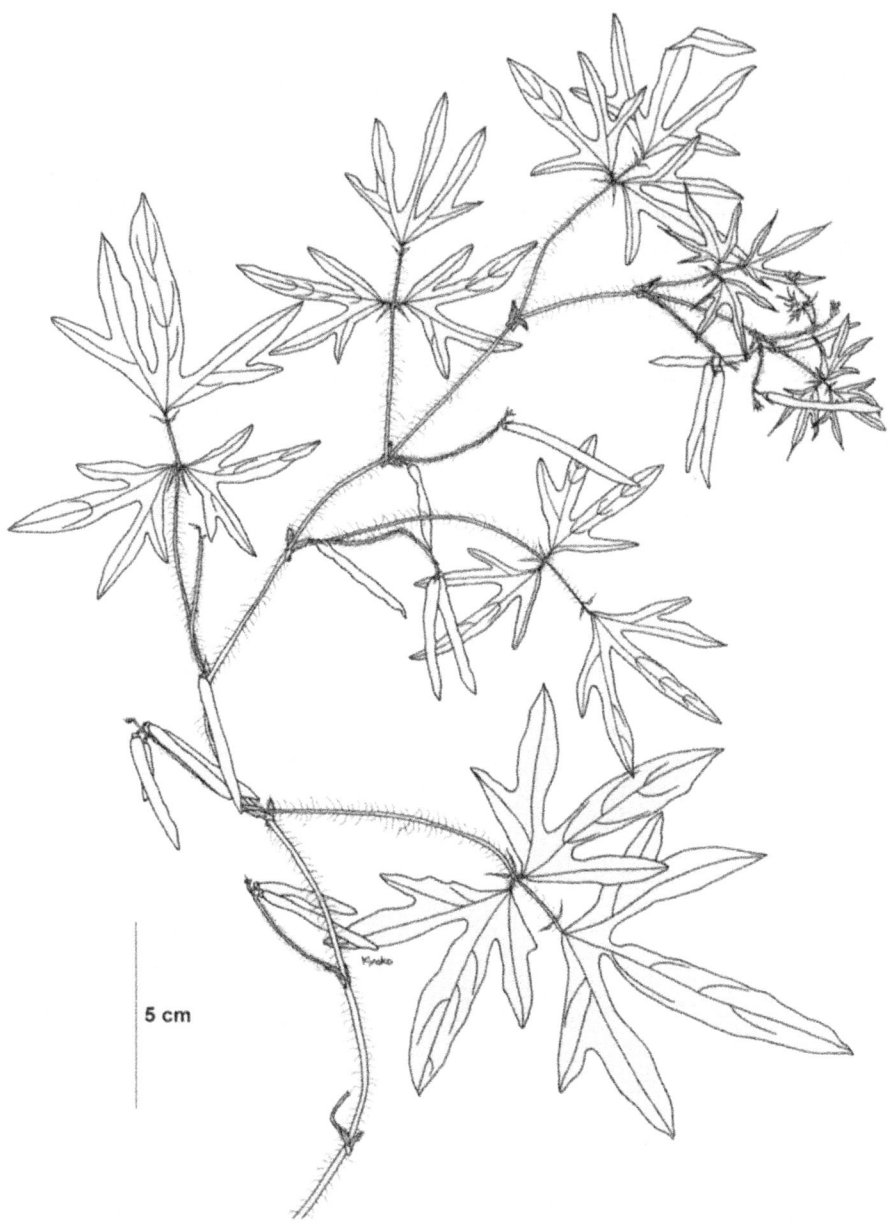

5 cm

Figure 4.55. Flowering and fruiting shoot of cultivated Vigna aconitifolia.
Drawn from JP105629 (India) by Kyoko Motoyoshi.

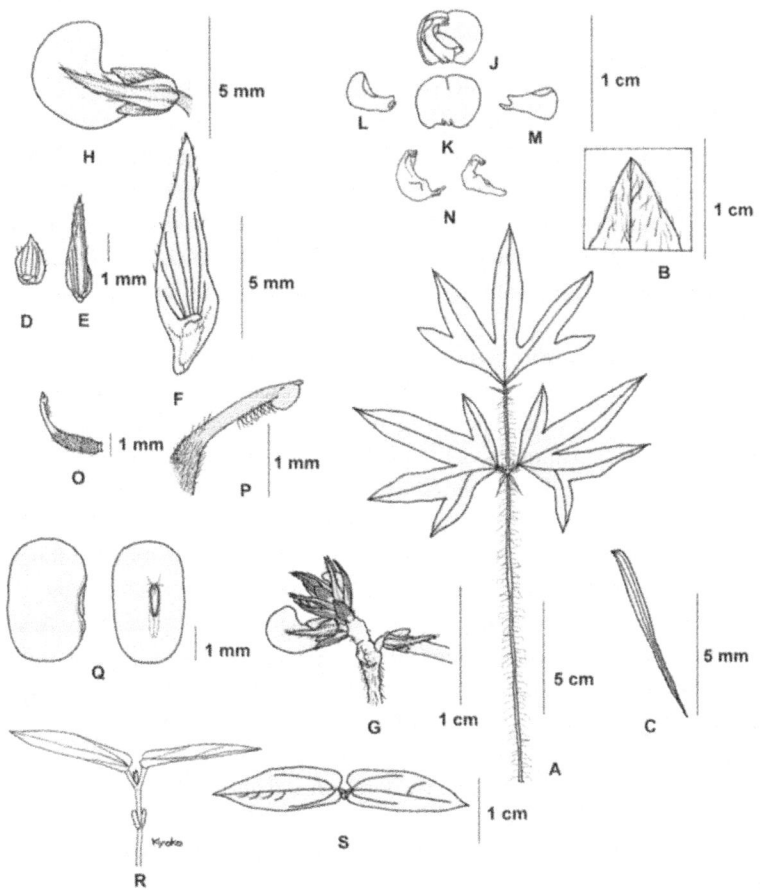

Figure 4.56. Plant parts of cultivated Vigna aconitifolia.
A: leaf; B: leaf apex; C: stipel; D: primary bract; E: secondary bract; F: stipule; G: inflorescence; H: flower bud and bracteole; J: flower; K: standard; L: right wing; M: left wing; N: keel; O:style; P: stigma and style beak; Q: seed; R & S: seedling shoot. Drawn from JP105629 (India) by Kyoko Motoyoshi.

Description: based on cultivated form, JP105629 from India
A semi-erect cultivated herb. *Stems* densely covered with very long (2 - 3 mm) spreading fine brown hairs. *Stipules* subulate, attenuate at the apex, acute at the base, 10 x 2 mm, sparsely ciliate with long (c. 1 mm) brown hairs on the margin. *Leaf petioles* 13.8 cm

long, densely covered with very long (1.5 - 2.5 mm) spreading brown hairs. *Leaflets* chartaceous, densely covered with very long (1.5 - 2.0 mm) spreading white hairs on both surfaces; *terminal leaflets* deeply 5 lobed, 7.3 x 8.0 cm; *lateral leaflets* deeply 4 lobed, 7.2 x 6.8 cm; *stipels* subulate, very long, 6.5 mm long, 0.6 mm wide.

Inflorescence axillary, 6 - 8 flowered; peduncles 3 - 4 cm long, densely covered with long (1.5 - 2 mm) retrorse brown hairs; rachis 0.5 - 1.0 cm long, covered with short (0.2 - 0.3 mm) white hairs. *Primary bract* ovate, acute at the apex, rounded at the base, 2.0 x 1.0 mm, sparsely ciliate with short hairs on upper margin. *Secondary bract* (pedicel bract) lanceolate, attenuate at the apex, obtuse or rounded at the base, 4.5 x 1.0 mm, very sparsely covered with short hairs. *Bracteoles* subulate or lanceolate, much longer than calyx, 4.5 x 1.2 mm, nearly glabrous. *Pedicels* ascending, 1 mm long in flower, 1.5 mm long in fruit.

Flowers bright yellow. *Calyx* campanulate, 2.6 mm long, tube 2.0 mm long. *Standard* asymmetrical, obliquely elliptic, 4.3 x 5.5 mm, emarginate at the apex, without an internal appendage. *Right wing* half concealing right side of keel-petal, obliquely narrowly obovate, lamina 3.7 x 2.5 mm, claws 1.0 mm long, auricle 0.5 mm long. *Left wing* spreading forwards, lamina 4.3 x 2.0 mm, claws 1.0 mm long, auricle 0.5 mm long. *Keel-petals* slightly incurved to the left, 5.8 mm long, left keel with a short pocket, the pocket 0.5 mm long. *Ovary* 1.9 mm long, covered with white hairs; style 3.7 mm long, prolonged beyond stigma to form very short style-beak, the beak 0.1 mm long.

Fruits spreading to pendulous, 3.0 - 3.5 cm long, sparsely covered with short hairs, brown when mature, 4 - 7 seeded. *Seed* elliptic, 4.1 x 2.5 x 2.3 mm, surface smooth, pale brown when mature, hilum short, 1.0 x 0.2 mm, linear, not protruding, aril not developed. Germination epigeal. *First and second leaves* with petiole, leaflets lanceolate, attenuate at the apex, rather cordate at the base, 1.8 x 0.6 cm.

Geographic distribution: India, mainly northern and northwestern parts. The wild form of *V. aconitifolia* occurs sporadically mainly in the northern and northwestern plains of the Deccan plateau (Arora and Nayar, 1984). Babu *et al.* (1985) state it is wild in Orissa, eastern India.

Altitude: 100 - 600 m, mean 333 m (3 records)

Vernacular names:
Bengali: kheri
English: mat bean, moth bean, dew bean
Hindi: mooth, moth, bhringga
German: mattenbohne
Gujarati: mut, math
French: haricot papillon
Kanda: madik
Malaysia: mittikelu
Maranti: math, matki
Nepali: kulthi
Punjabi: bhionji

Santal: birmung, moch, birmoch

Singhala (Sri Lanka): makushtha.

Tamil (Sri Lanka): kollu, tulkapavir

Tamil (India): kallupayaru, nari payaru, pani payaru

Telagu: kuncuma pesalu

Thai: matpe

Origin: It is believed *V. aconitifolia* was domesticated in India, Pakistan, Myanmar or Sri Lanka (Purseglove, 1974; Maréchal *et al.*, 1978).

Uses: In South Asia, the green pods are eaten as vegetables and the ripe seeds, whole or split, are eaten after being cooked (Purseglove, 1974). Seeds are sprouted and eaten with or without salt, or fried and salted (Jain and Mehra, 1980).

Ecology: It is grown as a hot-season crop in India from sea level to 1320m. It requires uniform high temperatures and is highly drought resistant. It remains alive on very little water and does best with an evenly distributed rainfall of 700 mm per annum. Heavy rain is harmful. This species grows best in dry light soils (Purseglove, 1974). The wild forms occur along field bunds and roadsides (Babu *et al.*, 1985).

2.3.2. V. aridicola *N. Tomooka & Maxted, Kew Bull. 57(3):613-624 (2002a)*

Chromosome number: 2n=2x=22

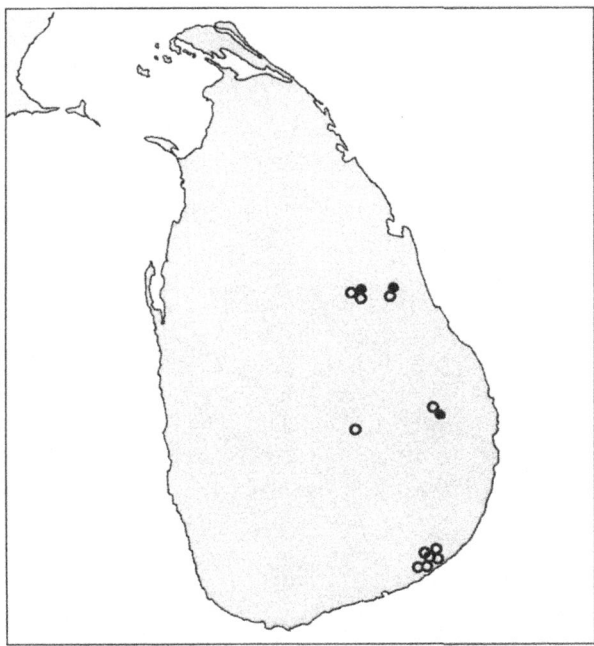

Figure 4.57. Distribution of Vigna aridicola *based on herbarium specimens (●) and direct germplasm collection (○) by author (N.T.).*

5 cm

*Figure 4.58. Lower plant parts and
flowering and fruiting shoot of* Vigna
aridicola. *Drawn from JP207977
(Sri Lanka, Type deposited at Kew,
as Tomooka CED2001SL28)
by Kyoko Motoyoshi.*

Taxonomic affinities and diagnosis: Based on molecular analysis, this species is closely related to *V. aconitifolia* within section *Aconitifoliae*. Useful characters to identify *V. aridicola* are stem with long spreading hairs; narrowly elliptic or oblong bracteole as long as calyx; pale yellow flower with purplish keel tip; brown or white bristle hairs on pod containing 5 - 7 seeds; mature seed having dull powder-like seed coat covering with protruding hilum; epigeal germination with first and second leaves without petiole.

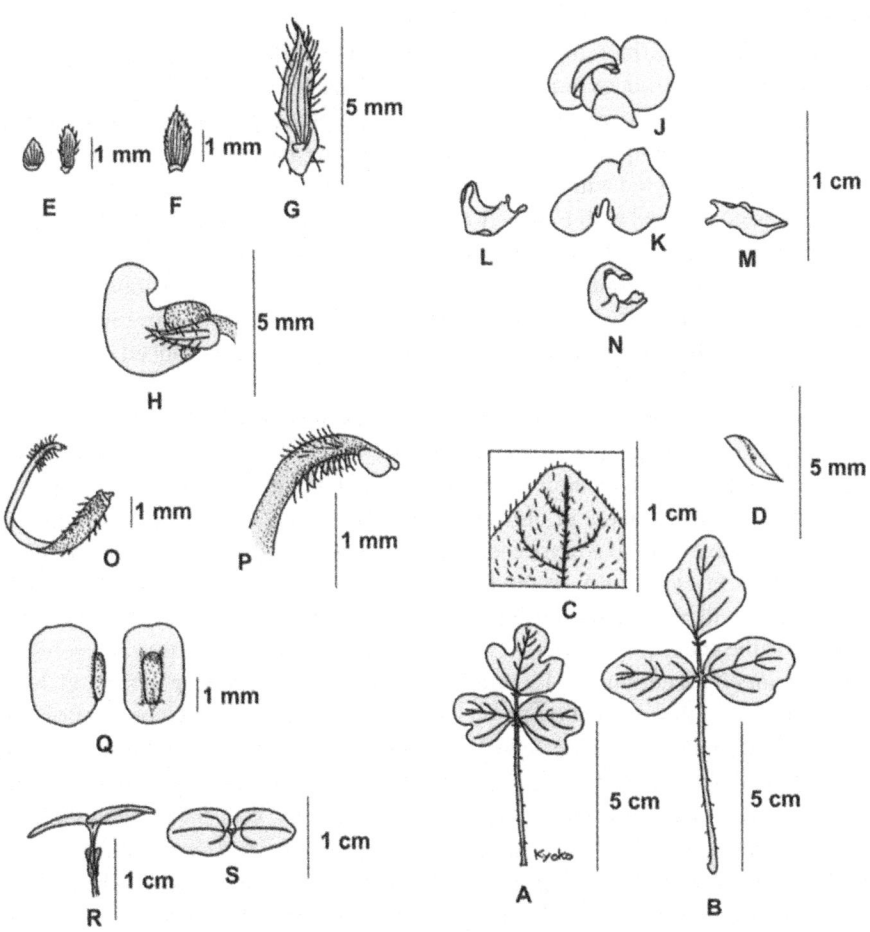

Figure 4.59. Plant parts of Vigna aridicola.
A & B: leaf; C: leaf apex; D: stipel; E: primary bract; F: secondary bract; G: stipule; H: flower bud and bracteole; J: flower; K: standard; L: right wing; M: left wing; N: keel; O: style; P: stigma and style beak; Q: seed; R & S: seedling shoot. Drawn from JP205896 (Sri Lanka) by Kyoko Motoyoshi.

Diagnostic characters which distinguish it from *V. aconitifolia* are entire to 3 lobed terminal leaflet; shorter bracteole; pale yellow flower; first and second leaves without petiole. *V. aridicola*, *V. stipulacea* and *V. trilobata* have been confused taxonomically because they often have similar leaflet morphology. However, compared with *V. aridicola*, *V. stipulacea* has nearly glabrous stem; larger ovate stipule; longer peduncle; shiny clear yellow flower; longer pod containing 13 - 14 seeds; hypogeal germination, whereas *V. trilobata* has orbicular to ovate stipule; golden yellow flower; smooth seed with orbicular protruding hilum having well developed aril.

Description: Based on JP205896 from Sri Lanka

A trailing herb, with a thick taproot. Initial stem internodes short, producing a basal rosette, secondary internodes long. *Stems* slender, densely covered with long (1 - 2 mm) spreading or retrorse yellowish white hairs. *Stipules* peltate, prolonged below the point of insertion, narrowly elliptic or oblong, acuminate at the apex, 3.5 - 5 x 0.8 - 1.5 mm, 5 - 9 nerved, sparsely covered with yellowish white hairs (0.5 - 1 mm). *Leaf petioles* 1.5 - 9 cm long, densely covered with long yellowish white hairs (1 - 2 mm) as stem, *rachis* 0.4 - 1.6 cm. *Leaflets* membranous, sparsely or densely covered with long (0.7 - 2 mm) spreading white hairs on both surfaces; *terminal leaflet* ovate to rhomboid, deeply to shallowly 3-lobed or nearly entire, acute or obtuse at the apex, obtuse or rounded at the base, 1.7 - 4.2 x 1.4 - 3.6 cm, *lateral leaflets* somewhat oblique, 2 - 3 lobed or entire, 1.1 - 4 x 1 - 3 cm; stipels narrowly elliptic, 2 - 2.6 mm long.

Inflorescence axillary, 5 - 10 flowered; peduncles slender, 5.5 - 15 cm long, sparsely to densely covered with long (1 - 1.3 mm) retrorse yellowish white hairs; rachis 2 - 9 mm long, sparsely covered with short (0.2 - 0.3 mm) white hairs. *Primary bract* peltate, ovate to broadly ovate, acute or obtuse at the apex, 1 - 1.5 x 0.6 - 1 mm, glabrous or sparsely white short hairy (0.2 mm), caducous. *Secondary bract* (pedicel bract) basifixed, narrowly ovate or cymbiform, acute at the apex, 1.8 - 2.7 x 0.8 - 1 mm, glabrous to sparsely ciliate with whitish short hairs (0.2 mm), caducous. *Bracteoles* narrowly elliptic, acuminate at the apex, 2 - 2.5 x 0.5 - 0.7 mm, as long as calyx, glabrous to sparsely hairy outside. *Pedicels* ascending, 1 - 1.2 mm long in flower, 1 - 2 mm long in fruit, glabrous.

Flowers pale yellow with purplish keel tip, 7 - 8 mm in diameter. *Calyx* campanulate, 1.5 - 2 mm long, glabrous outside, tube 0.9 - 1 mm long. *Standard* asymmetric, obliquely and broadly elliptic, 4.7 - 5.3 x 7.3 - 7.4 mm, emarginate at the apex, without an internal appendage. *Right wing* concealing the upper portion of the keel-petals, claws 0.9 - 1.1 mm long, lamina obliquely obovate, 3.5 x 2.6 - 3 mm, auricle 0.6 - 0.8 mm long. *Left wing* spreading, claws 0.8 - 0.9 mm long, lamina obliquely obovate or obliquely elliptic, 4 - 4.5 x 2.7 - 3.4 mm, auricle 0.6 - 0.8 mm long. *Keel-petals* spirally incurved to the left through 230° - 250°, 6 - 6.6 mm long, pocket on the left side petal 0.9 mm long. *Ovary* 1.9 mm long, covered with brown hairs (0.3 - 0.4 mm), 5 - 7 ovuled; style 6 - 7 mm long, shortly beaked beyond the stigma, the style beak 0.06 mm long.

Fruits spreading, linear, 1.5 - 2.8 cm long, 3 mm wide, with brown or white bristle-like upward pointing hairs (0.3 - 1 mm), blackish brown when mature, 5 - 7 seeded. *Seeds* ellipsoidal, 2.5 - 2.8 x 1.8 - 2.3 x 1.6 - 2 mm, brown or yellowish brown mottled with black spot when mature, with a dull white powdery like covering; hilum linear, 1.1 - 1.3 x 0.5 mm, more or less protruding, aril not developed. Germination epigeal; *first and second foliage leaves* simple, narrowly ovate (0.8 x 0.6 cm), sessile.

Geographical distribution: Dry zone of Sri Lanka.

Altitude: Sea level to 175 m, mean 70 m (13 records)

Phenology: Sri Lanka – February to March.

Ecology: This species grows in dry open exposed coastal and inland areas of Sri Lanka. It can be found growing in the same habitat as *V. stipulacea* and *V. trilobata*.

Soil type: Usually found in sandy soil.

Associated vegetation: Usually in grassland with species such as *Mimosa* sp. and *Atelosia* sp. Commonly found with *V. trilobata* but not on beaches. *V. aridicola* has broader ecological amplitude than *V. trilobata* in Sri Lanka since it can grow in turf grass inland. This species has been found growing adjacent to both *V. trilobata, V. stipulacea* and *V. radiata* var. *sublobata*.

Habitat: Usually grows in full sun or light shade. Adapted to disturbed habitats such as areas grazed by cattle or wild animals, or places where vegetation is regularly cut such as parkland.

Pests and diseases: Roots found heavily infested with nematodes at Welpallewela, Mahiyangana, Sri Lanka. Leaves and pods sometimes show signs of insect damage.

Habit: Usually trailing, ocassionally twining.

2.3.3. V. khandalensis *(Santapau) Raghavan & Wadhwa, Curr. Sci. Bangalore 41: 429 (1972)*

Chromosome number: Not reported

Taxonomic affinities and diagnosis: This species has been little studied but is considered as a member of section *Aconitifoliae*. Useful characters to identify *V. khandalensis* are erect stem; very large stipule; 3 lobed terminal leaflet; 2 lobed lateral leaflet; very large bracteole concealing flower bud. This species has very specific morphological characters and can easily be distinguished from other species.

Description: Based on Sedgwick & Bell 7953 from India, deposited at Kew

An erect herb. *Stems* angular, sparsely covered with retrorse brown hairs on angles. *Stipules* very large, 35.5 - 47.8 x 21.1 - 30.7 mm, broadly ovate, truncate at the base, sparsely covered with short hairs outside. *Leaf petioles* 4.8 cm long, sparsely covered with brown hairs. *Leaflets* sparsely covered with short yellowish hairs on both surfaces; *terminal leaflets* conspicuously 3 lobed, acute at the apex, obtuse at the base, 6.8 x 5.8 cm; *lateral leaflets* 2 lobed, obtuse at the apex, obtuse at the base, 4.3 x 3.5 cm; stipels lanceolate, 3.0 x 1.0 mm.

Inflorescence axillary, 4 - 10 flowered; peduncles 8.6 - 12.2 cm long, sparsely covered with long downward brown hairs, rachis 2.2 - 2.5 cm long, sparsely covered with short brown hairs. *Primary bract* ovate, 5 x 4 mm, sparsely covered with short hairs outside. *Secondary bract* (pedicel bract) cymbiform, obtuse at the apex, 10 x 5 mm, sparsely covered with short appressed hairs. *Bracteoles* very large, concealing the flower bud, cymbiform, obtuse at the apex, truncate at the base, 11 x 5 mm, ciliate with brown hairs on the margin. *Pedicels* ascending, 2.8 - 3.2 mm long in fruit.

Flowers yellow. *Calyx* campanulate, 3.2 mm long, tube 2.5 mm long. *Standard* asymmetrical, obliquely elliptic, 8 x 10 mm, without an internal appendage. *Right wing* obliquely obovate, lamina 7 x 5 mm, claws long elongated, 3 - 4 mm long, auricle 1.5 mm long. *Left wing* obovate, lamina 6.5 x 5.0 mm, claws elongated, 2.5 mm long, auricle 1.3 mm long. *Keel-petals* incurved to the left, left keel with a short keel pocket, the pocket 0.5 mm long. *Ovary* densely covered with white and brown hairs; style prolonged beyond stigma to form short style-beak, the beak 0.2 mm long.

Fruits ascending, linear, 5.0 - 6.3 x 0.3 - 0.4 cm, covered with long appressed brown hairs, 8 - 12 seeded. *Seed* oblong, 3.5 x 3.2 x 2.9 mm, blackish brown when mature, hilum linear, 1.5 x 0.4 mm, not protruding, aril not developed. Germination type unknown.

Geographical distribution: India

Ecology: A rainforest species.

Altitude: 150 m to 1000 m, mean 716 m (3 records).

Uses: Seed is harvested and is consumed in time of famine (Babu *et al.*, 1985).

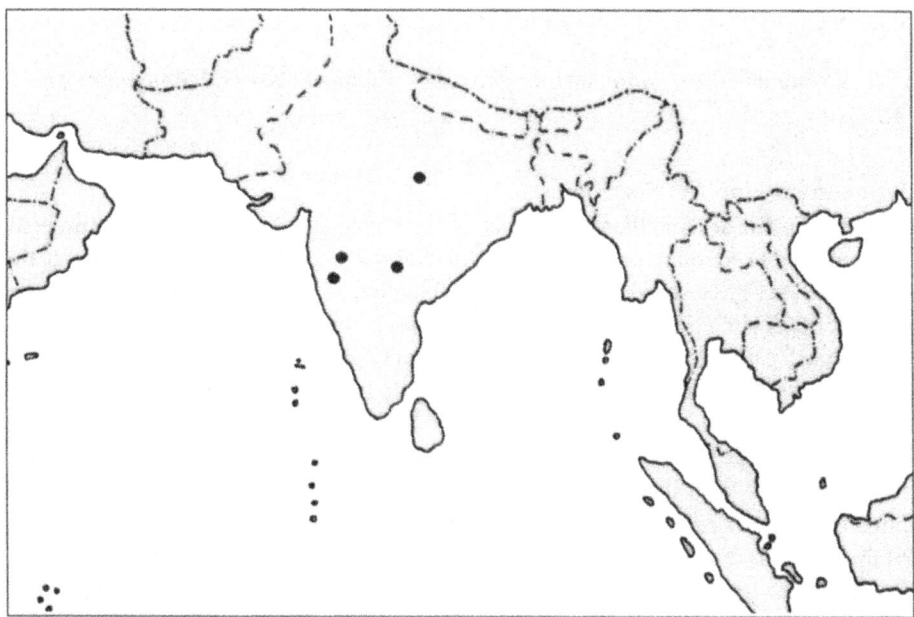

Figure 4.60. Distribution of Vigna khandalensis *based on herbarium specimens.*

5 cm

Figure 4.61. Flowering and fruiting shoot of Vigna khandalensis.
Drawn by Kyoko Motoyoshi from Sedgwick & Bell 7953 (India), Type, deposited at Kew.

2.3.4. V. stipulacea *Kuntze, Rev. Gen. 212 (1891)*

Chromosome number: 2n=2x=22

Taxonomic affinities and diagnosis: A clearly differentiated species in section *Aconitifoliae.* It appears most closely related to *V. trilobata* with which it has sometimes been confused. Useful characters to identify *V. stipulacea* are conspicuously large ovate stipule; long peduncle; shiny clear yellow flower; hairy mature pod containing 13 – 14 seeds; rough seed surface with slightly protruding hilum having slightly developed aril; hypogeal germination with petiolate first and second leaves. Diagnostic characters that distinguish it from *V. trilobata* are larger stipule; seed with less protruding oblong hilum having less developed aril; hypogeal germination.

Description: Based on JP 205892 from Sri Lanka

A trailing and twining herb. *Stems* angular, nearly glabrous, very sparsely covered with fine white hairs, 0.4 - 0.5 mm long. *Stipules* conspicuously large, ovate, acute at the apex, truncate or rather cordate at the base, 11.8 - 15.0 x 5.6 - 9.3 mm, surface glabrous, ciliate with short (0.3 mm) white hairs on the margin. *Leaf petioles* 5.1 - 7.6 cm long, sparsely covered with fine appressed white hairs, 0.4 - 0.6 mm long. *Leaflets* glossy, upper surface very rarely covered with white short (0.3 - 0.4 mm) hairs, lower surface sparsely covered with 0.5 - 0.6 mm long white hairs, 0.4 - 0.8 mm long whitish hairs on vein; *terminal leaflets* 3 lobed, obtuse at the apex, obtuse or rounded at the base, 1.7 - 2.6 x 1.8 - 2.6 cm; *lateral leaflets* oblique, rather shallowly 2 - 3 lobed, obtuse or rounded at the apex, rounded at the base, 1.6 - 2.5 x 1.2 - 2.4 cm; stipels ovate, acuminate at the apex, 1.6 x 1.0 mm.

Inflorescence axillary, 4 - 12 flowered; peduncles conspicuously long (22.7 - 29.6 cm), nearly glabrous to very sparsely covered with 0.4 - 0.5 mm long white hairs, rachis 3.7 - 4.7 mm long, glabrous. *Primary bract* ovate, acute at the apex, rounded or sometimes truncate at the base, 2.3 x 2.2 mm, glabrous. *Secondary bract* (pedicel bract) narrowly ovate, acuminate at the apex, obtuse at the base, 3.3 x 1.3 mm, surface glabrous, sparsely ciliate with short hairs on upper margin. *Bracteoles* subulate, attenuate at the apex, obtuse or truncate at the base, as long as calyx, 3.8 x 1.3 mm, glabrous. *Pedicels* ascending, 0.9 - 1.0 mm in flower, 1.7 mm in fruit.

Flowers shiny clear yellow. *Calyx* campanulate, 2.6 mm long, tube 2.2 mm long. *Standard* asymmetrical, obliquely elliptic, 8.1 x 12.3 mm, emarginate at the apex, an internal appendage absent. *Right wing* obliquely broadly obovate, concealing right side of keel-petal, lamina 5.6 x 6.3 mm, claws 1.8 mm long, auricle 0.9 mm long. *Left wing* spreading forward, obliquely obovate, lamina 6.5 x 4.8 mm, claws 1.4 mm long, auricle 0.5 mm long. *Keel-petals* spirally incurved to the left, 8.7 mm long, left keel-petal with a short mound-like pocket, the keel-pocket 1.3 mm long. *Ovary* 2.4 mm long, hairy; style 6.4 mm long, slightly prolonged beyond stigma to form style-beak, the beak 0.2 mm long.

Fruits spreading, linear, 4.5 - 5.5 x 0.2 - 0.3 cm, covered with 0.3 - 0.5 mm long brown hairs, blackish brown when mature, 13 - 14 seeded. *Seed* elliptic, 2.7 x 2.0 x 1.9 mm, brown

with small black mottle, surface rough, fine reticulate; hilum oblong, slightly protruding, aril slightly developed. Germination hypogeal. *First and second leaves* with petiole, simple, broadly elliptic, rather cordate at the apex and at the base, 1.1 x 1.0 cm.
Geographical distribution: The main center of distribution is South Asia. It is recorded from India, Bangladesh, Sri Lanka, Myanmar, Vietnam, Indonesia as far east as Irian Jaya, Madagascar and Yemen.
Altitude: 5 m to 700 m, mean 189 m (10 records)
Phenology: Sri Lanka - February to March; India – October to December; Myanmar - October to November.
Ecology: This species is a pioneer of secondary succession.
Soil type: Muddy.
Associated vegetation: Grasses.
Habitat: Found in open or lightly shaded habitats, particularly at the edge of paddy fields or in paddy fields that have been abandoned. Well adapted to habitats with high or intermediate disturbance.
Habit: Trailing.
Uses: Ethnic groups in parts of India apparently gather seeds of this species for food (Roxburgh, 1874; Maxwell, 1991; where this species is considered as part of *V. trilobata*).

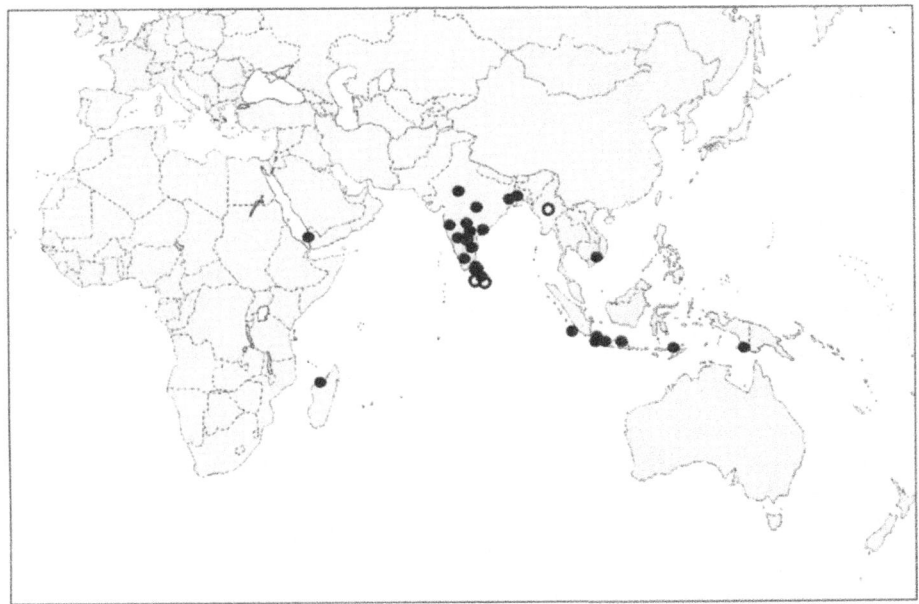

Figure 4.62. Distribution of Vigna stipulacea *based on herbarium specimens*
(●) and direct germplasm collection (○) by authors (N.T., D.A.V.).

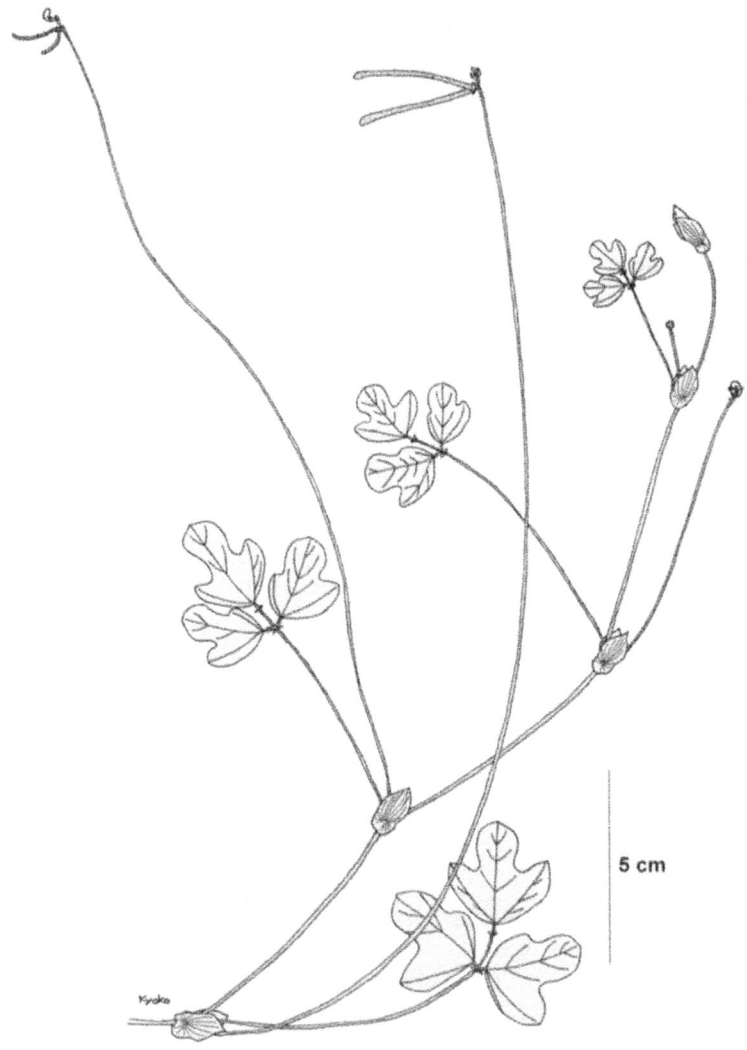

Figure 4.63. Flowering and fruiting shoot of Vigna stipulacea.
Drawn fron JP207976 (Sri Lanka) by Kyoko Motoyoshi.

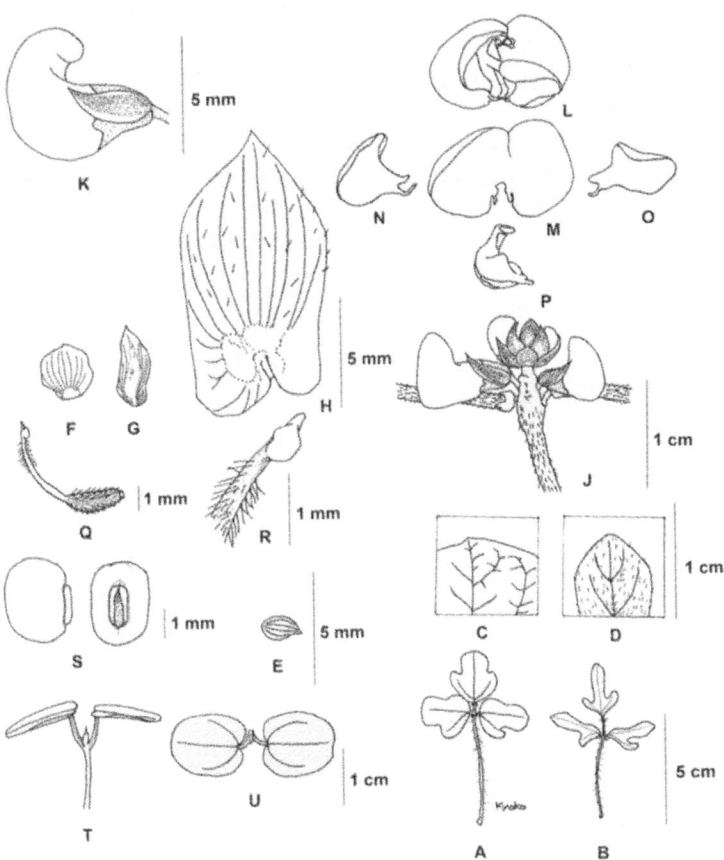

Figure 4.64. Plant parts of Vigna stipulacea.

A & B: leaf; C & D: leaf apex; E: stipel; F: primary bract; G: secondary bract; H: stipule; J: inflorescence: K: flower bud and bracteole; L: flower; M: standard; N: right wing; O: left wing; P: keel; Q: style; R: stigma and style beak; S: seed; T & U: seedling shoot. Drawn from JP205892 (Sri Lanka), except B & D from JP207975 (Sri Lanka) by Kyoko Motoyoshi.

2.3.5. V. trilobata *(L.)Verdcourt, in Taxon 17: 172 (1968)*

Chromosome number: 2n=2x=22

Taxonomic affinities and diagnosis: This species is most closely related to *V. stipulacea* within section *Aconitifoliae*. Useful characters to identify *V. trilobata* are orbicular to ovate stipule; golden yellow flower; glabrous mature pod; seed with protruding orbicular hilum having well developed aril. Diagnostic characters that distinguish it from *V. stipulacea* are smaller stipule; shorter peduncle; more protruding orbicular hilum with more developed aril; epigeal germination. Differs from *V. aridicola* by orbicular to ovate stipule; golden yellow flower; more protruding hilum with more developed aril; petiolate first and second leaves.

Description: Based on JP205895 from Sri Lanka

A trailing herb. *Stems* slender, glabrous to rather densely covered with fine yellowish hairs (1 - 2 mm long). *Stipules* orbicular to ovate, obtuse or acute at the apex, rounded at the base, 4.4 x 2.8 mm, ciliate with 0.3 - 0.4 mm long white hairs on the margin. *Leaf petioles* 1.5 - 7.0 cm long, glabrous to rather densely covered with white or yellowish hairs (0.3 - 1.5 mm long). *Leaflets* membranous, surface glabrous or sparsely covered

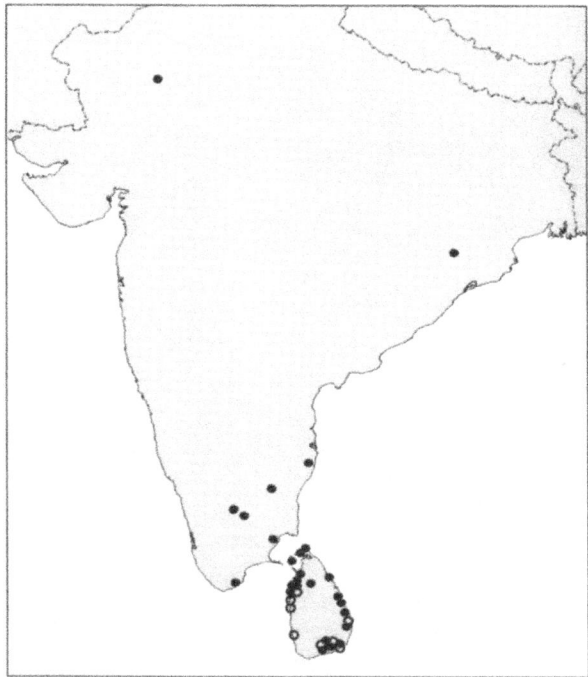

Figure 4.65. Distribution of Vigna trilobata *based on herbarium specimens(●) and direct germplasm collection (○) by author (N.T.).*

with short (0.3 mm) white hairs, ciliate with short (0.3 mm) white hairs on the margin; *terminal leaflets* ovate, entire or deeply 3 lobed, 1.5 - 2.3 x 1.6 - 2.1 cm; *lateral leaflets* ovate, entire or deeply 2 - 3 lobed, 1.3 - 2.1 x 1.2 - 2.0 cm; stipels ovate, 1.7 x 0.6 mm.

Inflorescence axillary, 2 - 6 flowered; peduncles glabrous or hairy as stems, 5.2 - 6.3 cm long; rachis 16.4 mm. *Primary bract* ovate or elliptic, obtuse or rounded at the apex, truncate at the base, 1.4 x 1.1 mm, glabrous. *Secondary bract* (pedicel bract) elliptic, acute or acuminate at the apex, truncate at the base, 2.1 x 1.1 mm, glabrous. *Bracteoles* lanceolate, acute or acuminate at the apex, truncate at the base, 2.2 x 1.2 mm, as long as or a little longer than calyx, glabrous or ciliate with short hairs on the upper margin. *Pedicels* ascending, 0.6 - 2.3 mm long in flower, 0.5 - 2.5 mm long in fruit.

Flowers golden yellow. *Calyx* campanulate, 2.5 mm long, tube 1.9 mm long, glabrous. *Standard* asymmetrical, obliquely elliptic, 7.6 x 11.1 mm, emarginate at the apex, an internal appendage absent. *Right wing* half concealing the right side of keel-petals, lamina broadly obovate, 6.2 x 5.5 mm, claws 1.8 mm long, auricle 1.0 mm long. *Left wing* spreading forward, lamina obliquely obovate, 7.0 x 5.7 mm, claws 1.7 mm long, auricle 1.2 mm long. *Keel-petals* spirally incurved to the left, 10.2 mm long, left keel-petal with a short horn-like pocket, the pocket 1.3 mm long. *Ovary* 2.6 mm long; style 10.2 mm long, prolonged beyond the stigma to form style-beak, the beak 0.4 mm long.

Fruits spreading, linear, 2.8 - 3.5 x 0.2 - 0.3 cm, short (0.1 - 0.2 mm long) white hairy when young, glabrous and brown when mature, 6 - 8 seeded. *Seed* elliptic, 2.3 x 1.8 x 1.7 mm, orange or brown when mature, surface smooth; hilum orbicular, 0.7 x 0.5 mm, protruding, aril well developed. Germination epigeal. *First and second leaves* cordate, 0.8 x 0.8 cm, with petiole.

Geographical distribution: In Sri Lanka confined to the coastline of both the wet and dry zone. It occurs in India mainly in Tamil Nadu state.

Altitude: Sea level to 500 m, mean 57 m (46 records)

Phenology: India – September to November; Sri Lanka – February to March.

Ecology: Two ecotypes has been found in Sri Lanka. Beach ecotype has glossy thick leaves often with entire leaflets and glabrous stems which has very thick and deep tap root (Fig. 4.67 A). Inland ecotype has deeply lobed rather thin leaflets and hairy stems (Fig. 4.67 C).

Soil type: Sandy soil.

Associated vegetation: Grasses, *Ipomea* sp., thorny bushes and *Canavalia* sp. On the upper reaches of sandy beaches *V. trilobata* may grow as isolated scattered plants or under thorny plants. It has been found growing in the same habitat as *V. aridicola* and *V. stipulacea* at Yala National Park, Sri Lanka.

Habitat: This species grows in Sri Lanka within about 5km of the coastline. It is commonly found on beaches where it may grow under thorny plants where animal grazing is common. It grows in open or lightly shaded habitats such as uder coconut plantation near the beach. It usually grows in disturbed habitats where it may be grazed by cattle or wild animals. Appears to be adapted to very dry conditions

Pests and diseases: Pod borers, stinkbug damage, nematode damage and galls on stem have been reported.

Habit: Trailing.

Uses: *V. trilobata* is reported to be cultivated in India as a cover crop and as a cattle fodder. Reported that the seeds of this species are gathered by tribal people (Roxburgh, 1874; Maxwell, 1991) may be due to confusion with *V. stipulacea* that produces much more easily gathered seeds in abundance.

Vernacular names:

English: jungli bean

Hindi: mugani

5 cm

Figure 4.66. Flowering and fruiting shoot of Vigna trilobata.
Drawn from JP207979 (Sri Lanka) by Kyoko Motoyoshi.

India: Phillipesara
Pakistan: mukni, jongli-math
Sinhala (Sri Lanka): bin-me, munwenna.
Tamil (Sri Lanka): navippayaru, pachapayaru, pani-payir

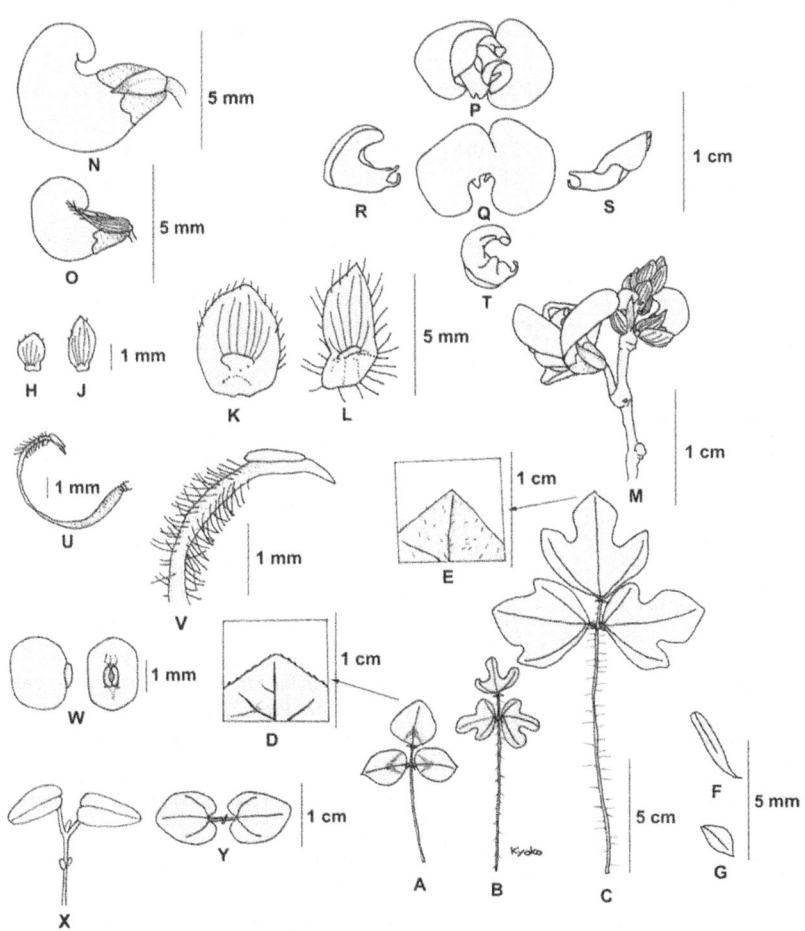

Figure 4.67. Plant parts of Vigna trilobata.
A, B & C: leaf; D & E: leaf apex; F & G: stipel; H: primary bract; J: secondary bract; K
& L: stipule; M: inflorescence: N & O: flower bud and bracteole; P: flower; Q: standard;
R: right wing; S: left wing; T: keel; U:style; V: stigma and style beak; W: seed; X & Y:
seedling shoot. Drawn from JP205895 (Sri Lanka) by Kyoko Motoyoshi.

COLOR PLATES 1 - 8 : Plant parts
Genebank reference number (Japan) and passport information for the accessions used in Plates

Plant parts (Plate 1 - 8)	JP/Col. no.	Passport information
V. aconitifolia	JP105629	local cultivar "Mukuni", Horti, 68km S of Solapur toward Bijapur, Karnataka, India, alt. 530m
V. angularis var. *angularis*	JP72985	local cultivar "Dainagon", Ategimori, Manba, Gunma, Japan
V. angularis var. *nipponensis*	JP107861	Sendai, Miyagi, Japan
V. aridicola	JP205896	road side, Gongala Wewa, Yala Nat. Park, Southern Prov., Sri Lanka, alt. 5m
V. exilis	JP205884	Kao Nov, Pub Pra, Ratchaburi, Thailand, alt. 150m
V. grandiflora	JP107862	road side, Ban Rakam,15km SW of Phitsanulok, Thailand, alt. 150m
V. hirtella	JP108851	road side, Kampong Lalok, 107km S of Kota Baharu, Kelantan, Malaysia
V. minima	JP205886	road side, 7km S of Boklua, Nan, Thailand, alt. 625m
V. mungo var. *mungo*	JP109668	local cultivar "Subsomotod", Phetchabun, Thailand
V. mungo var. *silvestris*	JP107874	India
V. nakashimae	JP107879	Ukushima, Goto islands, Nagasaki, Japan
V. nepalensis	JP107881	Arun valley, Nepal
V. radiata var. *radiata*	JP110830	released cultivar "Chai Nat 60", Thailand
V. radiata var. *sublobata*	JP107877	Madagascar
V. reflexo-pilosa var. *glabra*	JP109684	local cultivar, the Philippines
V. reflexo-pilosa var. *reflexo-pilosa*	JP108815	road side, Near Yonara bridge, Iriomote Is., Okinawa, Japan
V. riukiuensis	JP108810	open pasture, 2km NW of Kubura,Yonaguni Is., Okinawa, Japan
V. stipulacea	JP205892	beside tank embankment, Komawa Wewa, Yala Nat. Park, Southern Prov., Sri Lanka, alt. 5m
V. subramaniana	JP110836	Kalka-Dharampur rd., Himachal Pradesh, India, alt. 650m
V. tenuicaulis	JP109682	road side, Tambon Sansai, 6km S of Chiang Rai, Thailand
V. trilobata	JP205895	beach, Talawila, Puttalam Distr., Northern Prov., Sri Lanka, alt. 3m
V. trinervia var. *trinervia*	JP108840	road side, Kampong Batu Balai, 58km SE of Kuala Lipis, Pahang, Malaysia
V. trinervia var. *bourneae*	JP207981	Pallangi, Palni Hills, near Kodaikanal, Dindugal Distr., India, alt. 1670m
V. umbellata (cultivated)	JP100311	local cultivar "Ghore", Pangdwa, Pakhribas V.P., Nepal, alt. 1580m
V. umbellata (wild)	JP109675	road side, Fai Gae, 6km W of Nan, Thailand

COLOR PLATES 9 - 12 : Habitat
Genebank reference number (Japan) and passport information for the accessions used in Plates

Habitat (Plate 9 - 12)	JP/Col no.	Passport information
V. angularis var. nipponensis	JP90857	beside paddy, Gobo, Wakayama, Japan
V. aridicola	JP207977	open sandy grassland, Welpallewela, Mahiyangana, Kandy Distr., Central Prov., Sri Lanka, alt. 130m
V. dalzelliana	2001SL35	on wet slope, Kalpahana, Badulla Distr., Sri Lanka, alt. 790m
V. exilis	JP207983	growing on limestone, Ban Tum Ma Dear, A. Sai Yok., Kanchanaburi, Thailand, alt. 130m
V. grandiflora	JP108509	road side, 11km SW of Nakhon Sawan, Thailand, alt. 50m
V. hirtella	JP205885	Khao Yai Nat. Park, Pakchong, Nakorn Ratchasima, Thailand, alt. 750m
V. minima	JP210824	beside paddy, Mingala Ywathit, Pa-An, Kayin State, Myanmar, alt. 15m
V. nakashimae	JP110356	abondoned paddy, Narao, Goto Is., Nagasaki, Japan, alt. 5m
V. radiata var. sublobata	JP211874	near waterfall., Ahnesakan, Pyin-Oo-Lwin, Mandalay Div., Myanmar, alt. 855m
V. reflexo-pilosa var. reflexo-pilosa	JP201505	road side slope, Mandahara, Yonaguni Is., Okinawa, Japan
V. riukiuensis	JP201499	on cliff, Sanninu Dai, Yonaguni Is., Okinawa, Japan, alt. 40m
V. stipulacea	JP207975	abandoned paddy, 3km from Tabbowa, Puttalum Distr., Sri Lanka, alt. 1m
V. tenuicaulis	JP210817	beside paddy, border of Dawei and Thayetchaaung, Myanmar, alt. 15m
V. trilobata	JP210606	beach, near Patanangala Bangalow, Yala Nat. Park, Southern Prov., Sri Lanka
V. trinervia var. trinervia	JP207980	wet place, mountains of Hakagala Garden, Nuwara Eliya Distr., Sri Lanka, alt. 1630m
V. umbellata (wild)	JP108520	beside paddy, 13.5 km S of Tak, Thailand, alt. 155m

V. aconitifolia

V. angularis var. *angularis*

V. angularis
var. *nipponensis*

V. aridicola

V. exilis

V. grandiflora

V. hirtella

V. minima

V. mungo var. *mungo*

V. mungo
var. *silvestris*

V. nakashimae

V. nepalensis

1 mm

Plate 1. Seeds

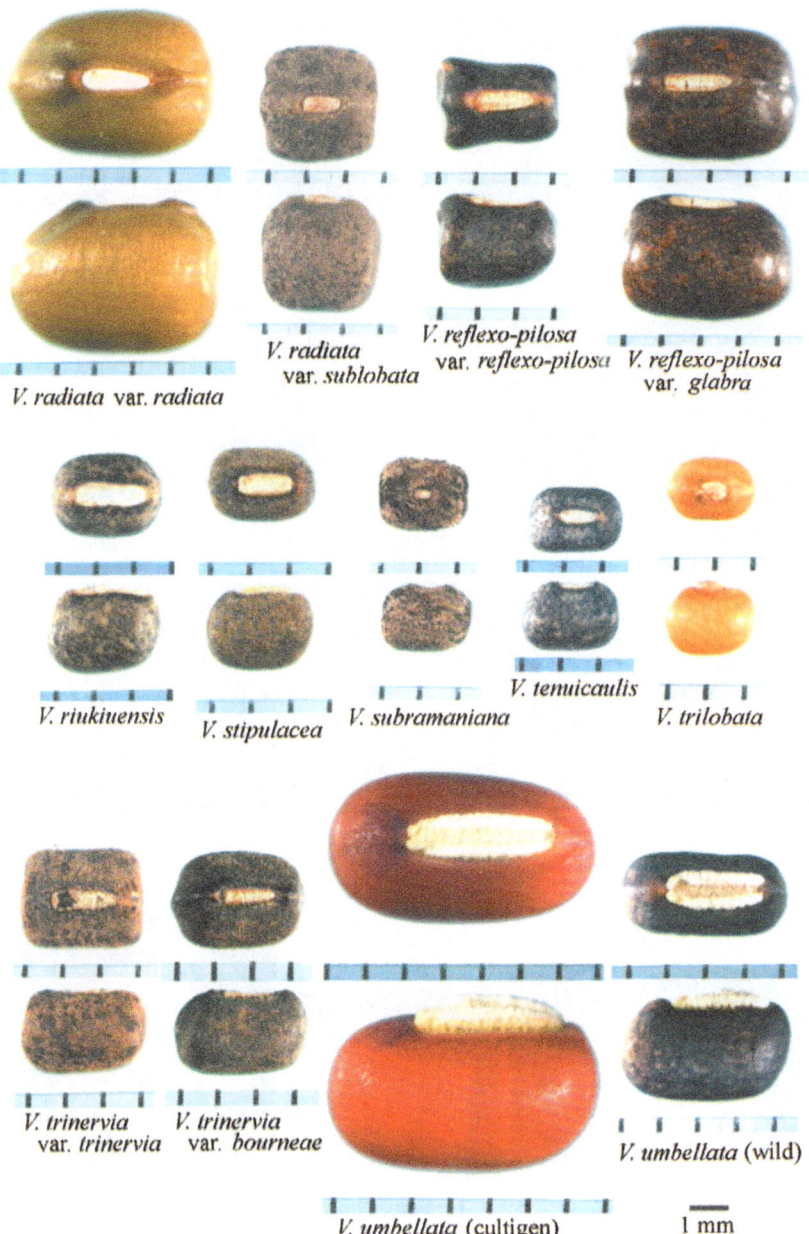

V. radiata var. *radiata*

V. radiata var. *sublobata*

V. reflexo-pilosa var. *reflexo-pilosa*

V. reflexo-pilosa var. *glabra*

V. riukiuensis

V. stipulacea

V. subramaniana

V. tenuicaulis

V. trilobata

V. trinervia var. *trinervia*

V. trinervia var. *bourneae*

V. umbellata (wild)

V. umbellata (cultigen)

1 mm

Plate 2. Seeds

154

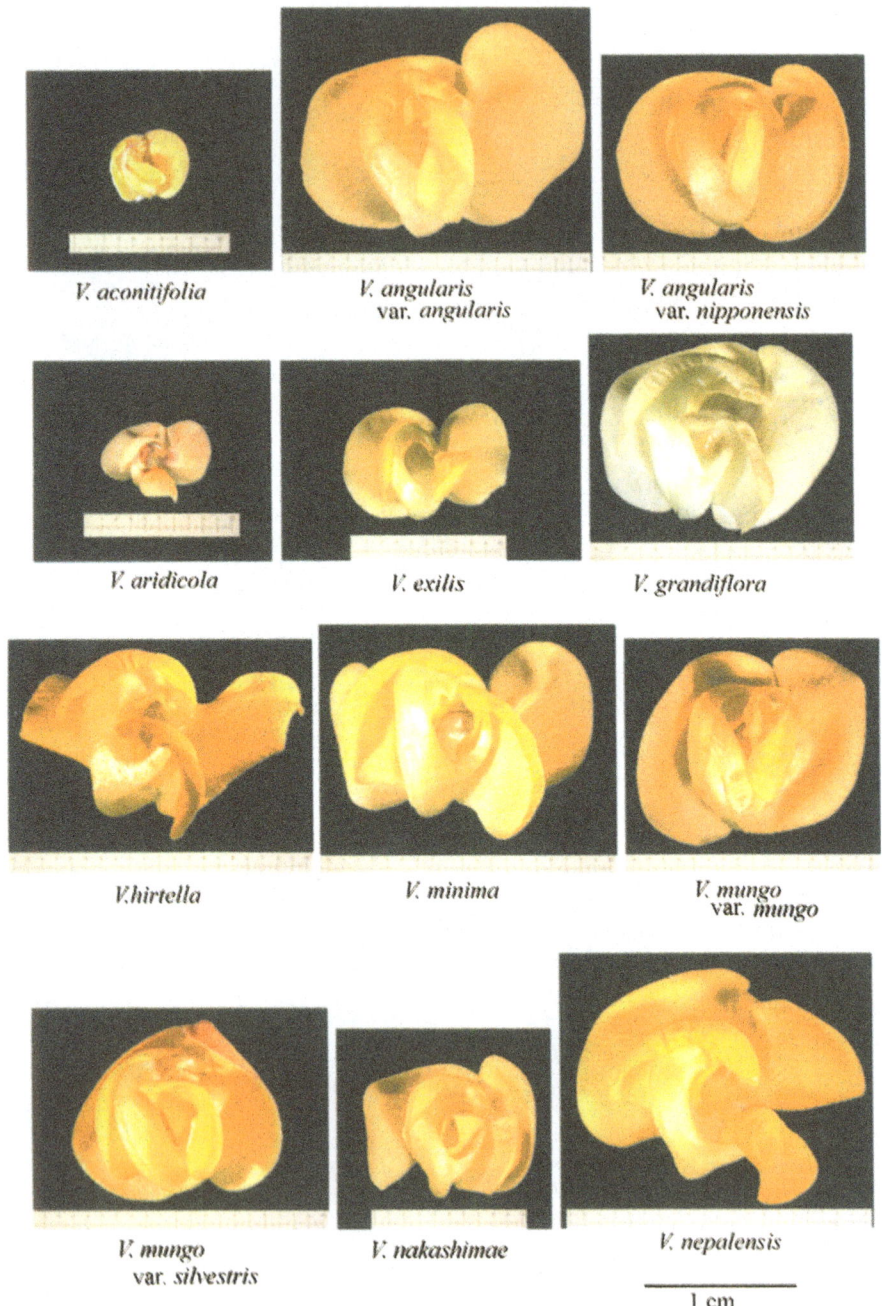

Plate 3. Flowers

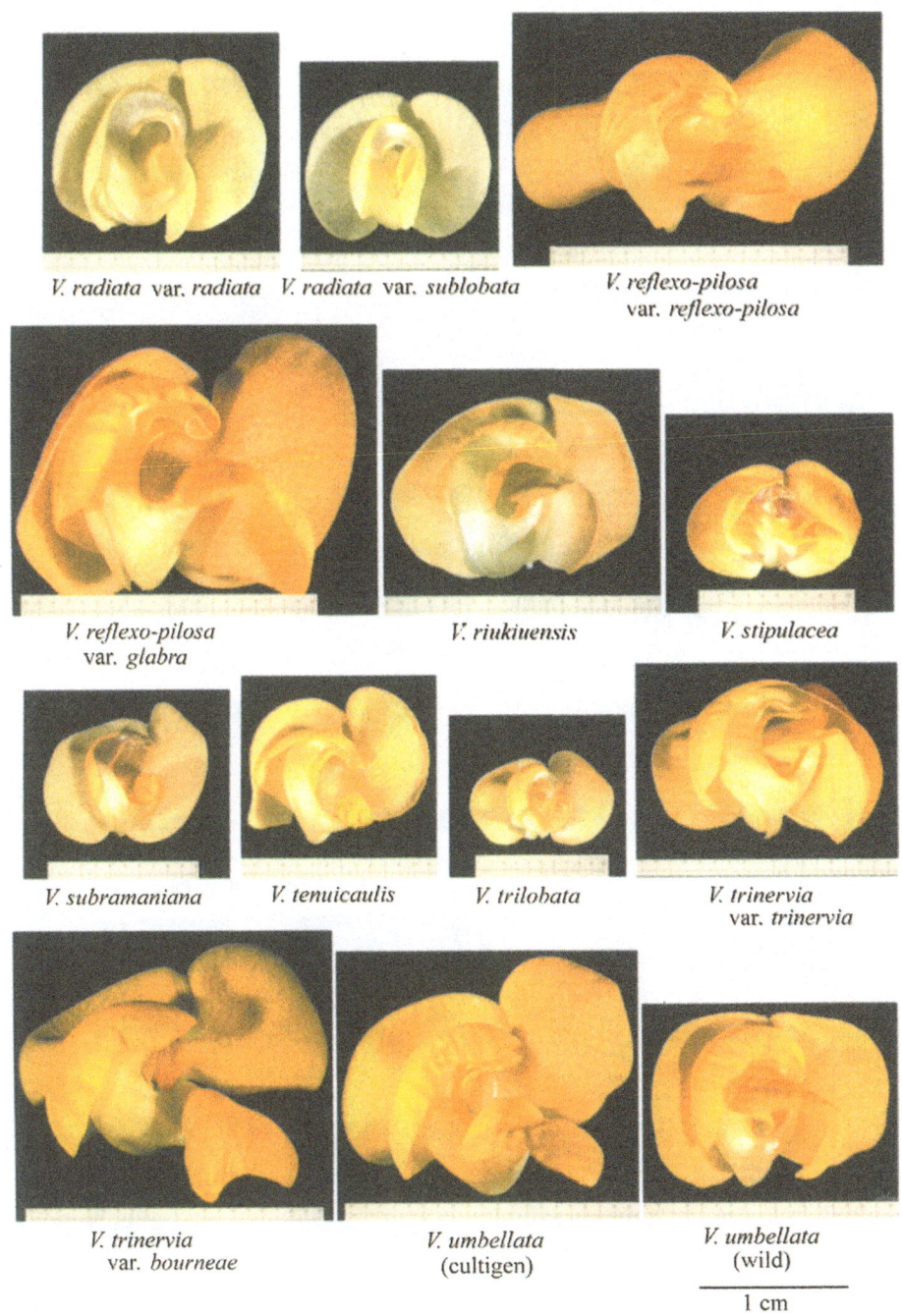

V. radiata var. radiata   V. radiata var. sublobata   V. reflexo-pilosa
var. reflexo-pilosa

V. reflexo-pilosa
var. glabra   V. riukiuensis   V. stipulacea

V. subramaniana   V. tenuicaulis   V. trilobata   V. trinervia
var. trinervia

V. trinervia
var. bourneae   V. umbellata
(cultigen)   V. umbellata
(wild)

1 cm

Plate 4. Flowers

156

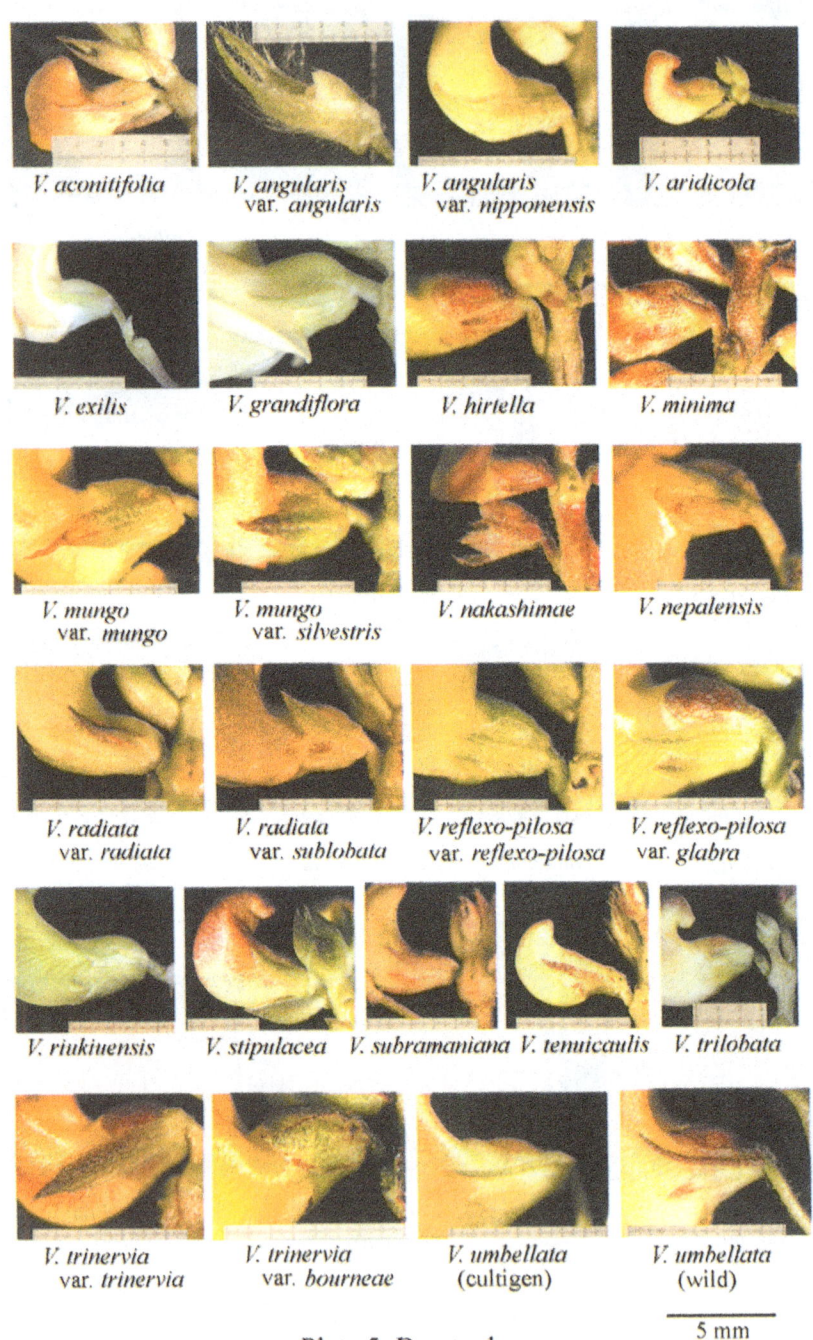

V. aconitifolia   *V. angularis*   *V. angularis*   *V. aridicola*
             var. *angularis*   var. *nipponensis*

V. exilis   *V. grandiflora*   *V. hirtella*   *V. minima*

V. mungo   *V. mungo*   *V. nakashimae*   *V. nepalensis*
var. *mungo*   var. *silvestris*

V. radiata   *V. radiata*   *V. reflexo-pilosa*   *V. reflexo-pilosa*
var. *radiata*   var. *sublobata*   var. *reflexo-pilosa*   var. *glabra*

V. riukiuensis   *V. stipulacea*   *V. subramaniana*   *V. tenuicaulis*   *V. trilobata*

V. trinervia   *V. trinervia*   *V. umbellata*   *V. umbellata*
var. *trinervia*   var. *bourneae*   (cultigen)   (wild)

5 mm

Plate 5. Bracteoles

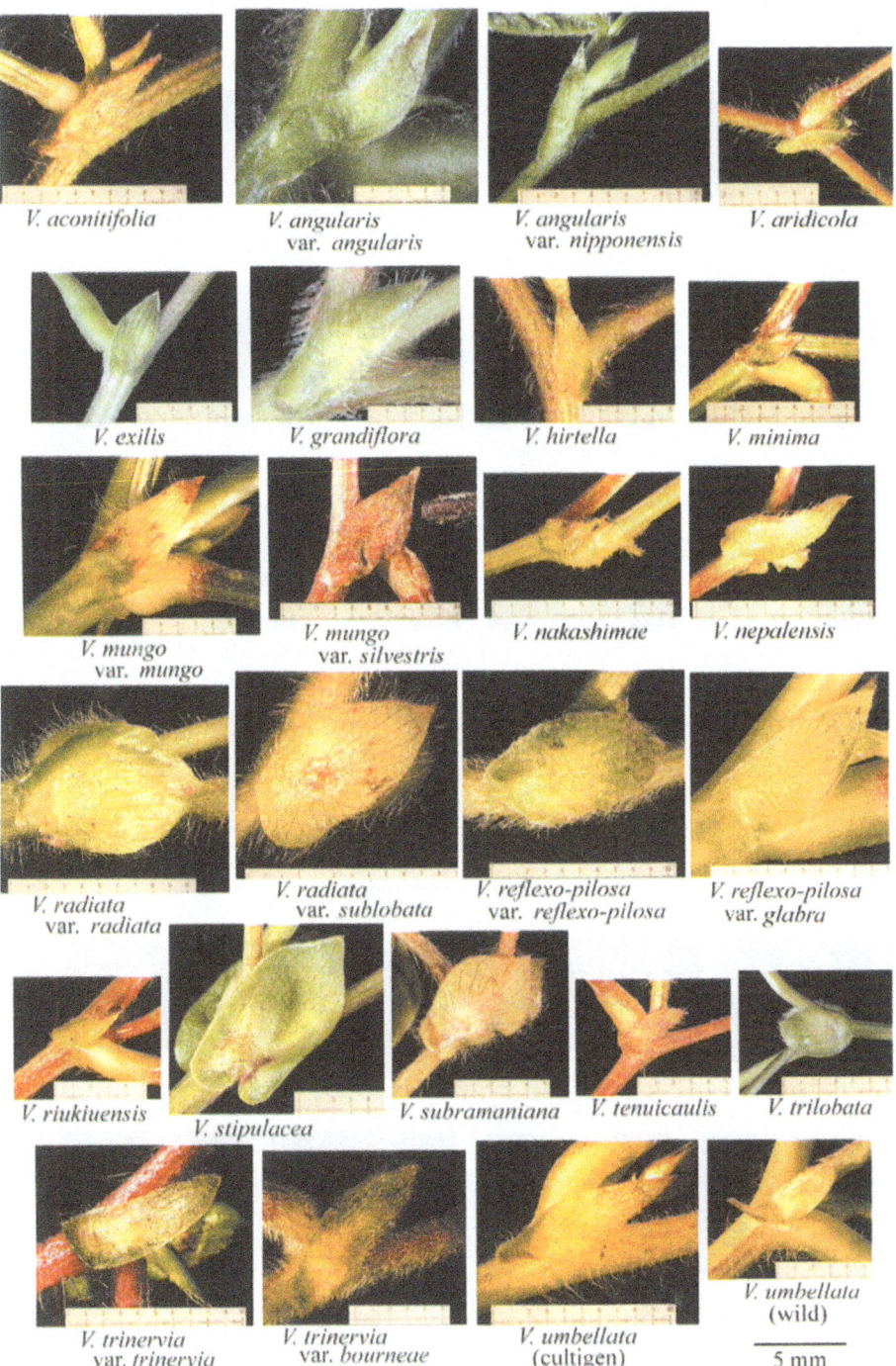

V. aconitifolia

V. angularis
var. *angularis*

V. angularis
var. *nipponensis*

V. aridicola

V. exilis

V. grandiflora

V. hirtella

V. minima

V. mungo
var. *mungo*

V. mungo
var. *silvestris*

V. nakashimae

V. nepalensis

V. radiata
var. *radiata*

V. radiata
var. *sublobata*

V. reflexo-pilosa
var. *reflexo-pilosa*

V. reflexo-pilosa
var. *glabra*

V. riukiuensis

V. stipulacea

V. subramaniana

V. tenuicaulis

V. trilobata

V. trinervia
var. *trinervia*

V. trinervia
var. *bourneae*

V. umbellata
(cultigen)

V. umbellata
(wild)

5 mm

Plate 6. Stipules

158

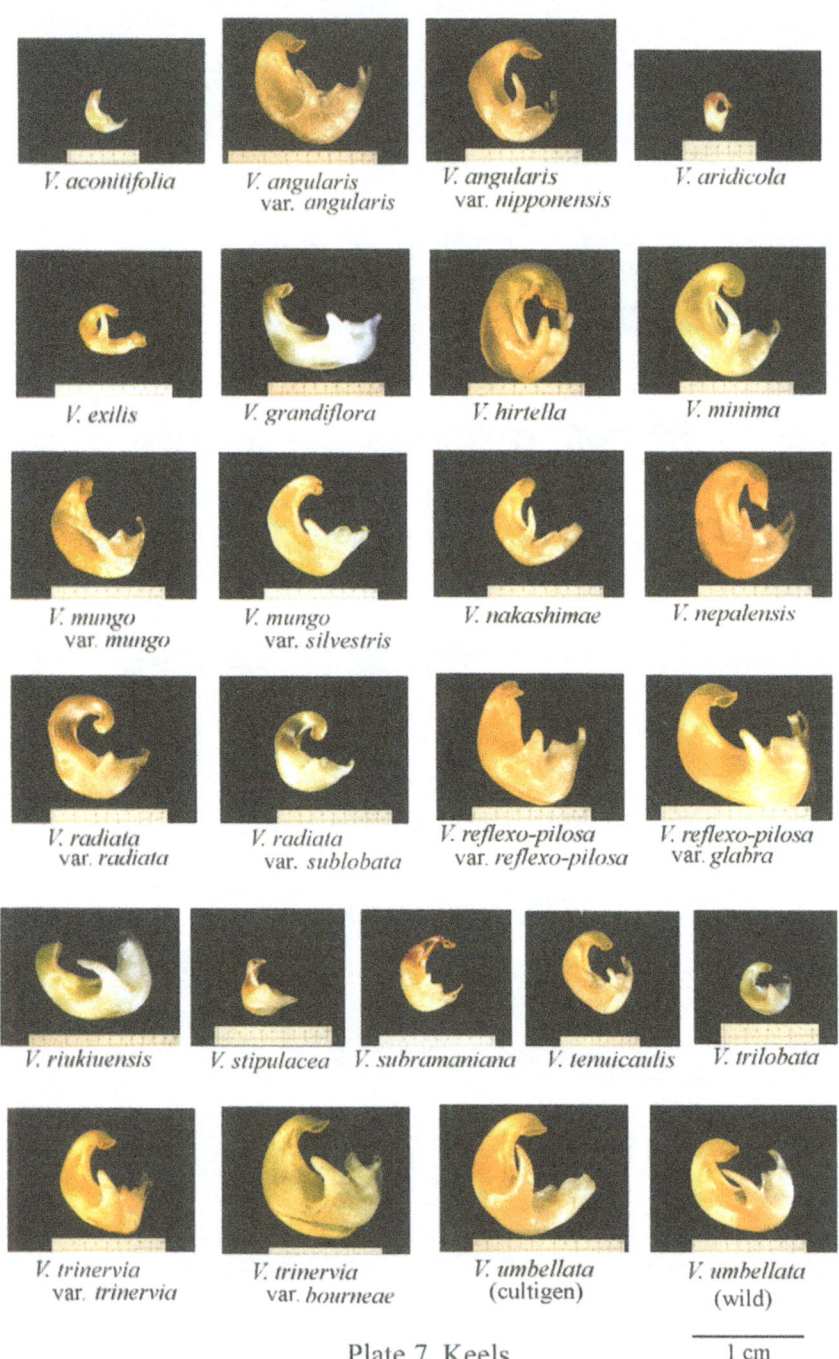

Plate 7. Keels

1 cm

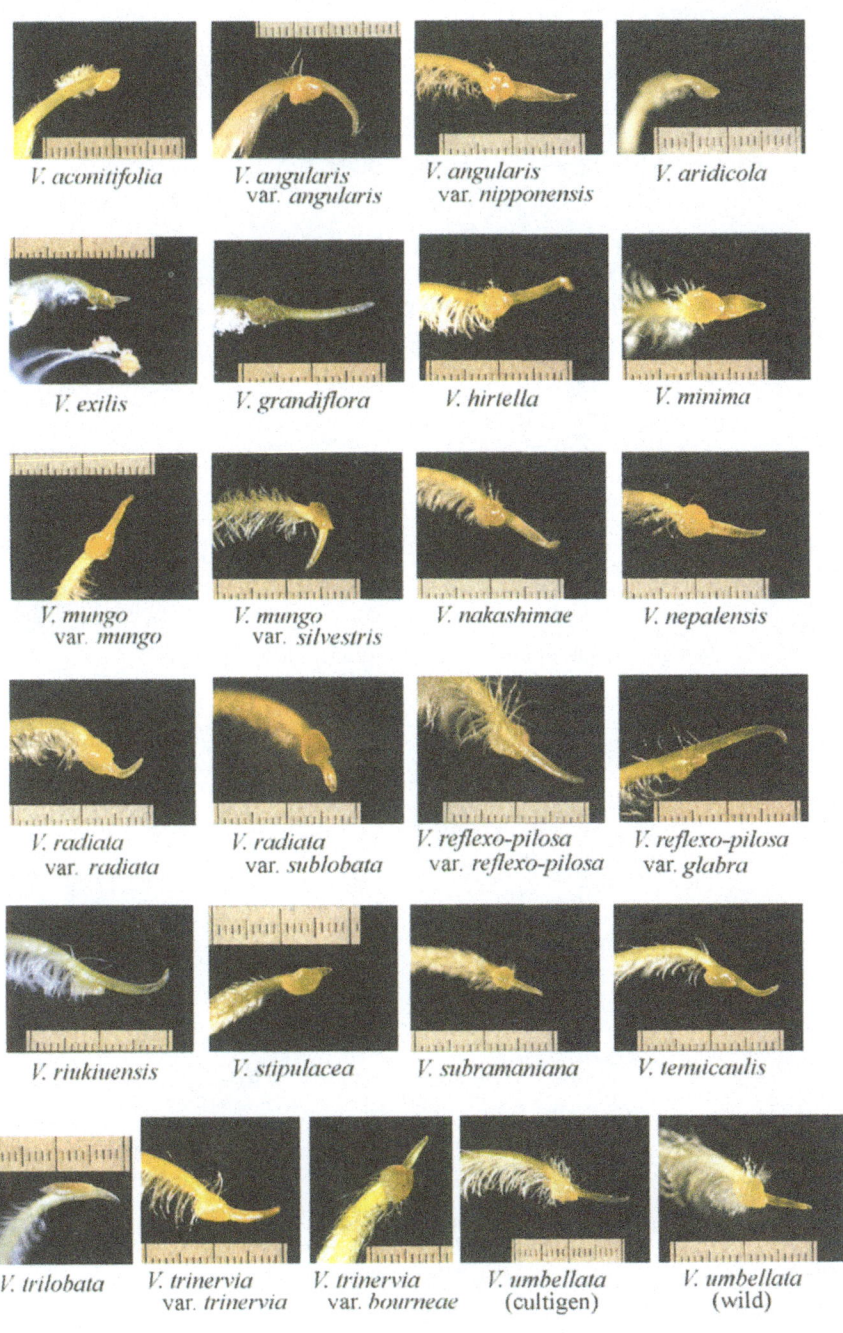

V. aconitifolia

V. angularis var. *angularis*

V. angularis var. *nipponensis*

V. aridicola

V. exilis

V. grandiflora

V. hirtella

V. minima

V. mungo var. *mungo*

V. mungo var. *silvestris*

V. nakashimae

V. nepalensis

V. radiata var. *radiata*

V. radiata var. *sublobata*

V. reflexo-pilosa var. *reflexo-pilosa*

V. reflexo-pilosa var. *glabra*

V. riukiuensis

V. stipulacea

V. subramaniana

V. tenuicaulis

V. trilobata

V. trinervia var. *trinervia*

V. trinervia var. *bourneae*

V. umbellata (cultigen)

V. umbellata (wild)

Plate 8. Style beaks

1 mm

160

V. angularis var. *nipponensis* (Japan, embankment, side of rice field)

V. aridicola (Sri Lanka, open sandy soil, grassland)

V. dalzelliana (Sri Lanka, wet roadside embankment)

V. exilis (Thailand, limestone rock)

Plate 9. Natural habitat and plant close up

V. grandiflora (Thailand, roadside vegetation)

V. hirtella (Thailand, edge of a forest)

V. minima (Myanmar, shade of trees beside rice field)

V. nakashimae (Japan, abandoned rice field)

Plate 10. Natural habitat and plant close up

162

V. radiata var. *sublobata* (Myanmar, roadside vegetation)

V. reflexo-pilosa var. *reflexo-pilosa* (Japan, embankment, side of rice field)

V. riukiuensis (Japan, seaside cliffs)

V. stipulacea (Sri Lanka, rice field)

Plate 11. Natural habitat and plant close up

V. tenuicaulis (Myanmar, side of rice field)

V. trilobata (Sri Lanka, sandy beach)

V. trinervia var. *trinervia* (Sri Lanka, hill country)

V. umbellata (Thailand, side of rice field near stream)

Plate 12. Natural habitat and plant close up

CHAPTER 5

ECO-GEOGRAPHIC ANALYSIS

Within the field of genetic resources science one area that is developing rapidly, with advances in computer hardware, software and associated databases, is the application of geographic information systems (GIS) to understand issues related to conservation, evaluation and use of plant genetic resources. The database developed for *Vigna* subgenus *Ceratotropis* consisting of herbarium and passport data, includes information on the location of populations. This geo-referenced data can be used as the basis for looking at commonality and differences among locations for a wide variety of cultural, geographic and ecological attributes such as altitude, climate, ethnobotany, soil and vegetation (Guarino *et al.*, 2002). Ethno-ecogeographic extrapolation requires care for the following reasons:

- Data on site location from herbarium specimens and passport data have varying levels of accuracy. Thus, while sites of recently collected germplasm may have been determined using a global positioning system (GPS), samples collected 15-20 years ago rarely have sites located so accurately.

- The data grids to which site location can be matched varies in size from region to region. Map grid size may be very fine, such as 1' square for some information but less fine for other information. In addition, map grids into which populations may be placed are fixed thus populations growing close together may fall into different grids. Asia is a difficult continent to analyze because it is non-contiguous and larger than South America and Africa.

- The date that samples are collected can be used to infer the pattern of flowering and seed production in different localities. The phenological interpretation of the data is dependent on the number of samples in the database and accuracy to which the phenological information can be derived from the time of collection. Sometimes phenological information is difficult to interpret from herbarium specimens, since herbarium samples are generally chosen to include flowers and fruits thus whether the whole population is primarily flowering or fruiting may not be clear.

The database for *Vigna* subgenus *Ceratotropis* provides an opportunity for eco-geographic and phenological interpretation. An initial synthesis and interpretation of this information is present for some attributes.

1. ECOLOGY

1.1. Altitude

Vigna subgenus *Ceratotropis* section *Angulares* species grow from sea level to 3500m in the Himalayas of Nepal. While some species, such as *V. riukiuensis,* are confined to low lying areas, other species, such as *V. dalzelliana,* grow from sea level to more than 2000m (Fig. 5.1).

Species in section *Aconitifoliae* are generally found growing at low altitude (Fig. 5.2). Only *V. khandalensis* that grows at a mean altitude of 716m, is an exception but based on only 5 site records. Of species in section *Aconitifoliae V. trilobata* grows at the lowest mean altitude of 57m. This reflects its usual habitat on coastal sand dunes.

Two species in section *Ceratotropis* grow at average altitudes above 500m, *V. mungo* (cultivated) and *V. subramaniana*, and two species grow below 500m, *V. radiata* var. *sublobata* and *V. grandiflora* (Fig. 5.3).

Species in section *Angulares* can be divided into three groups based on the altitude at which they have been found growing (Fig. 5.1). One group consists of species that grow at a mean altitude of over 500m in tropical regions. This group consists of *V. dalzelliana, V. hirtella, V. minima, V. nepalensis, V. tenuicaulis, V. trinervia* and *V. umbellata.*

The second group consists of those species that grow at low altitude in tropical and sub tropical regions. This group consists of *V. exilis, V. reflexo-pilosa* and *V. riukiuensis.* The third group consists of species that grow at low altitude in temperate regions. This group consists of *V. angularis* and *V. nakashimae. V. angularis* has a very wide distribution in temperate and sub-tropical regions. It also has a very wide longitudinal distribution from 85°E to 142°E (Fig. 4.2, chapter 4). In Japan (30 - 40°N) *V. angularis* grows below 500m. At lower latitudes (20 - 30°N) *V. angularis* grows at higher elevations where summer temperatures are comparable to those in Japan at lower elevation (Fig. 5.4).

1.2. Latitude

The species in the different sections of *Vigna* subgenus *Ceratotropis* are adapted to a different range of latitudes (Figs. 5.5 and 5.6). Species in section *Aconitifoliae* (tril, sti, kha, ari, aco) grow at latitudes of less than 30°N. Species in section *Ceratotropis* (ra-s,

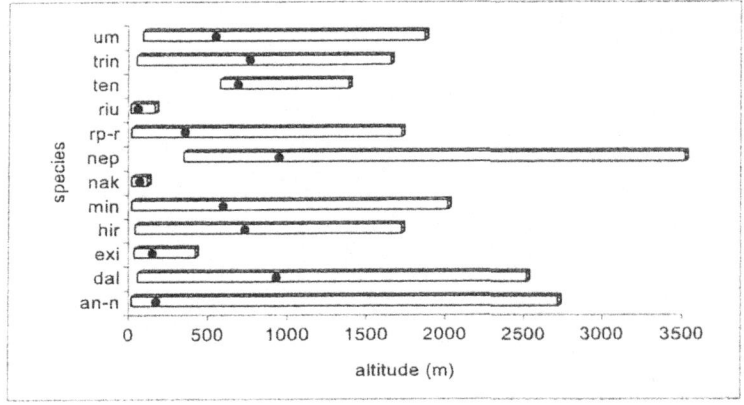

Figure 5.1. Mean altitude (●) and range that section Angulares *species grow.*

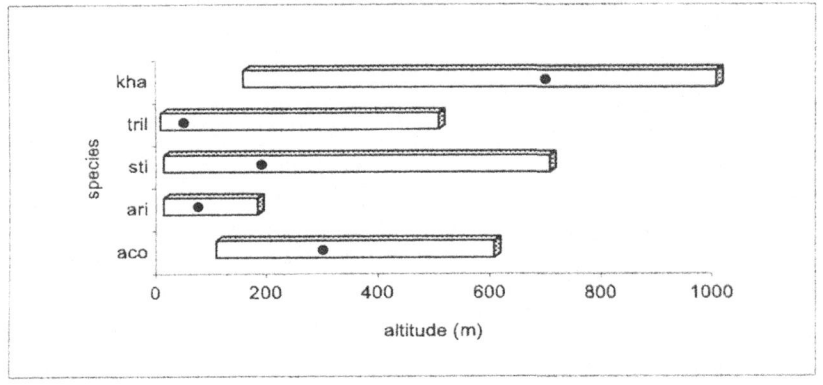

Figure 5.2. Mean altitude (●) and range that section Aconitifoliae *species grow.*

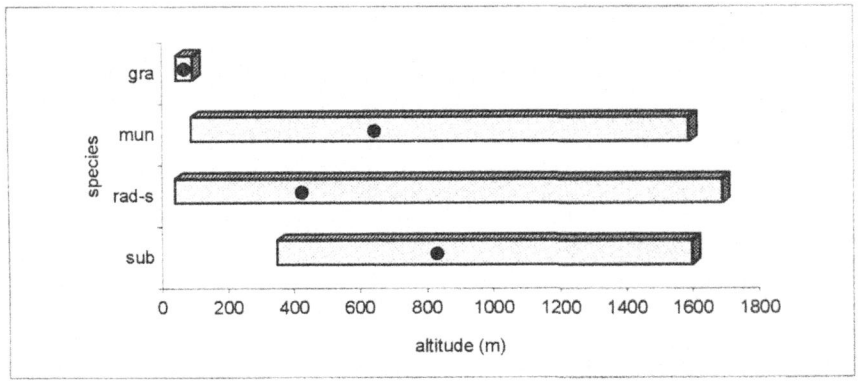

Figure 5.3. Mean altitude (●) and range that section Ceratotropis *species grow.*

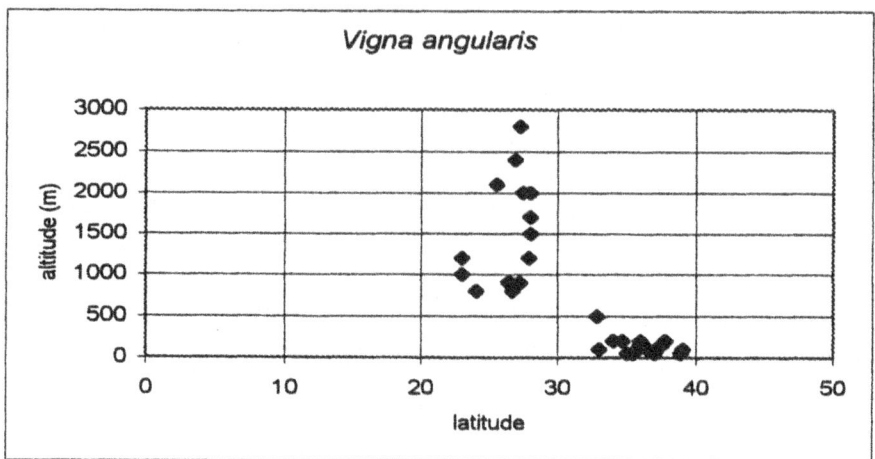

Figure 5.4. The altitude and latitude that wild, weedy and cultivated populations of V. angularis *grow.*

sub, sil, gra) grow at latitudes less than 33°N. Some species in section *Angulares* (um, trin, rp-r, min, ten, riu, nep, nak, hir, exi, dal, an-n) adapted to latitudes as far north as 40°N. The southern latitudes that species within section *Aconitifoliae, Ceratotropis* and *Angulares* grow are 16°S, 35°S and 19°S, respectively.

Section *Aconitifoliae* consists of tropical species with only *V. khandalensis* (1 site out of 6) and *V. trilobata* (3 sites out of 67) occasionally being found outside the tropics.

Section *Ceratotropis* consists of species that grow within and outside the tropics. The tropical species are *V. grandiflora* and *V. mungo* var. *silvestris*. The species that are found within and outside the tropics are *V. radiata* var. *sublobata* and *V. subramaniana*. *V. radiata* var. *sublobata* grows at a far greater range of latitudes than other species from 28°S to 33°N. In addition, its overall distribution is far wider than other wild species in the subgenus and this may reflect tolerance to varied environments.

Species in section *Angulares* consist of species that are only in the tropics (*V. exilis, V. tenuicaulis* and *V. trinervia*), species that occur inside and outside the tropics (*V. hirtella, V. minima, V. reflexo-pilosa, V. riukiuensis* and *V. umbellata*) and species that only occur outside the tropics (*V. angularis, V. nepalensis* and *V. nakashimae*).

1.3. Phenology

The months during which herbarium specimens of the wild species in the subgenus have been collected are indicative of the flowering and fruiting season of these species. For some species, there is sufficient information to determine whether the species is

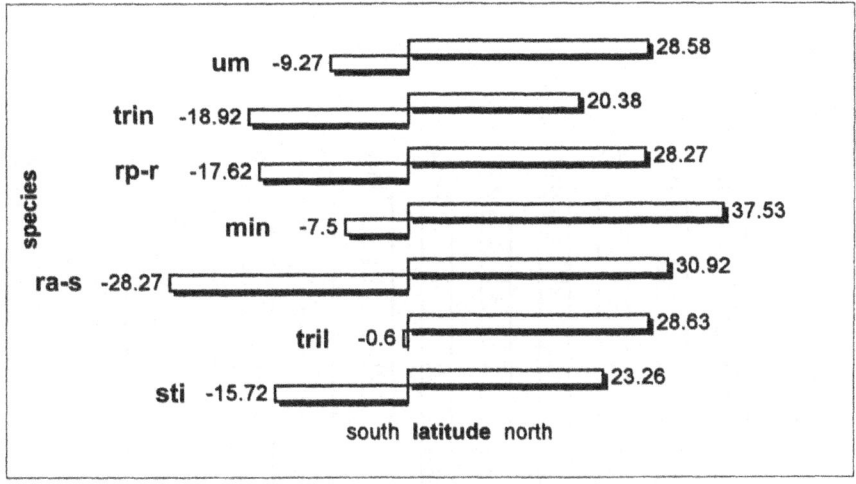

Figure 5.5. Latitude range for Vigna *subgenus* Ceratotropis *species that grow north and south of the equator.*

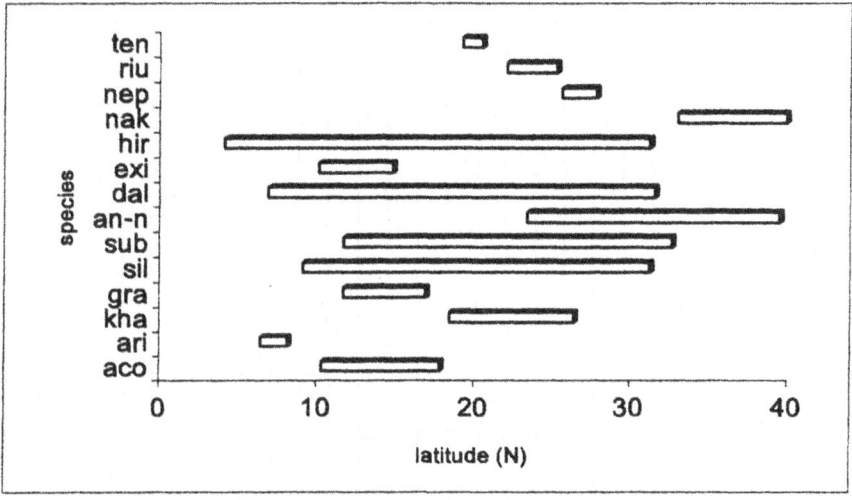

Figure 5.6. Latitude range for Vigna *subgenus* Ceratotropis *species that grow north of the equator.*

sensitive to changes in the seasons (Table 5.1).

V. angularis var. *nipponensis* and *V. nakashimae* both show clear seasonality to flowering with herbarium specimens only having been collected in a 4 or 5 month period during late summer and autumn. The seasonality for *V. angularis* var. *nipponensis* appears similar in India, Nepal and Myanmar, and East Asia. *V. aridicola, V. nepalensis, V. tenuicaulis* and *V. umbellata* have each only been collected in two consecutive

Table 5.1. Months of the year that Vigna subgenus Ceratotropis species have been collected as herbarium specimens or directly.

	Jan	Feb	Mar	April	May	June	July	Aug	Sept	Oct	Nov	Dec	Total[1] [months]
Section Angulares													
V. angularis var. *nipponensis*								3	18	47(14)	20(86)	3(17)	91(117)[5]
V. dalzelliana				1(3)			1	3	1		4	5	15(3)[6]
V. exilis	1					1	1	2			1	1(15)	8(15)[6]
V. hirtella		1(1)		1	1	1	1	1			5	3(14)	17(15)[8]
V. minima			1	1	1	1	8	3	6	15	23(1)	8(11)	88(12)[10]
V. nakashimae								1	10	12	1		24 [4]
V. nepalensis									2		2		4 [2]
V. riukiuensis	1	1(5)		5	1	6	1	3	7	3	7	10	45(5)[11]
V. reflexo-pilosa													
North[2]	2	2(10)	1	5	4	4	10	8	5	3	14	11	69(10)[12]
South[2]	2	2		4	4	3	3	4	6	2	1		31 [10]
V. tenuicaulis											(5)	(1)	(6)[2]
V. trinervia													
North[2]	1	1(27)		2	1	1	3		1	3	2	2	17(27)[9]
South[2]	2	3		4	2	6	1	1	1	1	3	3	27 [11]
V. umbellata (wild)							1				(37)	1(32)	2(69)[3]

Table 5.1. continued

	Jan	Feb	Mar	April	May	June	July	Aug	Sept	Oct	Nov	Dec	Total¹ [months]
Section Radiatae													
V. radiata var. *sublobata*													
North²	2		(1)		1	2	2		5	9	16	4	2 43(1)[10]
South²		7	9		4	4	3	3		2	2	1	35 [9]
V. subramaniana								1		4	6	1	1 13 [5]
V. grandiflora								2			2	(3)	4(3)[3]
V. mungo var *silvestris*									1	1	5		(1) 7(1)[4]
Section Aconitifoliae													
V. aridicola			1(3)	3(11)									4(14)[2]
V. stipulacea													
North²	2(1)	1(2)	3(2)				1	1	2	1	4		3 18(5)[9]
South²	1	1	3	1			2	1		1			10 [7]
V. trilobata	3(1)	2(7)	10(15)		1	2	2	1		4	3	8	6 42(23)[11]

¹ Numbers in bold are number of direct collections.
² North and South refers to collections made north or south of the equator.

months of the year. However, the data for these species, except *V. umbellata,* are rather limited. *V. radiata* var. *sublobata* is the most widely distributed species in the subgenus *Ceratotropis* and has been collected in 10 and 9 months of the year in the northern and southern hemispheres, respectively. However, most collections in the northern hemisphere are between August and November and in the southern hemisphere between February and May. This may reflect seasonality of this species when it grows in temperate regions.

Rather more species appear to flower year round. From the data presented *V. reflexo-pilosa* var. *reflexo-pilosa, V. riukiuensis, V. stipulacea, V. trilobata* and *V. trinervia* fall in this category. The data for *V. minima* is rather inconclusive and this may reflect the diversity of this species from which *V. nakashimae* (seasonal flowering) and *V. riukiuensis* (non-seasonal flowering) are presumed to have evolved.

It is probable that flowering of tropical *Vigna* species is closely related to rainfall. Thus, in Sri Lanka the optimum flowering time for dry zone *Vigna* species, *V. aridicola, V. stipulacea* and *V. trilobata,* is in the first half of the dry season from February to April. Highland species in the wet zone of Sri Lanka (*V. dalzelliana* and *V. trinervia*) grow in habitats that are not subject to the same extremes of climate of those in the *Vigna* species of the dry zone. Flowering of these highland species, however, also seems to be mainly in the months of February to April.

1.4. Analysis of population distribution and climate: an example using V. trinervia

After *V. radiata* var. *sublobata, V. trinervia* is the most widely distributed species in the subgenus *Ceratotropis*. It is a species that phylogenetically lies on the boundary of the three sections in the subgenus (Doi *et al.,* 2002; Tomooka *et al.,* 2002b). Analysis of the eco-climatic data from locations where *V. trinervia* has been collected suggests there are several groups within *V. trinervia* (Fig. 5.7). Most of the South Asian collections of this species cluster together. The collections from Southeast Asia form two groups. Based on the climate pixels there are differences in altitude among the clusters. However, average altitude for clusters do not reflect average collecting site altitude since most South Asian collections for which we have precise site data come from an average of 1,172m (12 records). This suggests that the climate pixel size is rather too large to obtain accurate relationship between collecting site and eco-climatic data.

There are now many ways to use Geographic Informations Systems (GIS) databases to analyze plant genetic resources (Guarino *et al.,* 2002). These methods are rapidly becoming more sophisticated. This initial analysis of *V. trinervia* suggests that greater precision is needed. With increasingly powerful computers and more refined (smaller grid size) climatic data for the Asian region, GIS analyses of the species in the subgenus *Ceratotropis* may reveal intra-specific variation not apparent using other methods.

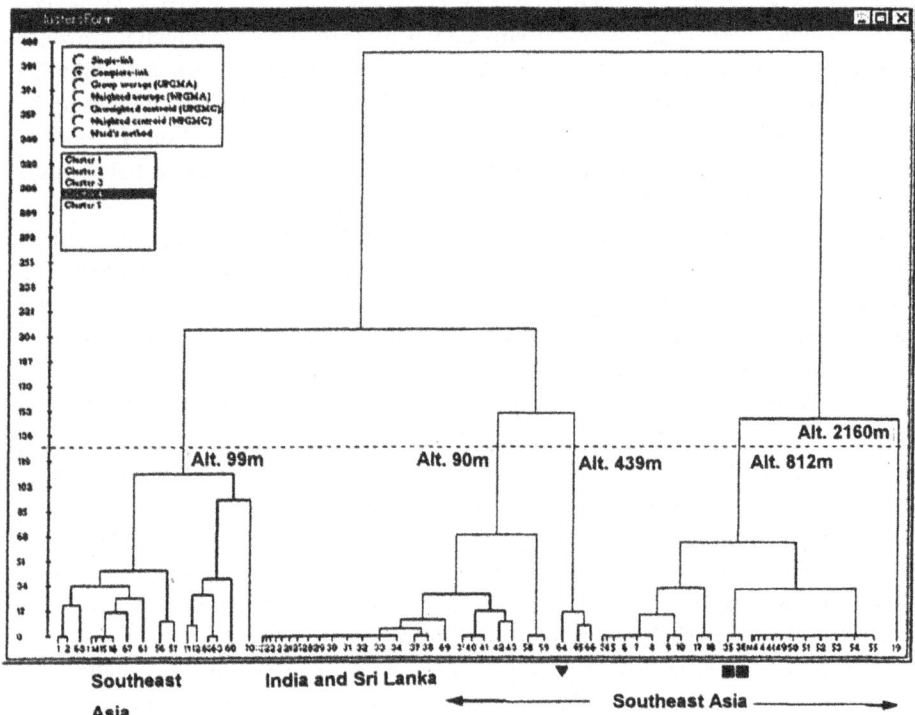

Figure 5.7. The relationship among 70 V. trinervia *populations based on climate and collection location in Asia and Papua New Guinea. Altitude refers to the average altitude for the climate pixels in which groups of samples were collected. Pixel size is 10 minutes square.* ▼ *is a sample from Papua New Guinea and* ■ *samples from Sri Lanka. This cluster diagram was generated by the complete link method using Flora Map. (information on this software can be found at http://www.floramap-ciat.org)*

2. GENETIC EROSION

The species in the genus *Vigna* subgenus *Ceratotropis* are best described as species adapted to disturbed habitats. Most species are found in land frequently disturbed by cutting vegetation such as roadside verges. Thus, many subgenus *Ceratotropis* species have the characteristics of pioneers of secondary succession, often associated with plants that would lend themselves easily to cultivation and domestication (Bunting, 1960; Harlan, 1992). Populations of such species are liable to be destroyed when development occurs in areas where they grow. For example, a population of *V. tenuicaulis* collected in northern Thailand was destroyed when the road next to which it grew was widened. Another characteristic of most species in the subgenus *Ceratotropis* that the authors have observed is that population size tends to be very small. Exceptions to this are occasional large populations of wild *V. umbellata* in Southeast Asia, *V. trinervia* that grows for long distances along the side of roads, and on large exposed outcrops of limestone populations of *V. exilis* can also be quite extensive.

While most species are found in open and disturbed habitats *V. minima* in northeast Thailand is an exception. This species grows in shaded habitats within dry deciduous forests where it twines up trees and is frequently found deep in forest shade away from forest paths. In Thailand the habitat of *V. minima* is largely restricted to National Parks since native dry deciduous forests have now been cut and replaced by agriculture.

To address genetic erosion of wild *Ceratotropis* it would be helpful to assess the levels of habitat disturbance that are optimum for each species. Some wild species may be well adapted to the level of disturbance associated with intensively farmed areas such as *V. stipulacea* that grows on paddy field surrounds in Sri Lanka. Others are adapted to the disturbance associated with orchard or plantation crops such as *V. trinervia* in Thailand and Malaysia. Outside the agricultural setting some species are adapted to parkland where grass is regularly cut such as *V. riukiuensis* in Okinawa. These examples suggest that addressing habitat conservation to prevent genetic erosion requires multiple approaches for subgenus *Ceratotropis* species. For many of the Asian *Vigna* we may never know the 'original' habitat of each species since we may be seeing today ecotypes adapted to man-made environments.

Land that is protected, such as within a National Park boundary, tends to have fewer highly aggressive weeds. This enables the natural flora and fauna to survive. One well-protected National Park is the Yala Park in Sri Lanka. Although populations of wild *Vigna* species can be found outside the Yala Park the number of populations increase sharply within the Park boundary. This is probably because aggressive weedy species that are not native to the area are less abundant in the Park.

3. PHYTOGEOGRAPHY AND CONSERVATION PRIORITIES

Based on distribution maps of different species presented in the conspectus (chapter 4) the region where most species in each section occur was determined (Fig. 5.8). This shows that section *Aconitifoliae* and *Ceratotropis* have their species diversity centers in South Asia whereas section *Angulares* has its center of species diversity in Southeast Asia.

Using the passport and herbarium databases, areas of greatest species diversity were determined using DIVA-GIS software that determines the number of species in a grid 1° square (http://www.cipotato.org/gis/tools/diva/htm). Three areas were highlighted

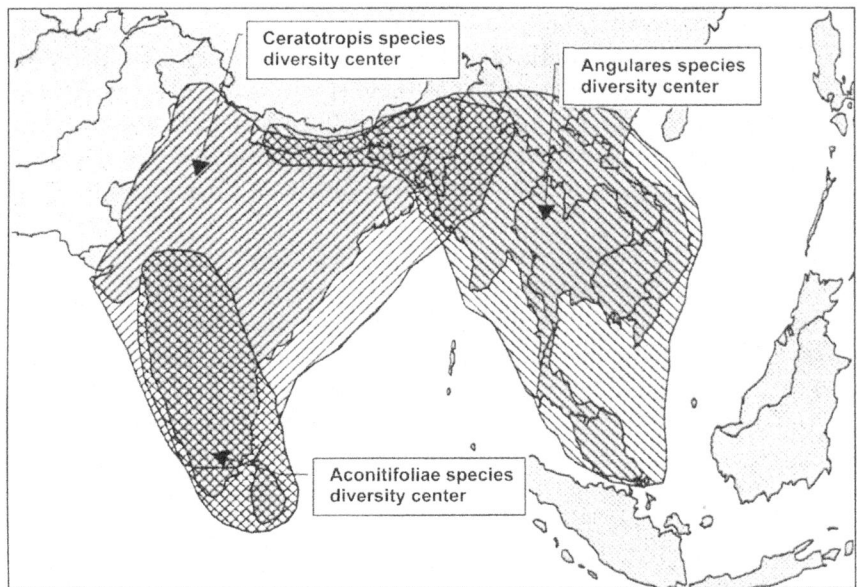

Figure 5.8. Main areas of species diversity for each section in subgenus Ceratotropis *based of individual species maps.*

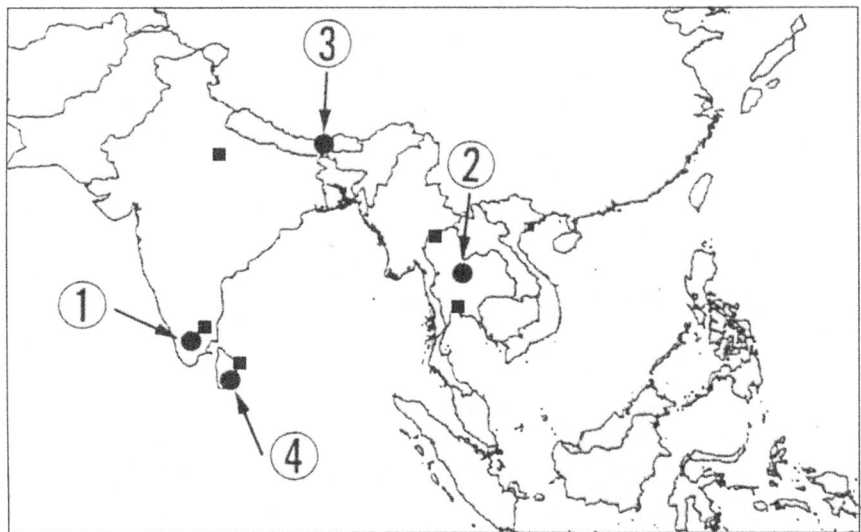

Figure 5.9. Areas of high species diversity determined applying DIVA-GIS software to the herbarium and passport database for Vigna *subgenus* Ceratotropis. *The grid size was one degree square. The positions 1, 2 and 3 were determined by DIVA-GIS software in order of species diversity. Position 4 was based on information of the authors. Secondary areas of species diversity are shown as black squares (■).*

as having the greatest level of species diversity (Fig. 5.9).

Area 1. This area is on the border of Tamil Nadu and Kerala States, India. In this area, eight *Ceratotropis* species have been found. These species are *Vigna aconitifolia, V. dalzelliana, V. mungo, V. radiata* var. *sublobata, V. stipulacea, V. subramaniana* and *V. trinervia* (including *V. trinervia* var. *bourneae*). This area thus has three species belonging to sections *Aconitifoliae* and *Ceratotropis* and 2 belonging to section *Angulares*.

Area 2. This area is in Thailand on the border with Laos between north and northeast Thailand. In this area, five wild subgenus *Ceratotropis* species grow *V. hirtella, V. minima, V. tenuicaulis, V. trinervia* and *V. umbellata*. Also in this area *V. radiata* is commonly grown as a crop. All the wild species in this area are members of section *Angulares* and of these wild species, four are not in area 1.

Area 3. This area is on the border of Nepal and Sikkim state, India. In this area, four *Ceratotropis* species grow. *V. angularis* var. *nipponensis, V. hirtella, V. mungo*, and *V. nepalensis*. All except *V. mungo* are in section *Angulares*. Three of the species in this center of diversity are not in diversity center 1 or 2.

The limitations of the software fail to detect a high level of species diversity in Sri Lanka due to different wild species that occur close to one another falling in different grid squares. In Sri Lanka, five species occur within one grid degree of one another based on our direct observations. These species are *V. aridicola, V. dalzelliana, V. stipulacea, V. trilobata* and *V. trinervia*. Of these species *V. dalzelliana* and *V. trinervia* belong to section *Angulares*, and the others to section *Aconitifoliae*. Thus, Sri Lanka can be considered a fourth area of high species diversity.

These four areas of *Vigna* species diversity are within biodiversity 'hotspots' identified by Myers *et al.* (2000). Thus these areas represent priority areas for conservation beyond their importance as *Vigna* species 'hotspots' (Ward, 2002).

These centers of species richness in *Vigna* subgenus *Ceratotropis* are areas that have complementary species diversity. Of these four areas, the richest in species and section representation is that in southern India in the Western Ghats. These four centers of species richness include 15 of the species in the subgenus only 6 are not included (*V. exilis, V. grandiflora, V. khandalensis, V. nakashimae, V. reflexo-pilosa,* and *V. riukiuensis*).

Two of the centers of species richness, the Thai center and Sri Lanka center, include protected areas. The Thai center includes the Phu Hin Rong Kha, Phu Luang and Phu Rua Reserves. The Sri Lanka center includes the Yala National Park. The center in southern India is close to but does not precisely coincide with protected areas, such as Periyar Lake Wildlife Sanctuary. The Nepal/Indian border center is close to the Koshi Tappu Wildlife Reserve and includes Khangchendzonga National Park. The later, however, is largely high mountains and elevations in this National Park begin at above 1800m. Thus, this Indian and Nepal/Indian border centers of *Vigna* species diversity should be the primary focus of conservation attention. Whether *Vigna* species occur in protected areas in or near these centers of species richness is not known except for the Sri Lankan center where *V. aridicola, V. trilobata* and *V. stipulacea* have been found in the Yala National Park. Detailed inventories of *Vigna* species in other protected areas in these centers of species diversity are warranted.

In addition to the main areas of species richness five other areas have been identified as having two or three species (Fig. 5.9, black squares). Of these those that include species not found in the four main areas of species richness should be considered as conservation targets. Some species do not occur in any of the highlighted areas of species richness such as *V. khandalensis*, *V. grandiflora*, *V. riukiuensis* and *V. nakashimae*. Of these species *V. khandalensis*, *V. grandiflora* and *V. nakashimae* are not reported to occur in protected areas (Table 3.2, Chapter 3). From our database only a very few populations of *V. khandalensis* and *V. grandiflora* are known therefore they should receive special conservation attention.

We end this chapter by providing a table that lists the 2000 IUCN conservation status (IUCN, http://www.redlist.org/info/categories_criteria .html) of each species based on information we have (Table 5. 2).

Table 5.2. *IUCN conservation status of taxa of the genus* Vigna *subgenus* Ceratotropis.

Section	Species	Conservation status (Redlist Categories - Criteric)
Angulares	*V. angularis* var. *angularis*[1]	Lower Risk, Least Concern
	V. angularis var. *nipponensis*	Lower Risk, Least Concern
	V. dalzelliana	Lower Risk, Least Concern
	V. exilis	Lower Risk, Least Concern
	V. hirtella	Lower Risk, Least Concern
	V. nepalensis	Data Defficient
	V. reflexo-pilosa var. *glabra*	Lower Risk, Near Threatened
	V. reflexo-pilosa var. *reflexo-pilosa*	Lower Risk, Least Concern
	V. tenuicaulis	Lower Risk, Least Concern
	V. trinervia var. *trinervia*	Lower Risk, Least Concern
	V. trinervia var. *bourneae*	Data Defficient
	V. umbellata (cultivated)[1]	Lower Risk, Least Concern
	V. umbellata (wild)	Lower Risk, Least Concern
	V. minima	Lower Risk, Least Concern
	V. nakashimae	Lower Risk, Least Concern
	V. riukiuensis	Lower Risk, Least Concern
Ceratotropis	*V. grandiflora*	Lower Risk, Near Threatened
	V. mungo var. *mungo*[1]	Lower Risk, Least Concern
	V. mungo var. *silvestris*	Lower Risk, Least Concern
	V. radiata var. *radiata*[1]	Lower Risk, Least Concern
	V. radiata var. *sublobata*	Lower Risk, Least Concern
	V. subramaniana	Data Defficient
Aconitifoliae	*V. aconitifolia*[1]	Lower Risk, Least Concern
	V. aridicola	Lower Risk, Least Concern
	V. khandalensis	Data Defficient
	V. stipulacea	Lower Risk, Least Concern
	V. trilobata	Lower Risk, Least Concern

[1]While these cultigens are not threatened at the species level, traditional land races are threatened with replacement in various regions. Thus land races would be classified as lower risk, near threatened.

CHAPTER 6

RESEARCH AND FUTURE PERSPECTIVES

1. RESEARCH

In this final chapter a series of research topics are discussed that are particularly relevant to the conservation, evaluation and use of the Asian *Vigna*. The final section of the chapter discusses future research perspectives for the Asian *Vigna*.

1.1. Cross compatibility studies

There are many reports of interspecific hybridization in the subgenus *Ceratotropis*, however, parental material germplasm cannot always be checked. Therefore, we review reports for which the parental materials used in the crosses are known. Broader coverage of the literature can be found in Dana and Karmakar (1990) and Fery (1980).

a. Section *Angulares* (Fig. 6.1)
The crosses among section *Angulares* species that result in fertile F_1 hybrids are shown (Fig. 6.1.a). Reciprocal crosses have been attempted with all the species shown in this figure. The species of the *Vigna minima* complex (*V. minima*, *V. nakashimae* and *V. riukiuensis*) and *V. umbellata* exhibit a different trend in inter specific cross compatibility. The species of the *V. minima* complex all produce fertile F_1 hybrids with other species of section *Angulares* that have been tested when they are the pollen (♂) parent (Fig. 6.1.b). In contrast, species of section *Angulares*, except *V. angularis*, can only be crossed with *V. umbellata* when *V. umbellata* is the seed (♀) parent (Fig. 6.1c). This suggests that at least two types of barriers to crossing have evolved in section *Angulares*. The relationship between particular species and groups of species in this section are discussed below.

V. angularis x *V. hirtella* (acc. JP109681)
This cross was successful producing a fertile F_1 progeny, when *V. hirtella* was used as the seed (♀) parent (Fig. 6.1a) (Tomooka *et al.*, 2000a).

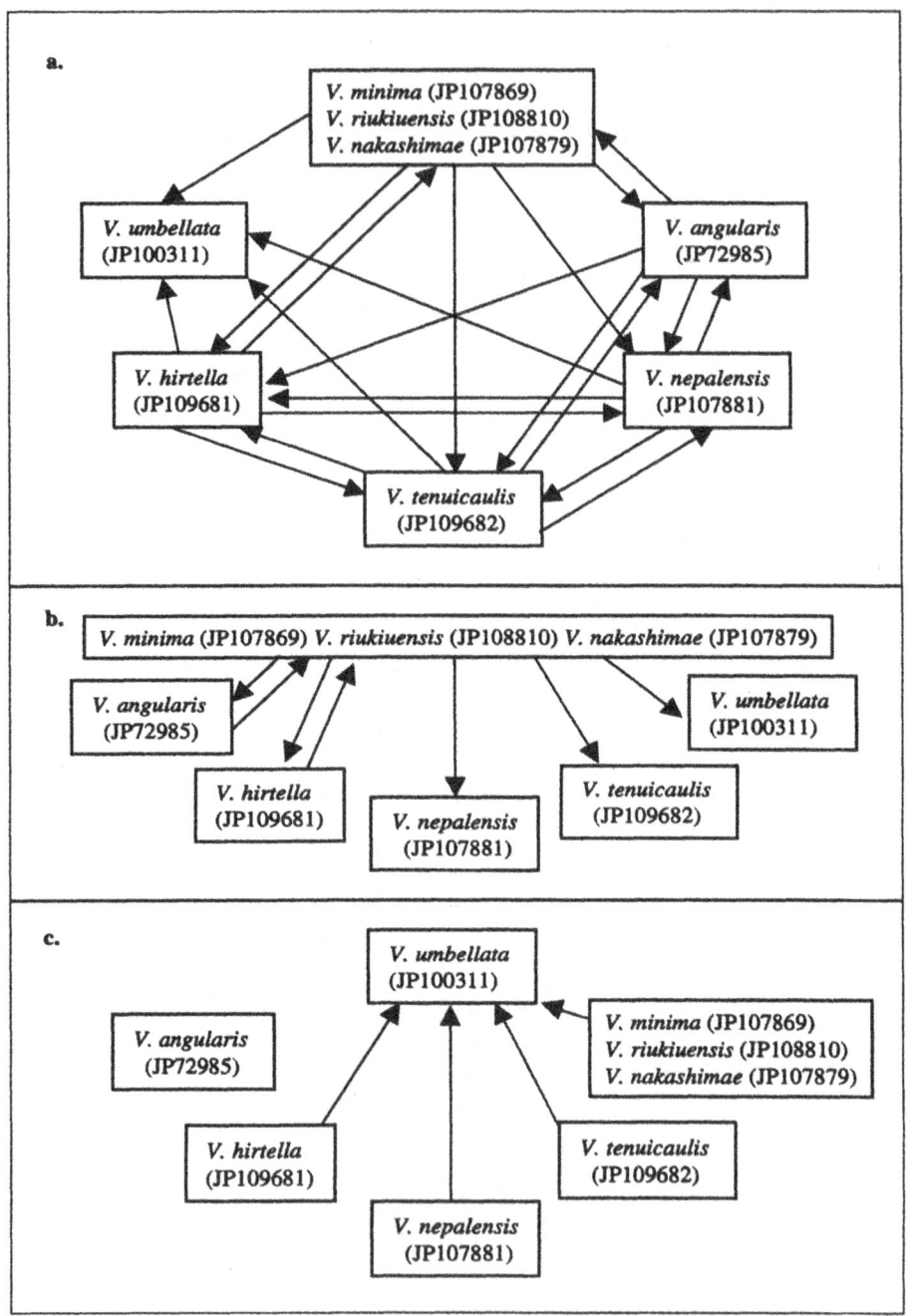

Figure 6.1.(opposite)

  a. Results of reciprocal crossing between eight section Angulares *species showing crosses with no barriers and that produce fertile F₁ plants.*

  b. Species of the V. minima *complex* (V. minima, V. riukiuensis *and* V. nepalensis) *when a pollen parent* (♂) *can produce fertile F₁ plants with* V. angularis, V. hirtella, V. nepalensis, V. tenuicaulis. *However, only* V. angularis (♂) *and* V. hirtella (♂) *produces fertile F₁ plants with species of the* V. minima *complex* (♀).

  c. V. umbellata (♀) *can produce fertile F₁ plants with pollen from* V. hirtella, V. nepalensis, V. riukiuensis *and* V. tenuicaulis *but not* V. angularis. *The reciprocal crosses do not produce viable F₁ seeds.*

V. angularis x [*V. minima, V. nakashimae, V. nepalensis, V. riukiuensis* and *V. tenuicaulis*] Crosses between *V. angularis* and this group of species show no major barriers to hybridization (Fig. 6.1.a & b). Fertile F₁ hybrids are produced when these species are used as female or male parents.

V. angularis x *V. umbellata*

  Hybrids can be obtained between these two cultigens only in the case when *V. umbellata* is used as the seed (♀) parent and by using embryo rescue (Ahn & Hartman, 1978; Chen *et al.*, 1983; Kaga *et al.*, 2000a). Pod set and pod development varies depending on the species that is pollen parent (Chen *et al.*, 1983). There is a high degree of chromosome homology in the hybrids and 11 bivalents are formed during meiosis. F_2 and subsequent generations grow normally. However, when a *V. umbellata* x *V. angularis* mapping population was developed, a high degree of segregation distortion was observed (Kaga *et al.*, 2000a).

V. umbellata x [*V. minima, V. nakashimae, V. nepalensis, V. riukiuensis, and V. tenuicaulis*]

  Crosses between *V. umbellata* and this group of species produces fertile F₁ hybrids only when *V. umbellata* is the seed parent (Fig. 6.1.c)

V. reflexo-pilosa (tetraploid)

  V. reflexo-pilosa var. *reflexo-pilosa* and var. *glabra* are cross compatible and produce fertile hybrids when crossed with each other (Tomooka *et al.*, 1991). Interspecific hybrids between *V. reflexo-pilosa* var. *glabra* (AVRDC V1160) and diploid cultigens (*V. mungo*, *V. radiata* and *V. umbellata*) require embryo rescue. In crosses between *V. reflexo-pilosa* var. *glabra* and *V. umbellata* 11 bivalents were observed in hybrids. In contrast, hybrids with *V. radiata* and *V. mungo* resulted in very low bivalent formation (Egawa *et al.*, 1988). This suggests that *V. reflexo-pilosa* is a member of section *Angulares* to which *V. umbellata* belongs.

b. Section *Ceratotropis*

V. radiata x *V. mungo*

Hybrids between *V. radiata* and *V. mungo* can be produced when *V. radiata* is the female parent (Chen *et al.*, 1983; Egawa, 1988). While there are no obvious barriers to getting hybrid seed between these two species most F_1 seeds are shrivelled and do not germinate. If F_1 plants are produced they show very low fertility and hybrids have irregular meiotic configurations with high frequency of univalents (Egawa, 1988; Egawa *et al.*, 1990a). For a range of crosses, F_1 hybrids between these two cultigens had less than 50% pollen stainability (Chen *et al.*, 1983).

V. radiata and *V. mungo* x their presumed wild progenitors

Intra-specific hybrids between *V. radiata* and *V. mungo* and their respective presumed wild progenitors, *V. radiata* var. *sublobata* and *V. mungo* var. *silvestris*, are normal, showing 94.0 - 99.0% F_1 pollen stainability (Miyazaki, 1982). However, some accessions showed reduced F_1 pollen stainability. The cross between *V. radiata* var. *radiata* and var. *sublobata* (JP107876) had 76.9 - 82.3% F_1 pollen stainability and the cross between *V. mungo* var. *mungo* and var. *silvestris* (JP107871) had 60.7 - 63.7% F_1 pollen stainability (Miyazaki, 1982). Cross combination variation in accessions of *V. radiata* var. *sublobata* from different regions suggests that ecogeographic differentiation of this variety requires further study (Miyazaki, 1982).

V. radiata x *V. grandiflora* (JP107862)

Crosses between these two species using various *V. radiata* cultivars as female parent resulted in 21% pod set (Egawa and Tomooka, 1994). Embryo rescue was necessary to obtain F_1 hybrids. However, lack of root development meant that no F_1 plants grew for more than a few weeks.

V. radiata and *V. mungo* x *V. subramaniana* (JP110836)

Cultivars of mungbean and blackgram crossed with *V. subramaniana* (♂) can result in F_1 plants but only when embryo rescue is used (Tomooka *et al.*, 2000a).

c. Intersectional crosses

Crosses between species in different sections generally fail completely. However, F_1 hybrids have been produced between *V. umbellata* (♂) and *V. radiata* (♀). The F_1 hybrid has low (about 5) bivalent formation and is sterile (Egawa *et al.*, 1990a). Hybrid plants between *V. radiata* and *V. reflexo-pilosa* var. *glabra* (♀) can be produced when embryo rescue is employed. Backcrossing F_1 plants to the diploid parent (*V. radiata*) have yielded viable seeds and fertile backcross plants (Chen *et al.*, 1989).

1.2. Gene pools

The total diversity within a crop species and related wild species that can potentially be used in conventional crop improvement is known as the crop gene pool. The gene pool concept was developed by Harlan and De Wet (1971) as follows:

- Primary gene pool (GP1) - the true biological species including all cultivated, wild and weedy forms of a crop species. Hybrids among these taxa are fertile and gene transfer to the crop is simple and direct.
- Secondary gene pool (GP2) - the group of species that can artificially hybridize with the crop, but where gene transfer is difficult. Hybrids may be weak or partially

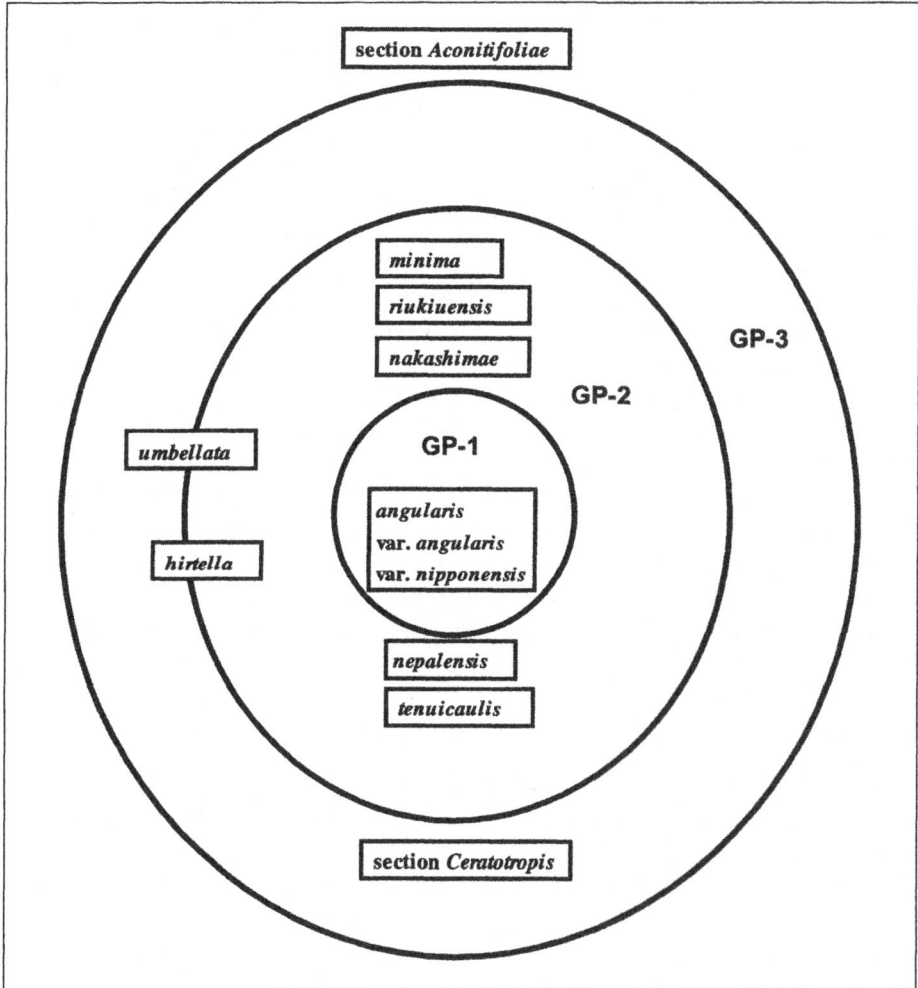

Figure 6.2. Gene pools of Vigna angularis.

sterile or chromosomes pair poorly.

- Tertiary gene pool (GP3) - including all species that can be crossed with difficulty (e.g. require in vitro hybrid embryo culture), and where gene transfer is impossible or requires radical techniques (e.g. radiation induced chromosome breakage).

Lawn (1995) proposed that the Asian *Vigna* consists of three more or less isolated gene pools based on cross compatibility studies. These gene pools correspond with sections as shown in Table 6.1.

Table 6.1. Relationship between gene pool and section

Gene pool (Lawn, 1995)	Section (Tomooka *et al.*, 2002a)
angularis - umbellata	*Angulares*
radiata - mungo	*Ceratotropis*
aconitifolia - trilobata	*Aconitifoliae*

From studies of cross compatibility between *V. angularis* and other taxa in section *Angulares,* a gene pool classification for azuki bean has been proposed (Tomooka *et al.*, 2000a) (Fig. 6.2) based on the concept of Harlan and De Wet (1971).

The primary gene pool of azuki bean consists of its wild, weedy and cultivated forms, *V. angularis* var. *angularis* and var. *nipponensis*. The secondary gene pool consists of two parts. Secondary genepool (a) consists of species that cross relatively easily both as seed parent and as pollen parent with *V. angularis* – *V. minima, V. nakashimae, V. nepalensis, V. riukiuensis* and *V. tenuicaulis*. Secondary gene pool (b) consists of *V. hirtella* [that cross only as female (seed) parent (*V. hirtella* JP109681)] and *V. umbellata* (that can only produce hybrids with the help of embryo rescue) and possibly other species in the *angularis-umbellata* gene pool. The tertiary gene pool would be taxa in the *radiata-mungo* gene pool.

1.3. Crop complexes: Vigna angularis *as a model*

Crop complexes are the biological species to which domesticated crops belong and include their wild and weedy relatives. Many crop complexes have been studied in depth because of the insights they give into domestication processes (eg. Beebe *et al.*, 1997; Oka and Morishima, 1971). Crop complexes are subject to both natural and artificial (human) selection and consequently they exhibit accelerated evolution.

Many crop species show similar patterns of variation between crops and their wild and weedy relatives. This parallelism in different crop complexes formed the basis of Vavilov's Law of Homologous Series (Vavilov, 1922). This parallel variation found

Table 6.2. Comparison of characteristics of wild, weedy and cultivated azuki.

Trait	Wild azuki	Weedy azuki	Cultivated azuki
Life cycle	Annual	Annual	Annual
Plant stature	Slender twining	Bushy to slightly climbing	Bushy
Pod dehiscence	Dehiscent	Dehiscent	Indehiscent
Pod color	Black	Black/straw	Black/straw/brown
Pod length (mm)[c]	29-91	29-97	84-111
Pod wall	Striated coriaceous	Striated coriaceous	Papyriferous, monoliform
Inflorescence[a]	Racemous, exerted from plant canopy	Racemous, not exerted from plant canopy	Racemous, not exerted from plant canopy
Branching[a]	2-4 lateral branches, many branchlets over one meter	Lacks lateral branches from lower axils	Lacks lateral branches from lower axils
Seedling[a]	Expands first leaf from dark purple (rarely green) main stem, extends lateral branches close to soil from axils of the first leaves and subsequent cauline leaves	Expands first leaves from green (rarely purple) main stem ca. 6-10cm above ground level and extends its first long branch from the forth or fifth node.	Expands first leaves from green (rarely purple) main stem well above ground level and extends its first long branch from the forth or fifth node
First leaf height (cm)[b]	2.05-3.74	6.19-13.1	8.05-15.4
First leaf length (cm)[b]	2.68-4.25	3.26-4.21	4.13-6.26
Seeds/pod[c]	3-13	4-11.8	6.4-9.8
Seed length (mm)[c]	3.65-4.59	4.26-5.78	5.2-8.39
Seed width (mm)	2.76-3.69	3.24-4.55	3.54-6.17
Seed thickness (mm)	2.37-3.3	2.79-4.3	3.03-5.76
100 seed wt. (g)[c]	0.9-4.9	1.2-9.3	12.2-21.7
Seed coat colors[c]	Black mottled on gray or green	Variable: yellow/brown, gray, yellow/green, brown mottled on light brown.	Usually red, also black, black on red, red/straw, or rarely same colors as weedy form.

Sources: a. Yamaguchi, 1992

      b. Yamaguchi and Nikuma, 1996

      c. Vaughan et al., 2000

among crop complexes which was initially recognized based on morphological characters, is similar to the parallel variation being documented today in molecular phylogeny and synteny studies.

In the subgenus *Ceratotropis*, there are crop complexes associated with each of the domesticated species. However, only the *Vigna angularis* complex (azuki bean complex) has been studied in depth. The azuki bean complex has a wide distribution from Nepal through China to Japan. Inter and intra-population studies have been conducted for populations in Japan of this complex. We present here results of research on the azuki bean complex since these studies provide information relevant to genetic diversity and conservation that may be applicable to other crop complexes of the subgenus *Ceratotropis*.

The *Vigna angularis* complex consists of two taxonomically recognized varieties var. *angularis* and its close wild relative and presumed progenitor, var. *nipponensis*. In addition, an intermediate type has been recognized and called a weedy type (Yamaguchi, 1992). The weedy type of azuki is variable in its morphological characters. Typical weedy azuki is recognized in the field by its bushy habit and often, greenish-yellow leaves. A table comparing the characteristics of wild, weedy, and cultivated azuki bean is shown (Table 6.2).

Populations over time

Genetic resources workers usually see populations at one moment in time and usually this is when seeds are near maturity. Thus, the impact of disturbance is not apparent.

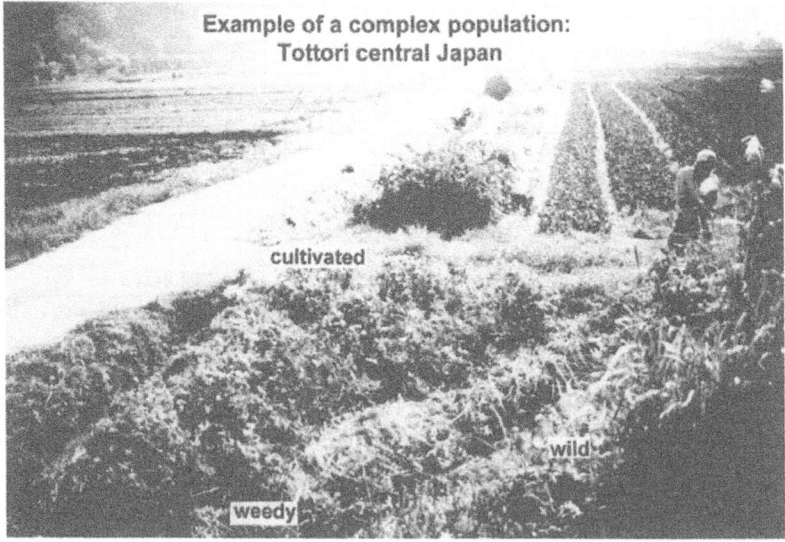

Figure 6.3. Part of a complex population and adjoining small plot of cultivated azuki bean at Koge, Tottori, Japan.

Monthly observations of wild and weedy azuki populations on Tsushima island, western Japan have revealed that of the 12 populations of wild, weedy or mixed wild/weedy populations observed monthly between early June and late October, all were disturbed at least once by either human activity or natural events. Six populations were disturbed once, 4 were disturbed twice and 2 were disturbed 3 times. The disturbances in order of frequency were weed killer (10), cutting (7), construction (1), farming (1) and flooding (1) (Yasuda and Yamaguchi, 1998a).

Studies of several population monitoring sites over several years in various parts of Japan have revealed the impact of disturbance and gene flow (Tomooka *et al.*, 2001b). Hybrid swarms are uncommon and maybe transient. Natural selection for the adaptive syndromes of either wild, weedy or cultivated types is considered to be strong and intermediates may be eliminated quickly (Harlan, 1965; Harlan & De Wet, 1965).

A different population type is the complex population. Complex populations tend to be large and have variable composition with wild and weedy plants scattered over a wide area and usually some fields in the vicinity are planted with azuki beans. A population at Koge village, Tottori, Japan is a good example (Figs. 6.3 and 6.4). This population has been studied over 5 years. Cultivated azuki beans are planted in different fields annually and weedy plants have been found in new locations each year.

The extent and role of gene flow among different components in the azuki bean complex is not known (Egawa *et al*, 1990b). However, insect visitors that accumulate pollen on their bodies have been recorded for different populations. This has shown that bees in the genera *Megachile* and *Xylocapa* are the most frequent azuki flower visitors (Fig. 6.5, Table 6.3). The complex flower structure, nectaries, large and bright colored flowers suggest that gene flow is likely to be mediated by insects. Since insects of the same genera visit flowers of wild and cultivated azuki, it is probable that gene flow among components of the azuki bean complex is facilitated by these insects. However, flower size differences may lower the rate of successful cross-pollination among crop complex components and result in gene flow being more likely in one direction. This would help explain, at least in part, what appears to be segregating sub-populations within complex populations (Fig. 6.4) (Wang *et al.*, 2002).

Field observations during one year and over several years reveal that populations of the azuki bean complex are changing constantly. Each population has its own particular characteristics of habitat disturbance and genetic make up. From a biosystematics perspective, wild, weedy and cultivated types of azuki bean can be recognized in the field at a moment in time. However, observations over time reveal a more dynamic picture.

Genetics of different population types

An analysis of genetic variation in the azuki bean complex from across most of its range has shown that wild *V. angularis* has a high level of genetic diversity in the Asian sub-tropical highlands of southwest China and the Himalayas (Mimura *et al.*, 2000). However, Mimura *et al.* (2000) hypothesized that cultivated azuki was most likely

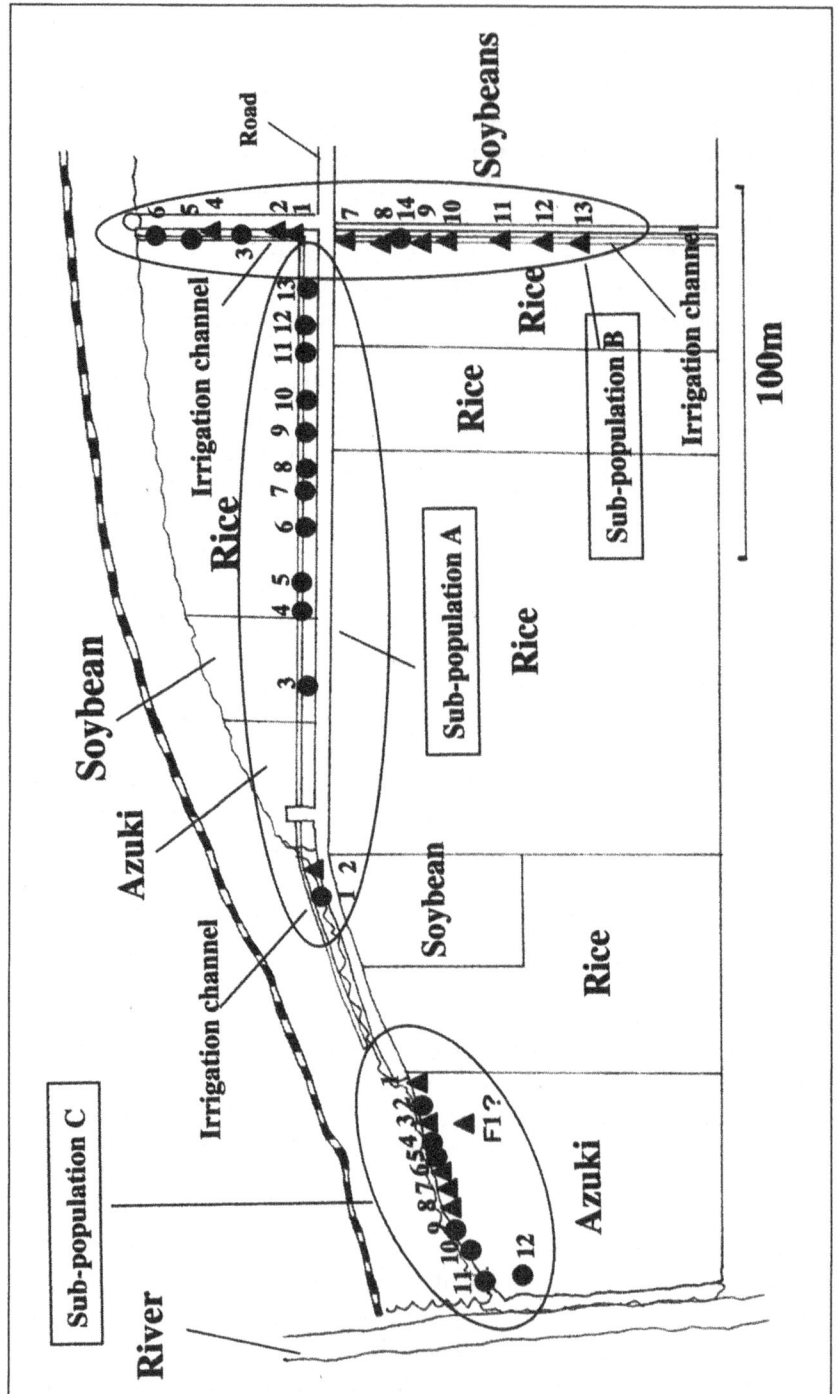

Figure 6.4. Distribution of wild (●) and weedy (▲) azuki bean plants in a complex population at Koge village, Tottori, Japan.

Table 6.3. Insect visitors to azuki bean populations [1]

Flower visitor		Wild azuki	Cultivated azuki
Family	Genus		
Megachilidae	*Megachile*	11	15
Anthophridae	*Xylocapa*	14	3
Apidae	*Bombus*	1	1
Halictidae	*Halictus*		1
Vespidae	*Polisters*		1
Vespidae	*Anterhynchium*		1
Syrphidae		1	1

[1] Based on observations at 4 populations from 7am to noon.

Figure 6.5. Kumabachi (carpenter bee), Xyclocapa appendiculata *foraging for nectar in flower of wild azuki bean* (Vigna angularis *var.* nipponensis).

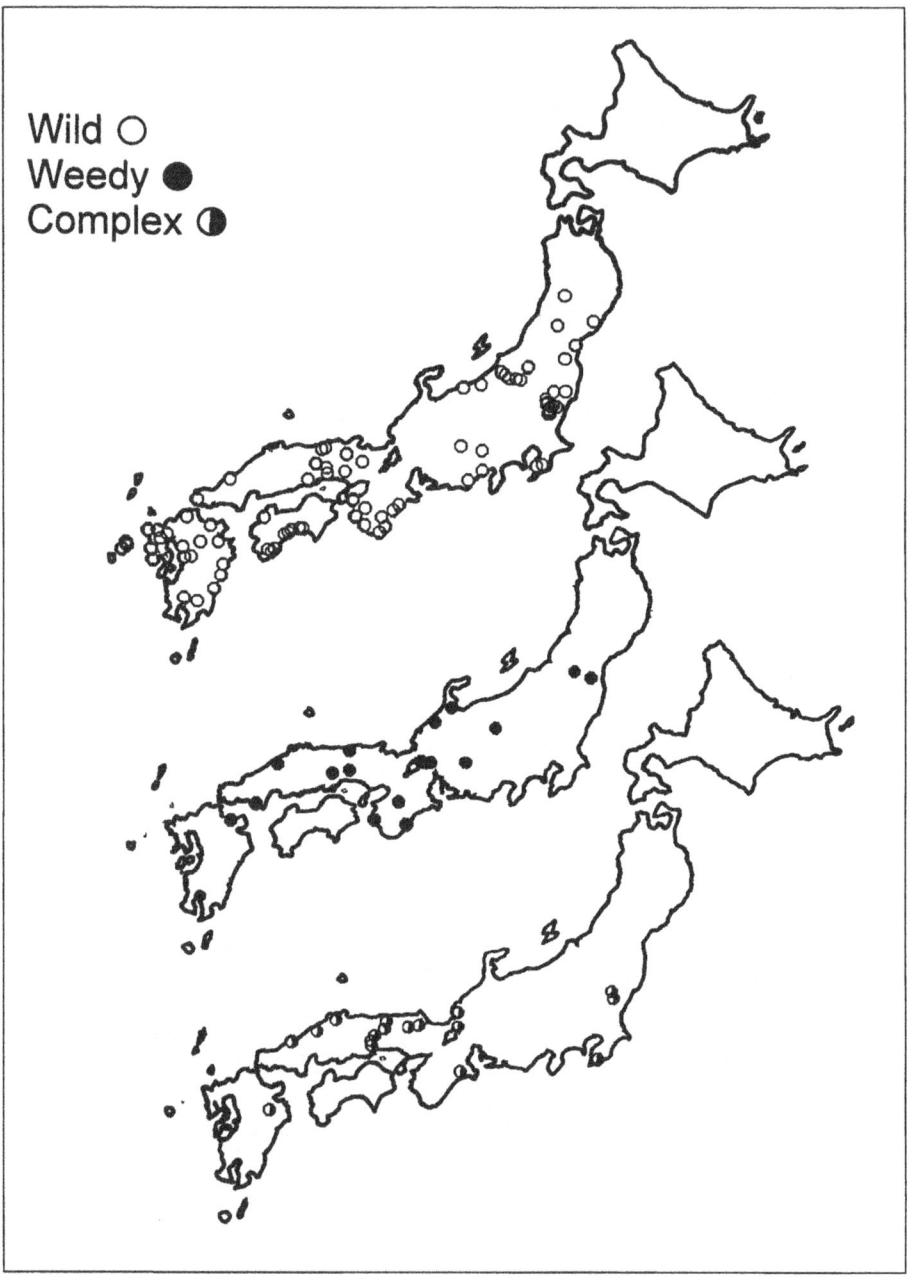

Figure 6.6. Distribution of Vigna angularis *complex population types in Japan based on direct collection by the authors.*

Table 6.4. Intra population variation based on RAPD and AFLP polymorphism detected in Vigna angularis *complex population groups.*

Population group	RAPD polymorphism[1]	AFLP polymorphism[1]
Cultigen	0.079	0.353
Weedy	0.124	0.561
Wild	0.132	1.191
Complex	0.152	-

[1] Based on Shannon's diversity index

domesticated from wild azuki bean in East Asia, Korea or Japan, probably through a stage like weedy azuki because wild and weedy azuki beans from East Asia show a high level of genetic similarity to cultivated azuki. At archaeological sites in Japan, beans have often been unearthed and these are presumed to be azuki bean (Yoshizaki, 1995, 1997). Others have speculated that azuki bean has its center of origin and center of domestication in China based on diversity of Chinese landrace azuki germplasm compared to other regions (Yee *et al.*, 2000). There are many areas within the region of distribution of wild and cultivated azuki bean for which germplasm has not been analyzed so questions related to diversity centers and origin of the azuki bean complex remains to be answered.

In Japan, there has been extensive exploration and genetic analysis of the azuki bean complex (Vaughan *et al.*, 2000; Xu *et al.*, 2000a, b; Yamaguchi, 1992). These studies give a clear insight into the relative genetic diversity of different components of the azuki bean complex. Maps of azuki bean complex population types across Japan reveal that the area where complex populations are most abundant is between 134°E and 137°E (Fig. 6.6). This may reflect the area where the environment is best suited to wild and weedy azuki in Japan. It may also reflect the history of Japan since this area encompasses the ancient capitals of Japan, Nara and Kyoto. In historic times to feed the population of Nara and Kyoto farmers would have grown azuki beans. The resulting human disturbance to wild populations and possibly gene flow may have had an impact on the genetic variation of the azuki bean complex we see in this region of Japan today.

Using the AFLP technique, Xu *et al.* (2000a) examined genetic diversity in populations across Japan. Wild populations showed clinal variation in AFLP polymorphism from west to east suggesting that wild populations are locally adapted. The local adaptation of weedy and wild azuki beans was also revealed by the relative inter and intra population genetic variation. About two thirds of the genetic variation was found between populations and one third within populations. Thus, for *ex situ* conservation of the azuki bean complex sampling many different populations will capture greater genetic diversity than sampling more individuals from few populations. Considering *in situ* conservation, conservation of widely scattered sites is commonly

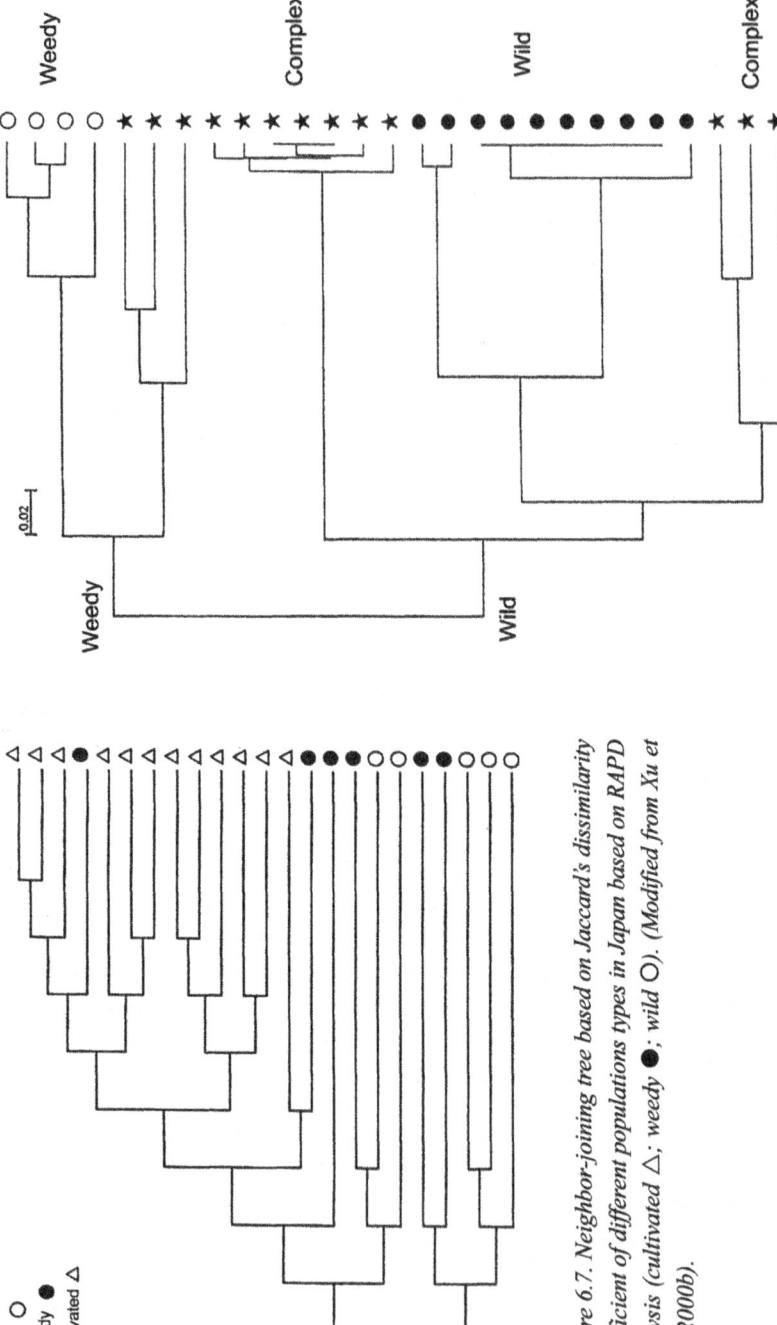

Figure 6.7. Neighbor-joining tree based on Jaccard's dissimilarity coefficient of different populations types in Japan based on RAPD analysis (cultivated △; weedy ●; wild ○). (Modified from Xu et al., 2000b).

Figure 6.8. UPGMA dendrogram, based on matrix generated from AFLP data, representing the association among individual plants from three different types of natural populations (wild ●; weedy ○; complex ★) belonging to the azuki bean species complex from Tottori prefecture. (Modified from Xu et al., 2000a).

preferable to *in situ* conservation of several sites in the same area (Maxted *et al.*, 1997).

Based on different molecular analyses, wild azuki bean has greater population variation than weedy and cultivated azuki bean (Table 6.4). However, the greatest population diversity is found in complex populations. Molecular analyses shed particular light on weedy and complex populations and they are discussed below.

Weedy populations

Three hypothesis have been proposed for the evolution of weedy races (De Wet and Harlan, 1975). These hypothesis are that weedy races are: a) Escapes from cultivation; (b) Hybrids between the wild type and the cultigen; (c) Directly evolved from the wild type.

Of these three hypothesis, DNA analysis results suggest that weedy populations of *V. angularis* may sometimes be the result of hybridization and sometimes be directly evolved from the wild type. Cluster analysis of AFLP data supports weedy populations resulting from both processes (Fig. 6.7). Two weedy populations cluster with cultivated azuki and three populations cluster with two groups of wild azuki. Evidence that supports the weedy type resulting from hybridization between wild and cultivated azuki includes:

• Seed coat color variation in some weedy populations resembles artificial hybrid populations;
• Populations that appear to be hybrid swarms have been observed (Vaughan and Kaga, 2000; Yamaguchi, 1992).

Evidence that supports weedy azuki having evolved directly from wild azuki includes:

• RAPD bands found in the wild and weedy types but not in the cultigen and no specific weedy RAPD bands (Xu *et al.*, 2000b);
• Higher genetic diversity in wild populations than weedy populations (Table 6.4);
• Field observations suggesting, in some locations, that weedy azuki is adapted to wetter habitats than nearby wild populations. For example, the sides of streams are particularly disturbed due to seasonal flooding and weedy azuki, which can be found in such habitats, may be specifically adapted to more highly disturbed ecological niches than the wild type. Larger seeds may be advantageous to support seedling growth through high level of silt cover and bushy habit may have been advantageous where flooding removes plants that support twining growth habit.

Complex populations

Field data and molecular analyses have enabled genetic characteristics of complex populations to be measured. Complex populations have been reported for other crop complexes e.g. *Vicia sativa* (Maxted, 1995), in the case of *Phaseolus vulgaris* they have been called inter-breeding complexes (Beebe *et al.*, 1997). AFLP methodology was used to analyze individual plants from three populations, wild, weedy and complex, growing within 6 km of each other in Tottori prefecture, Japan. All three populations analyzed were adjacent to farmland where small fields of cultivated azuki beans were being grown. The dendrogram resulting from this analysis is shown (Fig. 6.8). This

Table 6.5. Yields from collecting wild plants.

Species	Yield g/hr	Reference
Diploid wild wheat	1000g/hr clean seeds	Harlan, 1967
Tetraploid wheat	500g/hr	Ladazinsky, 1975
Lentil (wild)	10g/hr*	Ladzinsky, 1987
Lentil (wild in tilled field)	110g/hr*	Zohary, 1989
Azuki (**pod**) wild weed cultivated	735g/hr 1762g/hr 1980g/hr (control)	Yasuda and Yamaguchi, 1998 b
Azuki (**seed**) wild weed cultivated	187g/hr 170g/h 723g/hr (control)	Yasuda and Yamaguchi, 1998 b

*Estimated assuming seeds from 100 plants can be collected in one hour
from scattered wild populations and 200 plants in one tilled field.

shows that the complex population has plants that are similar to plants from both wild and weedy populations. In addition, wild like plants in the complex population have greater genetic diversity than the wild population and these plants form two groups. The high level of genetic diversity in complex populations suggests that complex populations should be a focus of attention for both *in situ* and *ex situ* conservation.

Domestication

Domestication of azuki beans occurred long ago but the process of domestication continues in breeding stations and farmers fields today. In common with other crops, a syndrome of characters have changed during domestication, such as, non- or reduced fruit (pod) shattering and increase in size of edible parts. For example, seed size of cultivated azuki is about 6 times the size of wild azuki. While domesticated azuki bean is non-dormant, compared to other crops the difference between dormancy in wild and domesticated azuki bean is not great.

In ancient times, would azuki bean have been a likely plant to domesticate? The yield of harvesting pods and seeds from wild and weedy populations of azuki in Japan has been determined (Yasuda and Yamaguchi, 1998b) (Table 6.5). These yields are compared with similar studies of wild plants in the Middle East. It is clear that quite large quantities of wild and weedy azuki can be quickly harvested. One pod of wild azuki will yield about three times as much weight of seeds as wild soybeans (Vaughan *et al.*, 2000). Thus, wild azuki may have appeared to be an attractive plant to harvest and then domesticate.

1.4. Analysis of representative germplasm sets

A core collection has been defined as a "limited set of accessions derived from existing germplasm collections, chosen to represent the genetic spectrum in the whole collection. The core should include as much as possible of its genetic diversity." (Brown, 1995). With a wild species germplasm collection, an initial representative (core) collection would include representatives of each species and for widely distributed species, incorporating accessions with known distribution, morphological, cytological and other attributes from widely scattered locations (Brown *et al.,* 1987). Representative (core) wild species collections have been shown to be an efficient and cost effective approach to evaluation (Kobayashi *et al.,* 1993; Maxted *et al.,* 1997, Tomooka *et al.,* 2000b).

The core collection concept at the species level has been used in diversity and evaluation studies of *Vigna* subgenus *Ceratotropis* germplasm. The aim of this section is to illustrate the potential of using core collections of *Vigna* spp. to help understand genetic diversity and find useful traits. As examples, we discuss characterization of a representative species collection for its various protease inhibitors (Konarev *et al.,* 2002) and evaluation for insect resistance (Tomooka *et al.,* 2000b).

1.4.1. Proteinase inhibitors

Proteinase inhibitors have a variety of functions in plants including involvement in plant resistance to insects and microorganisms (Ryan, 1990). These inhibitors also affect the nutritional quality of beans for humans, so efforts have been made to breed soybeans that lack trypsin inhibitors (Orf and Hymowitz, 1979).

The results of surveying proteinase inhibitors at the species level in the subgenus *Ceratotropis* are shown (Table 6.6) (Konarev *et al.,* 2002). These reveal the following variation in proteinase inhibitors in the subgenus *Ceratotropis*.

- Trypsin inhibitors (TI) are the most polymorphic protease inhibitors (Fig. 6.9) followed by chymotrypsin inhibitors (CI). Subtilisin inhibitors (SI) and cysteine proteinase inhibitors (CPI) are uniform across sections. CPI shows no variation among species in section *Ceratotropis* and *Aconitifoliae.*
- TI bands vary from 1 in *V. subramaniana* to 18 in the tetraploid *V. reflexo-pilosa.* Other species with abundant TI are *V. angularis* and *V. trinervia.*
- Several species have a considerable diversity of TI variation but absence or near absence of CI, such as *V. grandiflora, V. aconitifolia* and *V. stipulacea.*

Proteinase inhibitor variation revealed the following information concerning variation within and among species.
- Some species or species groups show very little intra-specific variation in TI and CI profile, such as, *V. minima, V. reflexo-pilosa, V. nakashimae, V. radiata* and *V. mungo.* Other species, such as *V. hirtella,* show considerable variation from accession to accession.

Table 6.6. Proteinase inhibitor polymorphism detected in Vigna subgenus Ceratotropis accessions (from Konarev et al., 2002)

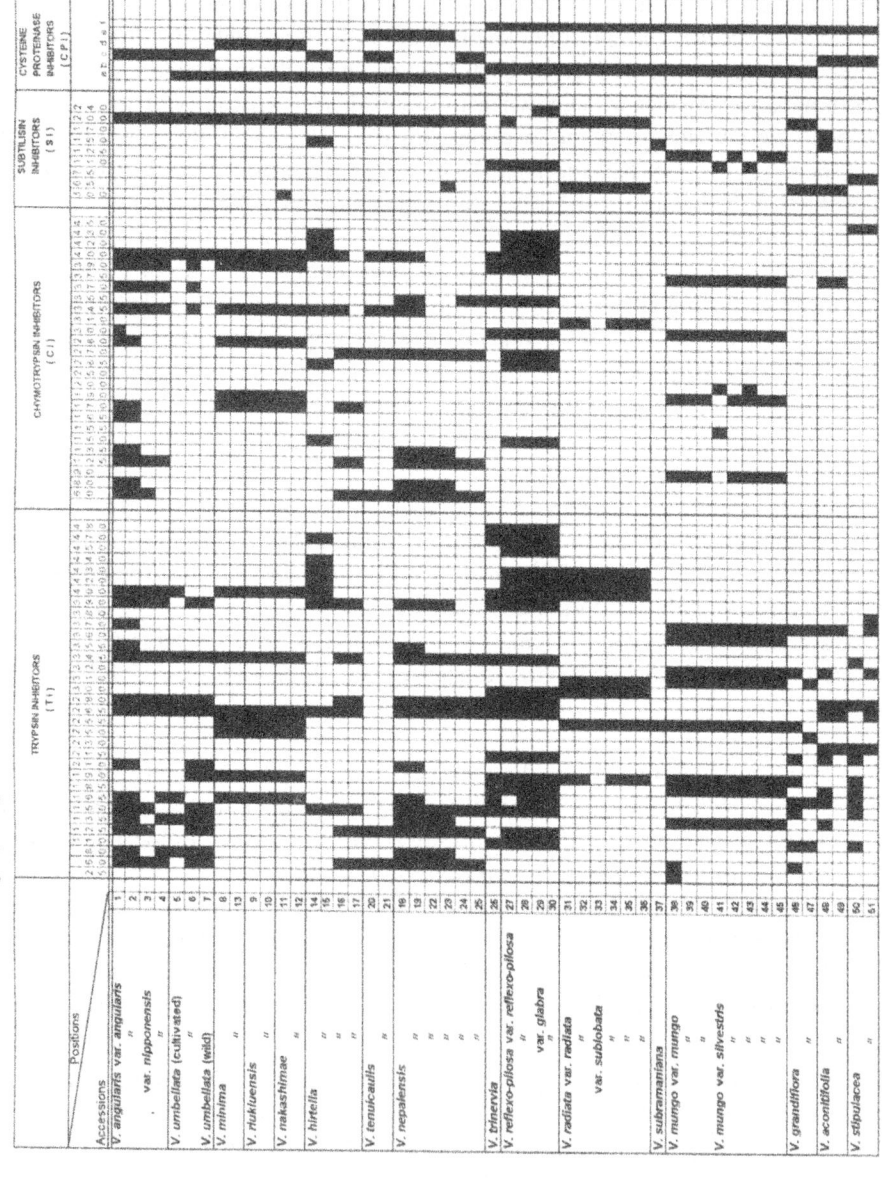

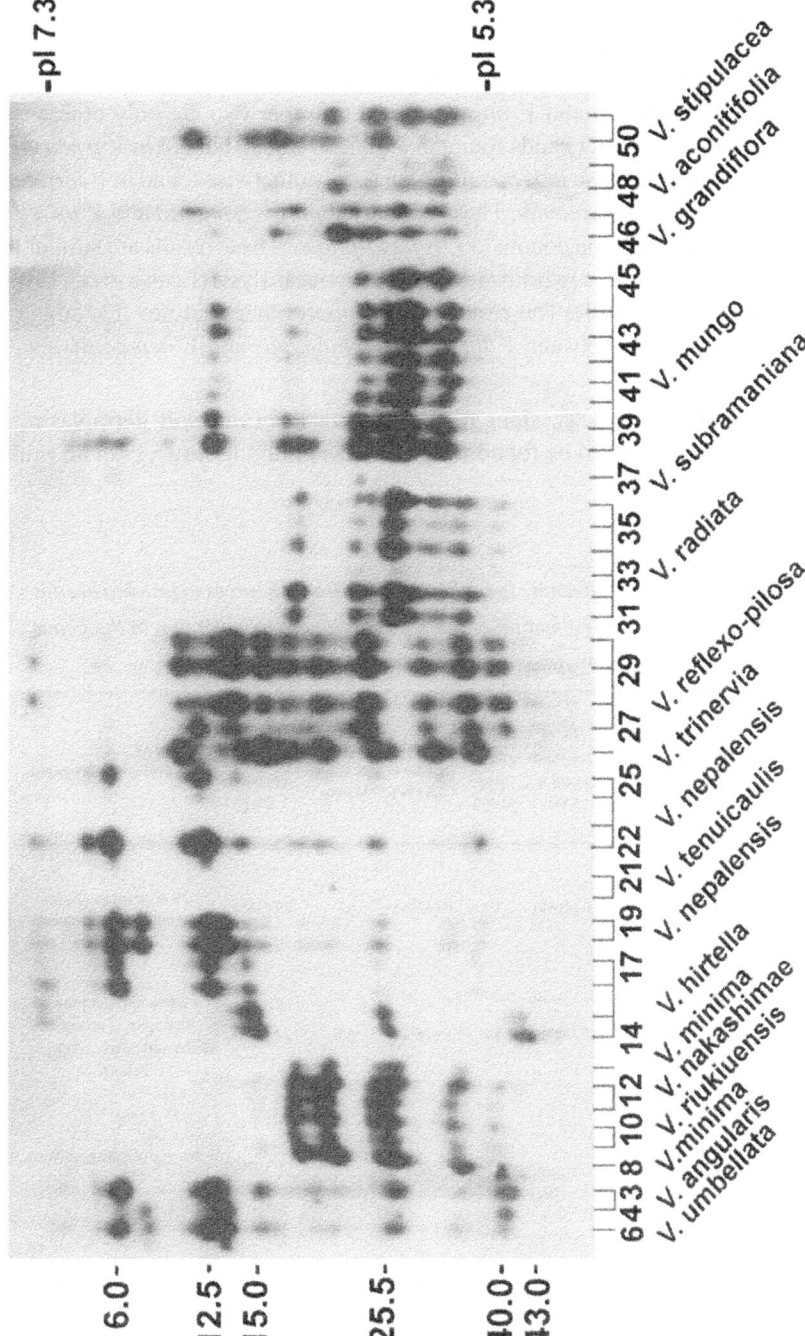

Figure 6.9. Trypsin inhibitor polymorphism in the genus Vigna *subgenus* Ceratotropis. *(from Konarev et al., 2002)*

- Results of analyzing proteinase inhibitors sheds light on the species that may have donated genomes to the tetraploid *V. reflexo-pilosa*. *V. trinervia* has 14 common TI and 4 common CI bands with *V. reflexo-pilosa*. Of TI bands in *V. reflexo-pilosa* not found in *V. trinervia*, one is found in *V. nepalensis*, *V. tenuicaulis*, *V. hirtella*, *V. umbellata* and *V. angularis* and a further two are only otherwise found in *V. hirtella*. Of CI bands found in *V. reflexo-pilosa* but not in *V. trinervia*, 4 are only found otherwise in *V. hirtella*. One is only otherwise found in *V. hirtella*, *V. nepalensis* and *V. tenuicaulis*. These results support *V. trinervia* and *V. hirtella* as the most likely genome donors to *V. reflexo-pilosa*. These results are similar to results from inter-specific hybridization and isozyme analysis (Egawa *et al.*, 1996; Jaaska and Jaaska, 1990). The results also support other analyses that suggest the close relationship between *V. hirtella*, *V. tenuicaulis* and *V. nepalensis*.

By using a few well-chosen accessions of the subgenus *Ceratotropis* germplasm, a broad picture of the variation to be found for proteinase inhibitors emerges. The value

Table 6.7. Comparison of effectiveness in finding new resistance sources to bruchid beetles between species level core collection and mungbean landrace collection in the genus Vigna subgenus Ceratotropis (modified from Tomooka et al., 2000b).

Plant materials	Bruchid species	No. of plant species tested	No. of resistant species (%)	No. of plant taxa tested	No. of resistant taxa (%)	No. of accessions tested	No. of resistant accessions (%)	Levels of resistance
Species level collection of the *Vigna*	*C. chinensis*	13	6 (46%)	19	8 (42%)	63	29 (46%)	29 accessions showed complete resistance
Species level collection of the *Vigna*	*C. maculatus*	13	5 (38%)	19	5 (26%)	63	24 (38%)	24 accessions showed complete, 2 accessions showed nearly complete resistance
Mungbean landraces	*C. chinensis*					426	0 (0%)	No significant resistance found
Mungbean landraces	*C. maculatus*					330	0 (0%)	No significant resistance found

of the results for crop improvement are that to reduce proteinase inhibitors that may have undesirable effects on legume nutritional quality, species with few proteinase inhibitors are available, such as *V. subramaniana* that has just one TI band and no CI. Evaluation of *V. subramaniana* resistance to bruchid insects has shown that despite low activity of these proteinase inhibitors, the same accession has complete resistance to *Callosobruchus chinensis* (azuki bean weevil) and *C. maculatus* (cowpea weevil) (Tomooka *et al.*, 2000b).

1.4.2. Insect resistance

Worldwide bruchid beetles are pests of legume seeds and can do major damage to legumes in storage. Using a representative species level germplasm collection, sources of resistance to this pest were sought, since evaluation of cultivated varieties and lines has revealed few useful sources of resistance (Epino & Morallo-Rejesus, 1982; Singh *et al.*, 1985). This representative germplasm collection was found to be effective in revealing new bruchid beetle resistance sources (Table 6.7). Eight taxa (*V. tenuicaulis*, *V. nepalensis*, *V. mungo* var. *mungo*, *V. mungo* var. *silvestris*, *V. subramaniana*, *V. reflexo-pilosa* var. *reflexo-pilosa*, *V. trinervia* and *V. umbellata*) consisting of 29 accessions have complete resistance to *C. chinensis*. Five taxa (*V. tenuicaulis*, *V. nepalensis*, *V. mungo* var. *silvestris*, *V. subramaniana*, and *V. umbellata*) consisting of 24 accessions showed complete resistance to *C. maculatus*. In addition, one accession of *V. nepalensis* and one accession of *V. minima* showed nearly complete resistance to *C. maculatus*. Various levels of resistance including delayed larval growth, low adult emergence and small size of emerged adult were also observed in other accessions. This suggests that the factors responsible for resistance are diverse across the subgenus *Ceratotropis*. Considering that bruchid strains or biotypes may overcome a single resistance gene, incorporation of multiple sources of resistance into *Vigna* crops is likely to be more durable (Credland, 1990).

A representative species level germplasm collection is an efficient means of revealing diversity for various traits. This approach to seeking novel genes can be highly efficient.

1.5. Biotechnology

A genetic resources monograph prepared in the 21st century inevitably needs to discuss information available at the genome and DNA level for the group under consideration. We present here a brief review of the main work that has been published related to genome maps, comparative genomics and gene mapping for *Vigna* subgenus *Ceratotropis*. This is a field that rapidly becomes out of date. Recent information related to this topic may be found from the internet at:
http://beangenes.cws.ndsu.nodak.edu
http:// www.ncbi.nlm.nih.gov/taxonomy

Table 6.8. Genome maps for Vigna subgenus Ceratotropis species.

Cross combination	Population analyzed	Markers used	Linkage groups resolved	Map distance	Level of distortion	Reference
V. angularis X V. nakashimae	F$_2$ population 80 plants	19 RFLP, 108 RAPD and 5 morphological markers	14	1250cM	19.7%	Kaga et al., 1996a
V. angularis X V. umbellata	F$_2$ population 86 plants	114 RFLP, 74 RAPD, 1 morphological marker	14	1702cM	29.8%	Kaga et al., 2000a
V. radiata var. radiata X V. radiata var. sublobata (from Madagascar)	F$_2$ population	151 RFLP, 20 cDNA and 1 pest locus	14	1570cM	12%	Menacio-Hautea et al., 1992
V. radiata var. radiata X V. radiata var. sublobata (from Australia)	F$_2$ population	52 RFLP, 56 RAPD, 2 morphological markers	12	758.3cM	14.5%	Lambrides et al., 2000
V. radiata var. radiata X V. radiata var. sublobata (from Australia)	Recombinant inbreed lines	113 RAPD, 2 morphological markers	12	691.7cM	24%	Lambrides et al., 2000

1.5.1. Genome maps

Genome maps are valuable in providing insights into genome organization, inheritance and linkage of traits. Genome maps are therefore helpful in determining breeding strategies for traits of interest. To date there have been five different genome maps developed for species within *Vigna* subgenus *Ceratotropis* (Table 6.8). Three have been developed for *Vigna radiata* (Menancio-Hautea *et al.*, 1992; Lambrides *et al.*, 2000) and two have been developed for *V. angularis* (Kaga *et al.*, 1996a, 2000a). A comparison between these maps is shown (Table 6.9). None of the maps developed so far have been able to resolve the 11 linkage groups, equivalent to the haploid chromosome number of both *V. radiata* and *V. angularis* (Sinha and Roy, 1979). This is probably due to the limited number of marker clones that are available for species in the subgenus *Ceratotropis*. To overcome this, the many DNA marker clones from related species, such as *Glycine max*, *Phaseolus vulgaris* and *V. unguiculata*, have been used to increase the saturation of the genome maps of *V. radiata* and *V. angularis* (Boutin *et al.*, 1995; Lambrides *et al.*, 2000; Kaga *et al.*, 2000a; Chaitieng *et al.*, 2002; Menacio-Hautea *et al.*, 1992).

The genome maps for *V. radiata* have been based on crosses between *V. radiata* var. *radiata* and *V. radiata* var. *sublobata*. In one genome map the *V. radiata* var. *sublobata* accession came from Madagascar, in the other two the accession came from Australia. In the resulting genome maps while the order of genome markers was similar, the level of distortion was higher in the cross that involved the Australian accession of var. *sublobata* and regions of distortion did not coinside with that produced with the Madagascar accession. This suggests that var. *sublobata* has considerable intra-specific genetic diversity and that the Australian form of var. *sublobata* is more distantly related to cultivated *V. radiata* than the Madagascar form (see section 1.1 of this chapter).

Of the genome maps reported, that between *V. angularis* and *V. umbellata* had the highest level of segregation distortion (29.8%) (Table 6.8). High levels of segregation distortion may not enable the trait(s) of interest to be found in segregating populations. When using distantly related species in crosses it is necessary to generate very large segregating populations, since distantly related species have less likelihood of recombination and thus the likelihood of getting the wanted trait in the desired background is low. However, large populations may not be easy to obtain as weak or sterile hybrid plants may limit seed production (Kaga *et al.*, 1996a). Ways to overcome this problem may include using bridging species to facilitate gene transfer (Tomooka *et al.*, 2001a).

To date only one genome map has employed recombinant inbred lines to generate the genome map (Lambrides *et al.*, 2000). In order to analyze traits controlled by many genes it is necessary to develop advanced generation recombinant inbred lines.

1.5.2. Comparative genome mapping

Since genome maps have been developed in cultigens of genera and subgenera related to subgenus *Ceratotropis* cultigens, there have been efforts to understand the

Table 6.9. Comparison of lengths of genomic blocks conserved among mungbean, common bean and soybean (Boutin et al., *1995)*

Species compared	Average length of conserved block (cM)	Standard deviation	Length of longest conserved block (cM)
Mungbean and common bean	36.6	27.6cM	103.5
Mungbean and soybean	12.2	9.4cM	37.8
Common bean and soybean	13.9	9.5cM	34.8

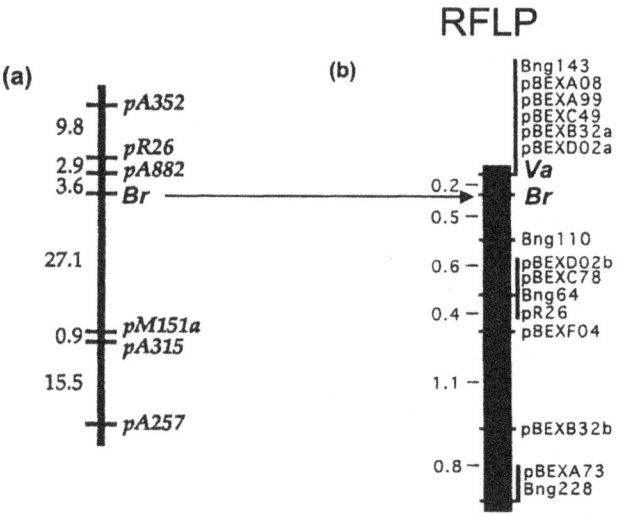

Figure 6.10. Progress in mapping the bruchid resistant gene Br on mungbean linkage group VIII using RFLP markers. Left side of each figure map distance in cM. [Composite from (a) Young et al. 1992 and (b) Kaga and Ishimoto, 1998. Copyright to both figures Springer-Verlag. Used with permission of Springer-Verlag and authors N. Young, University of Minnesota, USA and A. Kaga, National Institute of Agrobiological Sciences, Japan.]

comparative genome organization across these cultigens (Boutin *et al.*, 1995; Menacio-Hautea *et al.*, 1993). Comparisons of genome maps of *V. radiata* with *V. unguiculata* and *Phaseolus vulgaris* have revealed conserved blocks of considerable size, some containing loci for important traits. For example, the major QTL for seed weight in cowpea and mungbean span the same RFLP markers (Fatokun *et al.*, 1992). The comparison of *V. radiata* with *P. vulgaris* showed that average size of conserved blocks is about 36.6 cM with the longest of 103.5 cM (Table 6.9). There is therefore considerable scope for understanding genome organization in cultigens of subgenus *Ceratotropis* using probes from and comparison with better developed genome maps in other related species.

One of the best developed, genome maps among legumes is that of the soybean. Comparison of *V. radiata*, *Phaseolus vulgaris* and *Glycine max* revealed different types of genome organization. Conserved linkage blocks of *V. radiata* and *P. vulgaris* are smaller and highly scattered across the *Glycine max* genome. (Table 6.9, Boutin *et al.*, 1995). However, specific analysis of a genomic region influencing seed weight in soybean showed co-linearity of RFLP markers with mungbean (Maughan *et al.*, 1996).

1.5.3. Gene mapping

Much of the emphasis on understanding genome organization is as a tool in plant breeding to enable genes to be more readily transferred to elite breeding lines. To this end, several important traits have been mapped on the mungbean genome. Mapping in mungbean of resistance to bruchid beetle and powdery mildew provide contrasting examples of changes in mapping over the past 10 years.

By RFLP analysis bruchid resistance found in wild mungbean (Fujii & Miyazaki, 1987; Fujii *et al.*, 1989) was first mapped as a single major locus on linkage group VIII (Young *et al.*, 1992). The nearest RFLP marker was 3.6 cM distance from this locus (Fig. 6.10). Since this resistance gene, from *V. radiata* var. *sublobata* (TC1966), also has an inhibitory activity against stink bug (*Riptortus clavatus* Thunberg) and it was associated with novel cyclopeptide alkaloids, further efforts were made to map this gene (Kaga and Ishimoto, 1998). The resulting genetic map enabled the resistant dominant locus to be located to within 0.2cM of the nearest RFLP markers. This map distance may enable the gene to be cloned within a genomic library for eventual introduction into susceptible mungbean lines or other crops.

Powdery mildew resistance is a multi genic trait. The first attempt to map resistance in a breeding line identified 3 QTLs on three different linkage groups that accounted for 58% of the total variation (Young *et al.*, 1993). Using a different powdery mildew resistant line, 96 RFLP probes failed to identify any QTLs associated with resistance (Chaiteng *et al.*, 2002). Subsequently 100 AFLP primer pair combinations were tested and 4 out of more than 5000 polymorphic bands were found to be associated with resistance. The main QTL associated with resistance was found on a new linkage group and accounted for 68% of the total variation.

A system has been developed to regenerate azuki and mungbean from organogenic calli (Yamada *et al.*, 2001, Jaiwal *et al.,* 2001). This opens new possibilities in genetic manipulation among *Vigna* subgenus *Ceratotropis* species. The genome of mungbean ranges from 470 - 560Mb about 50% bigger than the small genome of rice and about half that of soybean (Murray *et al.*, 1979; Arumuganathan and Earle, 1991). Therefore mungbean and other Asian legume crops are good candidates for more in depth genome analysis in the future. The results of such research would likely have an impact beyond the Asian legumes, as the discussion above on comparative genome mapping suggests.

2. FUTURE PERSPECTIVES

Much progress has been made over the last decade in studies on the Asian *Vigna*, however, there remain many areas of research for which the Asian *Vigna* have hardly been investigated. Below we discuss research areas where future studies may be beneficial in relation to conservation and use of the Asian *Vigna*.

2.1. Genetics and cytology

In many crops the genetics of domestication has been been shown to involve a syndrome of characteristics, many of them being closely linked (Doebley and Stec, 1993; Koinange *et al.*, 1996; Xiong *et al.*, 1999). The Asian *Vigna* that are used in agriculture represent varying steps on the way to domestication. *V. trinervia* is cultivated as a cover crop but retains wild characteristics. *V. umbellata* may be best described as a semi-domesticate, whereas *V. mungo* is fully domesticated. In Thailand one problem to growing mungbean (*V. radiata*) is asynchronous maturity. A better understanding of genetics related to the control of flowering and other traits associated with domestication may help in manipulating the traits most important to adaptation to various agricultural systems.

Reports of cross compatibility studies are numerous (see section 1.1 of this chapter). However, studies to determine the nature of inter sectional incompatibility and whether this can be overcome requires investigation. There have been no attempts to identify the different genomes within Asian *Vigna* as has been done with crops such as wheat and rice. DNA hybridization may enable the broad genome groups to be determined, while FISH (fluorescent *in situ* hybridization) and related methodologies may enable the details of chromosome structure and rearrangements to be elucidated.

New molecular markers for use in *Vigna* mapping studies are becoming available (Li *et al.*, 2001; Wang *et al.*, 2002). These will lead to increasingly detailed genome maps. The next decade may witness a quantitative improvement in the genome maps of the Asian *Vigna* that may assist in molecular breeding. However, at the same time the development of recombinant inbred lines need to parallel this progress so that appropriate populations are available for genetic analysis. Standard genome mapping populations that are internationally recognized need to be developed. There will become a need for a mechanism, such as a *Vigna* genetic committee, to ensure *Vigna* germplasm, genetics and genomics has an established international basis as has been done for other crops.

Transformation methodology has been developed for *V. angularis* and *V. radiata* (Yamada *et al.*, 2001, Jaiwal *et al.*, 2001). The next step will be to develop methodology to transform the other Asian *Vigna*.

At the population level, newly developed co-dominant micro-satellite (SSR) markers are revealing in more detail population genetic structure (Wang *et al.*, 2002). Population studies in centers of *Vigna* species diversity may help in understanding the evolutionary dynamics and speciation processes of Asian *Vigna*.

Population genetic studies conducted with attention to ecological factors would enable the full context of a population to be better understood. Such studies over several years would assist in development of *in situ* conservation strategies and may provide information relevant for ecologically balanced agriculture. Since the cultivated Asian *Vigna* are seriously affected by a variety of insect pests, wild *Vigna* populations may provide research sites to rapidly find natural enemies of these pests.

2.2. Germplasm

There remain many areas where biodiversity of the Asian *Vigna* is poorly understood, these include:

- the intra-specific diversity in *V. hirtella*, *V. minima* and *V. tenuicaulis*;
- *V. dalzelliana*, the wild form of *V. aconitifolia* and *V. khandalensis* have not been studied;
- crop complex studies for the Asian *Vigna* except *V. angularis* have not yet been conducted.

Large areas of Asia have been poorly explored for *Vigna* species. Particularly fruitful areas for future exploration will be:

- The Yunnan/Shichuan biodiversity hotspot and surrounding areas in Myanmar, northeast India and Bhutan (Morell, 2002);
- Indonesia, particularly Sumatra, Kalimantan and Irian Jaya;
- Laos, Cambodia and Vietnam.

2.3. Conservation

Research into conservation of the Asian *Vigna* needs to address ways to conserve "weedy" species. We have not discussed in this book on-farm conservation since literature on the topic specifically related to the Asian *Vigna* is lacking. However, as one approach to comprehensive conservation this topic needs attention. Recently, the outline for a methodology for on-farm conservation has been elaborated (Maxted *et al.*, 2002). Particularly useful will be detailed diversity studies in the four regions of *Vigna* species diversity highlighted in chapter 5 - South India, Thailand, eastern Nepal and neighboring India and Sri Lanka. Surveys in these areas that combine specialists of different plant groups to map, analyze and monitor crop relatives in several major protected areas may provide a stimulus to *in situ* conservation efforts.

Most *Vigna* species produce seeds in sufficient quantity to make *ex situ* conservation, a conservation method of choice where rapid habitat destruction is ongoing. Emergency actions, as have been undertaken to collect germplasm for *ex situ* conservation in areas affected by the Five Gorges Dam project in China, may need to be undertaken in other areas. Country wide specific surveys of wild *Vigna* for *ex situ* conservation may be particularly rewarding and fill a gap in current national genebank collections. In Japan such an approach and subsequent analysis of the germplasm has revealed useful sources of insect resistance (Kashiwaba *et al.*, 2002).

2.4. Characterization

Improving the nutritional quality of crops is always a research priority. Basic information on the chemical composition of the main *Vigna* cultigens is known (Duke, 1981). For *V. radiata* improving protein quality is focus of breeding programs in Thailand. The broad genetic diversity in the Asian *Vigna* gene pool may contain particular nutritional factors that can enhance nutritional quality of the cultigens. Conversely within the Asian *Vigna* gene pool adverse nutritional factors may be absent in some germplasm as has been shown for the near absence of trypsin inhibitors in *V. subramaniana* and *V. tenuicaulis*. Such germplasm may be useful for improving nutritional quality (Konarev *et al.*, 2002).

2.5. Evaluation

Efforts to evaluate the Asian *Vigna* have focused on screening germplasm. Future evaluation research requires an improved understanding of what is being evaluated. For example, testing for bruchid resistance has not considered biotypes within bruchid species that may affect results. Evaluation of *V. umbellata* for resistance to strains of *Callosbruchus chinensis* in Japan gave a different response when evaluated using Thai strains of *C. chinensis* (Tomooka unpublished data). Similarly, evaluation for resistance to powdery mildew has indicated that strains from different countries give different reactions in the same germplasm (Chaitieng *et al.*, 2002).

While wild species germplasm maybe most useful as a source of major genes for pest or disease resistance, some wild species may have useful genes for specific environmental niches such as *V. aridicola* for hot, dry areas and *V. exilis* for highly calcareous soils.

2.6. Last comments

The Asian *Vigna* in agriculture will always be ranked as "secondary crops" on a global scale compared to rice, wheat, maize and soybeans. Hence support for research related to these crops will also be "secondary" in scale. The close evolutionary relationship between cultivated Asian *Vigna*, African *Vigna* and *Phaseolus* of the New World suggest that international collaboration, perhaps within the framework of a global network, would raise all these crops to the level, they hold together, of a "primary" crop group on a global scale. Research strength from formal partnerships across the world would accelerate research since, for example, genetic markers that are time consuming, technically difficult and expensive to produce, may be shared among different related species. In addition, global partnership may enable seemingly obscure but potentially valuable research to be addressed such as genetic diversity in the subgenus *Sigmoidotropis* that appears to be central to the diversification of the *Vigna-Phaseolus* complex (Maréchal *et al.*, 1978).

At the start of the 21st century 4 new Asian *Vigna* have been described, *V. aridicola*, *V. exilis*, *V. nepalensis* and *V. tenuicaulis*. Three of these new species appear to be closely related to cultigens. Each of these four species consists of a unique genetic constitution with genes of potential use in agriculture. If 4 new species have been described so recently, how many Asian *Vigna* have yet to be found and described? The pace of habitat destruction and parallel genetic erosion in Asia is faster than previously realized (Jepson *et al.*, 2001). Thus the need for multiple approaches to conserve and study genetic resources of Asian *Vigna* will be a fundamental research priority in the 21st century.

APPENDIX I

Synonymised list of *Vigna* subgenus *Ceratotropis* taxa

A =Azukia D=Dolichos P=Phaseolus R=Rudua V=Vigna

Species name	Synonym	Ref
V. aconitifolia (Jacq.) Maréchal	*P. aconitifolius* Jacq.	3
	P. palmatus Forskal	3
	D. dissectus Lam.	3
V. angularis (Willd.) Ohwi & Ohashi	*D. angularis* Willd.	3
	P. angularis (Willd.) Wight	3
	A. angularis (Willd.) Ohwi	3
	Adzuki subtrilobata (Franchet & Savatier) Takahashi	4
	Abrus precatorius L.	4
	P. chrysanthos Savi	4
	P. mungo L.	4
	P.mungo L. var *radiatus*	4
	P. mungo var. *sublobata*	4
	P. mungo var. *subtrilobata* (Franchet & Savatier) Matsumura	4
	P. radiatus L.	4
	P. radiatus L. var. *subtrilobata* Franchet & Savatier	4
V. angularis var. *nipponensis* (Ohwi) Ohwi & Ohashi	*P. nipponensis* Ohwi	3
	P. trilobus Ait. *sensu* auctt. Jap.	1
	P. angularis f. nipponensis (Ohwi) Kitamura	1
	A. angularis var. *nipponensis* (Ohwi) Ohwi	1
	P. angularis var. *nipponensis* (Ohwi) Ohwi	1
V. aridicola N.Tomooka & Maxted	*V. trilobata* (L.) Verdcourt *sensu* Maxwell	8
V. dalzelliana (Kuntze) Verdcourt	*P. pauciflorus* Dalzell	3
	P. dalzellianus O. Kuntze	3
	P. dalzellii Cook	3
V. exilis Tateishi & Maxted	*P. pauciflorus* Dalzell *sensu* Gagnepain	6
	P. dalzellianus O. Kuntze *sensu* Craib	6
	V. dalzelliana (Kuntze) Verdcourt *sensu* Maréchal, Mascherpa & Stainier	6
V. grandiflora (Prain) Tateishi & Maxted	*P. sublobatus* (Roxb.) var. *grandiflorus* Prain	2
	V. radiata var. *grandiflora* (Prain) Niyomdham	2
	P. sublobata var. *grandiflorus*	1
V. hirtella Ridley		1
V. khandalensis (Santapau) Raghavan & Wadhwa	*P. grandis* Dalzell & Gibson	3
	P. khandalensis Santapau	3
	V. grandis (Dalzell & Gibson) Verdcourt	3
V. minima (Roxb.) Ohwi & Ohashi	*P. calcaratus* Roxb. var. *gracilis* Prain	5
	P. calcaratus Roxb. var *glabra* (Roxb.) Prain	5
	P. heterophyllus Hayata	5
	P. minimus Roxb.	3
	A. minima (Roxb.) Ohwi	3
	P. minimus f. *typicus* Hosogawa	5
	P. minimus f. *linearis* Hosogawa	5
	V. minima f. *heterophylla* (Hosokawa) Ohwi & Ohashi	5
	P. gracilicaulis Ohwi	5
	A. heterophylla (Hosogawa) Masamune	5
	V. gracilicaulis (Ohwi) Ohwi & Ohashi	5
	V. minima f. *linearis* (Hosokawa) Huang & Ohashi	5

Species name	Synonym	Ref
V. mungo (L.) Hepper	*P. mungo* L.	3
	P. viridissimus Ten.	1
	P. hernandezii Savi	3
	A. mungo (L.) Masamume	3
V. mungo var. *silvestris* Lukoki, Maréchal & Otoul	*V. radiata* var. *sublobata* (Roxb.) Verdccourt	5
V. nakashimae (Ohwi)Ohwi & Ohashi	*P. nakashimae* Ohwi	1
	A. nakashimae (Ohwi) Ohwi	1
V. nepalensis Tateishi & Maxted	*Vigna* sp. *sunsu* Tateishi	6
V. radiata (L.) Wilczek	*P. radiatus* L.	3
	P. hirtus Retz.	3
	P. abyssinicus Savi	3
	P aureus Roxb.	3
	A. radiata (L.) Ohwi	3
	R. aurea (Roxb.) Maekawa	3
	P. chanetii (Levl.) Levl.	2
	P. chanetii Levl.	2
	P. radiatus var. *typica* Matsum.	1
	Vigna aureus (Roxb.)Hepper	1
	Vigna mungo sensu Hepper, non L.	1
	P. mungo sensu F.B.I. non L.	1
V. radiata var. *sublobata* (Roxb.) Verdcourt	*P. sublobatus* Roxb.	3
	P. trinervius Wight	3
	P. trinervius Wight & Arn.	1
	V. brachycarpa Kurz.	1
	P. radiatus auct. mult. non *l. sensu stricto*	1
	V. opisotrichia Richard	3
	V. perrieriana Viguier	3
V. reflexo-pilosa Hayata	*P. reflexo-pilosus* (Hayata) Ohwi	1
	P. mungo L. *sensu* Forbes & Hemsley	1
	P. neocaledonicus Baker f.	1
	A. reflexo-pilosa (Hayata) Ohwi	1
	V. glabrescens Maréchal, Mascherpa & Stainier	1
	V. Catjang Endl. var. *sinensis* King *sensu* Matsum.	1
	V. minima var. *minor* (Matsum.) Tateishi	1
V. reflexo-pilosa var. *glabra* (Maréchal, Mascherpa & Stainier) N.Tomooka & Maxted	*P. glaber* Roxb.	3
	P. glabrescens Steudel	3
	P. mungo L. var. *glaber* (Roxb.) Baker	3
	V. glabrescens Maréchal, Mascherpa & Stainier	3
	V. radiata (L.) Wilczek var. *glabra* (Roxb.) Verdcourt	3
	V. reflexo-pilosa Hayata	1
	P. calcaratus var. *glaber* (Roxb.)Prain	1
	V. mungo var. *glabra* (Roxb.) Baker	1
V. riukiuensis (Ohwi) Ohwi & Ohashi	*V. lutea* Gray var. *minor* Matsum.	1
	P. mimima (Roxb.) Ohwi & Ohashi *sensu* Walker	1
	P. rotundifolius Hayata	5
	V. marina (Burm.) Merr. var. *minor* (Matsum.) Masamune	5
	P. minimus f. *rotundifolius* Hosokawa	5
	P. riukiuensis Ohwi	5
	A. riukiuensis (Ohwi) Ohwi	5
	V. minima var. *minor* Tateishi	1

Species name	Synonym	Ref
V. stipulacea (Lam.) Kuntze	*V. trilobata* (L.) Verdcourt	5
	D. stipulaceus Lamark	5
V. subramaniana (Babu ex Raizada)	*P. wightianus* Grah	5
M. Sharma	*P. subramanianus* Babu ex Raizada	5
V. tenuicaulis N.Tomooka & Maxted	*Vigna* sp. *sensu* Tomooka *et al.*	8
V. trilobata (L.) Verdcourt	*D. trilobatus* L.	3
	D. trilobata L.	1
	P. trilobatus (L.) Schreber	3
	P. trilobus sensu Aiton et auct. mult.	3
V. trinervia (Heyne ex Wight & Arnott)	*P. trinervius* Heyne	5
Tateishi & Maxted	*V. radiata* var. *sublobata* (Roxb.) Verdccourt	5
V. trinervia var. *bourneae*	*V. bourneae* Gamble	5
(Gamble) Tateishi & Maxted		
V. umbellata (Thunb.) Ohwi & Ohashi	*D. umbellatus* Thunb.	3
(cultigen)	*P. pubescens* Blume	3
	P. calcaratus (Roxb.)Kurz.	3
	P. calcaratus Roxb.	1
	P. chrysanthus Savi.	1
	P. torous Roxb.	3
	P. torosus Roxb.	1
	P. ricciardianus Tenora	1
	P. riccardianus Tenore	3
	V. calcarata (Roxb.) Kurz.	3
	P. calcaratus var. *major* Prain	1
	P. calcaratus var. *rumbaiya* Prain	1
	A. umbellata (Thunb.) Ohwi	3
	V. bourneae Gamble	7
V. umbellata (Thunb.) Ohwi & Ohashi	*P. calcaratus* Roxb. var. *gracilis* Prain	3
(wild form)	*P. gracilicaulis* Ohwi -written on manuscript	3

References

1. Lumpkin and McClary, 1994
2. Lock and Heald, 1994
3. Maréchal *et al*., 1978
4. Sacks, 1977
5. Tateishi, 1984
6. Tateishi and Maxted, 2002
7. Tomooka, Egawa and Kaga, 2000
8. Tomooka and Maxted, 2002
9. Tateishi and Ohashi, 1990

APPENDIX II

Appendix II contains the herbarium data gathered during the course of preparing this book. Many herbarium specimens are duplicated we indicate here only one herbarium that was observed. The list of herbaria visited is given below. Passport data that was another source of information for this book is not included, however, this information is accessible via the internet at http:// www.gene.affrc.go.jp/plant/.

BM - British Museum (Natural History), Cromwell Road, London, SW7 5BD, UK;

BO - Herbarium Bogoriense, Jalan Raya Juanda 22-24, Bogor Java, Indonesia;

BR - Herbarium, National Botanic Garden of Belgium, B-1860 Meise, Belgium;

BSIS - Herbarium, Botanical Survey of India, Industrial Section, Indian Museum, Calcutta 700 016, West Bengal, India;

CAL - Central National Herbarium, P.O. Botanic Garden, Howrah, Calcutta, 711-103, West Bengal, India;

DD - Dhera Dun, Herbarium, Systematic Botany Branch, Forest Research Institute, New Forest, Dehra Dun 248 006, Uttar Pradesh, India;

DUH - Herbarium, Botany Department, University of Delhi, New Delhi, 110-007, Deli Union Territory, India;

FRI - Australian National Herbarium, Division of Forestry and Forest Products, CSIRO, P.O.Box 4008, Queen Victoria Terrace, Canberra, A.C.T. 2600, Australia;

K - Herbarium, Royal Botanic Gardens, Kew, Richmond, Surrey, TW9 3AB, United Kingdom;

KYO - Kyoto University, Kitashirakawa Oiwake-cho, Sakyo-ku, Kyoto 606-8502, Japan;

L - Rijksherbarium, Van Steenisgebouw, Einsteinweg 2, 2300 RA Leiden, the Netherlands;

NBPGR - National Bureau of Plant Genetic Resources, P.O. Pusa, New Dehli 11012, India;

P- Natural History Museum, 61 rue Buffon 75231 PARIS CEDEX 05, France;

PDA - Botanic Gardens, Peradeniya, Sri Lanka;

TUS - Herbarium, Tohoku University, Aoba-ku, Sendai, Miyagi 980-8578, Japan;

Appendix II. Data on *Vigna* subgenus *Ceratotropis* species from herbarium specimens.

Species	Status	Locality	Cty	Longitude	Latitude	Alt.m	Collector	Coll. no	Herb.
aconitifolia		Chota Nagpore	IND	78 08E	11 38N	100	Clarke CB	20402a	BM
aconitifolia		Salem, Nammakal,Taluk,2km before Nammakal	IND	78E	10 23N		Matthew, Jayaseelan	19635	K
aconitifolia		Dindigul, Tandigudi road	IND	78E		300	Matthew KM	52205	K
aconitifolia		ICRISAT site, 30 km NW of Hyderabad, A.P.	IND	78 05E	17 40N	600	van der Maesen	3794	K
aconitifolia		Baji, Ad Dali Rd.	YEM	44 44E	13 41N		McKilligan SN		BM
angularis	weed	Shichikuike, Kamiyada, Kameoka, Kyoto	JPN	135 35 01E	34 59 43N		K Mimoro	2218	KYO
angularis	weed	Fukuchiyama, Kyoto	JPN	135 04 30E	35 18 30N		Y Araki	14486	KYO
angularis	weed	Ohtsuki, Ohtsuki, Yamanashi	JPN	138 56 46E	35 36 29N	400	Y Tateishi	457	TUS
angularis	weed	Yamanashi pref., Kowase, Nigioka, Ohtsuki, Yamanashi	JPN	138 57 35E	35 36 53N	350	Y Tateishi	1524	TUS
angularis	weed	Ohkubo, Hannou, Saitama	JPN	139 18 35E	35 51 25N	250	Y Tateishi	514B	TUS
angularis		Yase, Kyoto	JPN	134 31E	35 27N		Koidzumi	sn	BKF
angularis		Unzen, swamp nr bathing pool	JPN	130 02E	32 14N		Rogers A	N99/37	K
angularis angularis	cult.	Ye Luk Tau, Tung Koo Shen, Tapu Dist., Kaytung	CHN				WT Barg	21679	KYO
angularis angularis	cult.	Ridinea, Naga Hills	IND	94 07E	25 40N	1700	Bor N L	6250	K
angularis angularis	cult.	Tosa, Kouchi	JPN	133 24 11E	33 27 45N		Doi et al.		KYO
angularis angularis	cult.	Ujidawara mura, Chifukudani, Tsuzuki gun, Kyoto	JPN	135 51 35E	34 51 18N	300	G Murata	72454	KYO
angularis angularis	cult.	Tango, Kyoto	JPN	135 10 30E	35 40 40N		K Horie	22	KYO
angularis angularis	cult.	Aomori	JPN	140 45E	40 38 40N		N Kinashi	611	KYO
angularis angularis	cult.	Kyoto University Botanic Garden, Kyoto	JPN	135 47 05E	35 01 26N		S Kitamura		KYO
angularis angularis	cult.	Mt. Hayachine, Iwate	JPN	141 27 11E	39 32 55N		T Makino		KYO
angularis angularis	cult.	Fukuchiyama, Kyoto	JPN	135 07 15E	35 17 24N		Y Araki	13839	KYO
angularis angularis	cult.	Sasayama, Taki-gun, Hyogo	JPN	135 14 17E	35 06 08N		Y Araki	14661	KYO
angularis angularis	cult.	Shiroyama, Oosumi, Kagoshima	JPN	130 59 49E	31 35 49N		Z Tashiro		KYO
angularis angularis	cult.	Fukuchiyama, Kyoto	JPN	135 04 30E	35 18 30N			13842	KYO

Species	Status	Locality	Cty	Longitude	Latitude	Alt.m	Collector	Coll no	Herb.
angularis nipponensis	wild	North Triangle	BUR	97 42E	27 45N		Kingdon-Ward		BM
angularis nipponensis	wild	Moshuan-shan, Zhejiang Province	CHN				K Kimura		KYO
angularis nipponensis	wild	Oo Chi Shan, near Llam Uk Vill., Lungnan Dist.	CHN	114 47E	24 54N		Lau SK	4746	BM
angularis nipponensis	wild	Chantong Yingfou, North China	CHN				Licent E	6355	BM
angularis nipponensis	wild	Moshuan-shan, Zhejiang Province	CHN						KYO
angularis nipponensis	wild	Kohima, Naga Hills	IND	94 07E	25 40N	1600	Bor N L	6234	K
angularis nipponensis	wild	Nungklor 3500, Khasia	IND	91 30E	25 30N		Clarke CB	44819	BM
angularis nipponensis	wild	Sikkim	IND	88 25E	27 42N		Clarke CB	25461B	K
angularis nipponensis	wild	Jheflen, Simla	IND	77 10E	31 06N	2000	Gamble JS		K
angularis nipponensis	wild	Simla	IND	77 10E	31 06N		Henry Collet	379	K
angularis nipponensis	wild	The Glen, Himalya	IND				Rich H H	311	K
angularis nipponensis	wild	9 km to Solan, 54 km to Kalka, Simla, Himachal Pradesh	IND	77 07E	30 58N	1447	van der Maesen LJ	2954	K
angularis nipponensis	wild	Yokokawa, Iwaki, Fukushima	JPN	140 39 22E	36 55 27N		A Kimura et al.		TUS
angularis nipponensis	wild	Shinhukurogawa, Yoshinari, Tottori	JPN	134 13 17E	35 29 16N		A Tanaka	14116	KYO
angularis nipponensis	wild	Okamasu, Kokufu, Iwami-gun, Tottori	JPN	134 15 53E	35 28 05N		A Tanaka	21127	KYO
angularis nipponensis	wild	Kimizudukuri, Tomiya, Kurokawa-gun, Miyagi	JPN	140 55 56E	38 24N		A Yokota	682	TUS
angularis nipponensis	wild	Bettcyo, Ohtsu, Shiga	JPN	135 51 17E	35 00 51N		C Hashimoto	2311	KYO
angularis nipponensis	wild	behind Ginkaku temple, Sakyo-ku, Kyoto	JPN	135 48 12E	35 01 21N		Doi		KYO
angularis nipponensis	wild	Yamashiroen, Yase, Kyoto	JPN	135 49 25E	35 04 47N		G Koidzumi		KYO
angularis nipponensis	wild	Okukenbayashi mura, Ikaruga gun, Tanba, Kyoto	JPN	135 24 56E	35 09 15N		G Murata	10276	KYO
angularis nipponensis	wild	Miyanokuchi, Oko-mura, Kochi	JPN	133 43 31E	33 37 14N		G Murata	11058	KYO
angularis nipponensis	wild	Katsura, Kyoto	JPN	135 42 52E	34 58 40N		G Murata	13490	KYO
angularis nipponensis	wild	Akasaka yama, Makino, Takashima-gun, Shiga	JPN	136 01 24E	35 30 59N	150	G Murata	18908	KYO
angularis nipponensis	wild	Tokura, Haga, Shiso-gun, Hyogo	JPN	134 32 41E	35 17 12N	600	G Murata	20326	KYO
angularis nipponensis	wild	Matsuo, Ukyo-ku, Kyoto	JPN	135 44 06E	34 59 57N		G Murata	22277	KYO
angularis nipponensis	wild	Heta-bashi, Kishida gawa, Heta, Hamasaka, Mikata, Hyogo	JPN	134 27 28E	35 36 29N	10	G Murata	56401	KYO

Species	Status	Locality	Cty	Longitude	Latitude	Alt.m	Collector	Coll. no	Herb.
angularis nipponensis	wild	Niyodomura near Matsubara, Takaoka-gun, Kochi	JPN	133 06 24E	33 29 41N	700	G Murata		KYO
angularis nipponensis	wild	btwn, Manba-Mt. Higashimikubu, Manba, Tano-gun, Gunma	JPN	138 54 59E	36 07 33N	500	G Murata et al.	990	KYO
angularis nipponensis	wild	Hiwasa, Kaifu-gun, Tokushima	JPN	134 32 17E	33 43 53N		G Nakai	4040	KYO
angularis nipponensis	wild	Mizusawa, Iwate	JPN	141 10 19E	39 09 30N		H Iwabuchi	6379	KYO
angularis nipponensis	wild	Higashiusuki-gun, Miyazaki	JPN	131 40 54E	32 28 53N		H Koyama	7523	KYO
angularis nipponensis	wild	Oishida, Kitamurayama-gun, Yamagata	JPN	140 22 28E	38 35 34N	100	H Ohashi et al.	10756	TUS
angularis nipponensis	wild	Riverside of Karasugawa, Nakajima, Fujioka, Gunma	JPN	139 04 55E	36 17 08N		H Ohba	74904	KYO
angularis nipponensis	wild	Koharu, Mizushitani, Nango, Higashiusuki-gun, Miyazaki	JPN	131 22 41E	32 23 14N	400	H Takahashi	8408	KYO
angularis nipponensis	wild	Abukuma Mts., Mafune Pass, Miyagi	JPN	140 34 44E	37 30 57N		J Ikeisu	930	TUS
angularis nipponensis	wild	Nakazato-mura, Tano-gun, Gunma	JPN	138 50 25E	36 05 23N	625	J Murata	2543	KYO
angularis nipponensis	wild	Koshigaya, Saitama	JPN	139 47 08E	35 53 10N		J Ohwi	85	KYO
angularis nipponensis	wild	Harima Nozato, Himeji, Hyogo	JPN	134 41 30E	34 49 42N		K Fukuda	2417	KYO
angularis nipponensis	wild	Goma, Hiyoshi, Funaii-gun, Kyoto	JPN	135 28 26E	35 11 38N		K Nagai	25328	KYO
angularis nipponensis	wild	Hozukyo, Ukyo-ku, Kyoto	JPN	135 44 06E	34 59 57N	500	K Nagai	25553	KYO
angularis nipponensis	wild	Hiigawa, Jyonan-ku, Fukuoka	JPN	130 23 07E	33 32 53N		K Nakajima		KYO
angularis nipponensis	wild	Abe river, Ashikubo, Shikiji, Shizuoka, Shizuoka	JPN	138 24 51E	34 56 26N		K Nakayama et al.	114	KYO
angularis nipponensis	wild	Nishiya, Niihama, Ehime	JPN	133 17 58E	33 56 36N		K Ochi		KYO
angularis nipponensis	wild	Syuyoue, Tokuyama, Yamaguchi	JPN	131 49 50E	34 02 45N		K Oka	32116	KYO
angularis nipponensis	wild	Kawauchi Ninomaru, Aoba-ku, Sendai, Mayagi	JPN	140 51 19E	38 15 28N		K Saito		KYO
angularis nipponensis	wild	Hachiman, Gujo-gun, Gifu	JPN	136 56 35E	35 43 18N		K Shiota		KYO
angularis nipponensis	wild	Foot of Mt. Mitake, Nishitama-gun, Oume, Tokyo	JPN	139 09 11E	35 46 51N	350	M Mizushima	15003	KYO
angularis nipponensis	wild	Hukahori, Matsukawa, Higashiyama, Higashiiwai-gun, Iwate	JPN	141 15 23E	38 56 21N		M Suzuki		TUS
angularis nipponensis	wild	Nariai to Kawakubo, north of Takatsuki, Osaka	JPN	135 37 30E	34 53 54N		M Tagawa	6567	KYO
angularis nipponensis	wild	Shoshayama, Himeji, Hyogo	JPN	134 41 30E	34 49 42N		M Yoshizawa	13609	KYO
angularis nipponensis	wild	northern Morikuni, Toyonagasabushi, Niimi, Okayama	JPN	133 45 54E	34 57 42N		N Fukuoka et al.	1865	KYO
angularis nipponensis	wild	SE ft of Mt. Yokayama dake, Suino, Kinomoto, Ika-gun, Shiga	JPN	136 13 38E	35 30 21N		N Fukuoka et al.	6127	KYO

Species	Status	Locality	Cty	Longitude	Latitude	Alt.m	Collector	Coll. no	Herb.
angularis nipponensis	wild	Kamigamo, Kyoto	JPN	135 45 31E	35 04 05N		N Kinashi		KYO
angularis nipponensis	wild	btwn. Yajougahara-Moroyose, Hamasaka, Mikata, Hyogo	JPN	134 26 04E	35 36 53N	50	N Kurosaki	13701	KYO
angularis nipponensis	wild	Nagasaki	JPN	129 53E	32 44N		Oldman R	376	K
angularis nipponensis	wild	Minoo dam, Minooi, Osaka	JPN	135 28 51E	34 51 26N	300	S Fujii	2566	KYO
angularis nipponensis	wild	Takachihokyo, Miyazaki	JPN	130 57 14E	31 52 31N	250	S Hatushima et al.	22401	KYO
angularis nipponensis	wild	Higami, Higami-gun, Hyogo	JPN	135 02 58E	35 10 26N		S Hosomi	5998	KYO
angularis nipponensis	wild	Higami, Higami-gun, Hyogo	JPN	135 02 58E	35 10 26N		S Hosomi	7549	KYO
angularis nipponensis	wild	Higami, Higami-gun, Hyogo	JPN	135 02 58E	35 10 26N		S Hosomi		KYO
angularis nipponensis	wild	Mt. Hiei, Kyoto	JPN	135 49 33E	35 03 20N		S Kitamura	HL4	KYO
angularis nipponensis	wild	Tokyo Uni. campus, Hongo, Bunkyo-ku, Tokyo	JPN	139 45 49E	35 42 52N		S Matsuda		KYO
angularis nipponensis	wild	Mukaijima, Fushimi-ku, Kyoto	JPN	135 45 31E	34 54 50N		S Miki		KYO
angularis nipponensis	wild	Irifune, Sakaide, Kagawa	JPN	133 51 19E	34 19 15N		S Sakaguchi	10179	KYO
angularis nipponensis	wild	Hirata, Ohfunato, Iwate	JPN	141 43 21E	39 07 18N		S Sasamura		TUS
angularis nipponensis	wild	Shirakawa, Fukushima	JPN	140 12 53E	37 07 47N		S Suzuki		TUS
angularis nipponensis	wild	Nakamura, Muki, Kaifu-gun, Tokushima	JPN	134 24 57E	33 39 51N		S Takato	243	KYO
angularis nipponensis	wild	Akano, Nishiora-chiku, Maizuru, Kyoto	JPN	135 23 28E	35 31 10N	60	S Tsugara et al.	19433	KYO
angularis nipponensis	wild	Yamatogawa, Asaka, Sumiyoshi-ku, Osaka	JPN	135 30 48E	34 35 18N	5	T Fujii	1256	KYO
angularis nipponensis	wild	near Tarumizu dam, Natori, Miyagi	JPN	140 51 51E	38 10 36N		T Kurosawa	1051	TUS
angularis nipponensis	wild	Uchi, Kakuda, Oyama, Miyagi	JPN	140 50 23E	37 57 01N	40	T Mori	5392	TUS
angularis nipponensis	wild	Houzawa, Izuki-ku, Sendai, Miyagi	JPN	140 47 51E	38 21 59N		T Naito et al.		TUS
angularis nipponensis	wild	btwn. Otsuki-Sakaine, Watarimura, Tama-gun, Kumamoto	JPN	130 41 57E	32 14 03N		T Shimizu	372	KYO
angularis nipponensis	wild	Sakamoto, Oohara no ishi zukuri, Nishi-ku, Kyoto	JPN	135 39 14E	34 56 54N		T Yamazaki	1783	KYO
angularis nipponensis	wild	Ashiodani, Yotsuya, Hiyoshi, Funai-gun, Kyoto	JPN	135 32 45E	35 12 24N	240	TTsugara et al.	20810	KYO
angularis nipponensis	wild	Ewa, Miyama, Kitakuwada-gun, Kyoto	JPN	135 39 37E	35 19 03N	280	TTsugaru et al.	21149	KYO
angularis nipponensis	wild	Shosha yama, Himeji, Hyogo	JPN	134 39 29E	34 53 15N		Tagawa	2180	KYO
angularis nipponensis	wild	Suyama, Kuwata-gun, Kyoto	JPN	135 24 56E	35 09 15N		Y Araki	13795	KYO

Species	Status	Locality	Cty	Longitude	Latitude	Alt.m	Collector	Coll. no	Herb.
angularis nipponensis	wild	Mt. Akiwa, Shizuoka	JPN	137 51 59E	34 58 42N		Y Kurosawa		KYO
angularis nipponensis	wild	Sotogawara, Shiroishi, Miyagi	JPN	140 37 49E	38 00 43N	30	Y Murakami	11	TUS
angularis nipponensis	wild	Kumamoto castle, Kumamoto	JPN	130 42 21E	32 48 14N		Y Shimada	7812	KYO
angularis nipponensis	wild	Izumi, Shimomashiki-gun, Kumamoto	JPN	130 44 04E	32 46 53N		Y Shimada		KYO
angularis nipponensis	wild	Asagaya, Kanazawa, Ishikawa	JPN	136 43 33E	36 30 12N		Y Sugie	1291	KYO
angularis nipponensis	wild	Nakanomi, Kawachi, Ishikawa-gun, Ishikawa	JPN	136 38 34E	36 23 06N		Y Sujie	1291	KYO
angularis nipponensis	wild	Aobayama, Sendai, Miyagi	JPN	140 51 17E	38 15 08N	100	Y Tateishi	16658	TUS
angularis nipponensis	wild	Ooya, Shizuoka, Shizuoka	JPN	138 26 16E	34 57 02N	50	Y Tateishi et al.	16148	TUS
angularis nipponensis	wild	near Tanuki-ko lake, Fujinomiya, Shizuoka	JPN	138 34 53E	35 20 47N	650	Y Tateishi et al.	16159	TUS
angularis nipponensis	wild	Osawa, Natori, Miyagi	JPN	140 48 38E	38 11 47N		Y Tateishi et al.	11070	TUS
angularis nipponensis	wild	Wakayama	JPN	135 09 51E	34 13 17N				KYO
angularis nipponensis	wild	Kwangrung, Kyonggi-do	KOR				G Koidzumi		KYO
angularis nipponensis	wild	Mt. Taeduk, Taemyong-dong, Namgu, Taegu, Kyongsangpuk	KOR	127 22E	36 56N	200	Oh Soo-Young	0326-85-89	TUS
angularis nipponensis	wild	Genzan, Kanran	KOR				S Kitamura	H.L.67	KYO
angularis nipponensis	wild	Mt. Jiri, from Machon to Baekshick	KOR				S Okamoto		KYO
angularis nipponensis	wild	Mt. Jiri, Rhogodan	KOR	127 43E	35 20N		T and F Yamazaki	3218	KYO
angularis nipponensis	wild	Mt. Seolmsan, Sunam-ri, Dong-myeon, Chonwon-gun, Chungchongnam-do	KOR	126 52E	36 27N	235	T Nemoto et al.	2857	
angularis nipponensis	wild	Chong near Cibrikat	NEP	82 45E	29 01N	2700	O Polunin	3311	BM
angularis nipponensis	wild	near Humseem	NEP			2000	Stainton et al.	4061	BM
angularis nipponensis	wild	Dhawalagiri, Myagdi Dist. Boghara (2010m)-Lipshe(1840 m)-Lapche Kharka(2060 m)- Dobang(2360 m), Myagdi	NEP	84 23E	28 35N	2090	M Mikage et al.	9685219	TUS
angularis nipponensis	wild	Hejjo and Vicinity	PKR	125 44E	39 01N		PDorsett, WMorse	6222	K
angularis nipponensis	wild	Kaohsiung Co. Tengchu	TWN			1650	H Ohashi et al.	12783	TUS
angularis nipponensis	wild	Kaohsiung Co., Tengchu	TWN			1650	H Ohashi et al.	13089	TUS
angularis nipponensis	wild	Kaohsiung Co.,Meishan, Tenchi, Southern Cross Rd	TWN	120 55E	23 16N	1500	H Ohashi et al.	24137	TUS
angularis nipponensis	wild	Shihcho, Chiayi Co.	TWN	120 51E	23 42N	1000	TC Huang	9961	TUS

Species	Status	Locality	Cty	Longitude	Latitude	Alt.m	Collector	Coll. no	Herb.
angularis nipponensis	wild	Chiayi Co: Mt. Ali-Shan , Shihcho, Fushan	TWN	120 48E	23 30N	1450	Y Tateishi et al.	21526	TUS
angularis nipponensis	wild	Chiayi Co: Mt. Ali-Shan , Fushan	TWN	120 48E	23 30N	1400	Y Tateishi et al.	17934	TUS
angularis nipponensis	wild	Kaohsiung Co: Meishan, Southern Cross Road,	TWN	120 55E	24 47N	1300	Y Tateishi et al.	25170	TUS
aridicola	wild	Btw. Kaikudeh-Polonnaruwa, marker 59, Polonnaruwa Dist., N. Central Prov.	LKA	81 23E	8 01N		VE Rudd et al.	3154	PDA
aridicola	wild	Btw. Amparai-Maha Oya, mile marker 19 ~20, Amparai District, Eastern Prov.	LKA	81 39E	7 26N		VE Rudd et al.	3226	PDA
aridicola	wild	Btw. Amparai-Maha Oya, markers 19/20, Ampari, E. Prov.	LKA	81 38E	7 27N	175	VE Rudd et al.	3227	K
aridicola	wild	open exposed dry area, Polonnaruwa, Sacred Area	LKA	81 00E	7 57N	61	Wolfgang Dittus	7003.2602	PDA
dalzelliana	wild	Simla area	IND	77 10E	31 07N	1000		33350	FRI
dalzelliana	wild	Karwar, Tamil Nadu	IND				Ambo	6587	K
dalzelliana	wild	Katgal - N. Kanara, Bombay	IND	72 52E	19 07N	170	Ambo	6852	K
dalzelliana	wild	Pulney (Palni) Hills, Shembagnur, Tamil Nadu	IND	77 18E	10 10N	2000	Auglade L	2136	K
dalzelliana	wild	Kerala, Jirchur	IND	76 35E	10 18N	59	Babu CR	59	K
dalzelliana	wild	Central India, Bastar	IND	81 56E	19 12N	59	Babu CR	59	K
dalzelliana	wild	Jolpad, south Canara, Tamil Nadu	IND				Barber CA	2341	K
dalzelliana	wild	Anamallags	IND	79 46E	11 25N	40	Beddome	2233	BM
dalzelliana	wild	Tena district,	IND	72 55E	26 23N	320	Biswas	1435	CAL
dalzelliana	wild	Vilpatti Valey, Pulneys, Tamil Nadu	IND	77 16E	10 12N		Bourne	2001	K
dalzelliana	wild	Machur Path Shola, Pulneys, Tamil Nadu	IND	77 16E	10 12N		Bourne	2002	K
dalzelliana	wild	Vilpatti Valley, Pulneys, Tamil Nadu	IND	77 16E	10 12N		Bourne	2003	K
dalzelliana	wild	Poombari Valley, Pulneys, Tamil Nadu	IND	77 16E	10 12N		Bourne	2568	K
dalzelliana	wild	Tamil Nadu, Coimbatore, Konalar, Anamolai Hills	IND	79 11E	12 10N	1925	Chandrabox	57787	CAL
dalzelliana	wild	Bengal, Burakur, Burdwan	IND	80 30E	22 47N	400	Clarke CB	25270a	K
dalzelliana	wild	Punjab	IND	76 04E	31 25N		Drummond JR	1529	K
dalzelliana	wild	Niigiri, Tamil Nadu	IND	76 38E	11 20N	1700	Gamble JS	14915	K

Species	Status	Locality	Cty	Longitude	Latitude	Alt.m	Collector	Coll. no	Herb.
dalzelliana	wild	Nelahola, Nilgiris, Tamil Nadu	IND	76 41E	11 23N	1000	Gamble JS	15667	K
dalzelliana	wild	Nilgiri, Nelakota, Tamil Nadu	IND	76 25E	11 33N	1000	Gamble JS	15668	K
dalzelliana	wild	Wadakonichery,Trichur,Kerala	IND	76 13E	10 31N	50	Gopinathan	sn	DUH
dalzelliana	wild	Annandale, Simla	IND	77 10E	31 07N	1000	Gov of India	9550	BSIS
dalzelliana	wild	Dindigul, Kodaikanal, Mahilkundram, Palni (Pulney) Hills, Tamil Nadu	IND	77 30E	10 12N	2200	KM Matthew et al.	50664	K
dalzelliana	wild		IND			1500	KM Matthew et al.	51452	K
dalzelliana	wild	Kallar,Trivandrum district, Kerala	IND	77 07E	8 42N	350	Mohanon	65163	CAL
dalzelliana	wild	Madras,Travancore district	IND	80 16E	13 05N	<50	Narayanaswami	1480	CAL
dalzelliana	wild	Maharashtra, nr Ambaune, Mulshi	IND	73 22E	18 37N	500	Neodi	99477	CAL
dalzelliana	wild	Panorama point, Bisle Ghat, Hassan Dist. Mysore	IND	76 05E	13 04N	260	Nicolson DH et al.	2308	K
dalzelliana	wild	Poona ,Mysore, Shimoga district, Hulical Ghat	IND	76 39E	12 18N	700	Raghavan	83078	CAL
dalzelliana	wild	Kerala, Cannanore district,Kannoth Raf.	IND	76 09E	16 52N	160	Ramachandran	58269	CAL
dalzelliana	wild	Khandala	IND	74 47E	20 01N	1200	Sharma	sn	DUH
dalzelliana	wild	Caucan	IND				Stock et al.		K
dalzelliana	wild	Karnatak, North Canara, Yellapore	IND	74 42E	14 58N	500	Talbot	710b	CAL
dalzelliana	wild	36 km to Munnar, Cardamum estate, 5 km to Poopara, Kerala	IND	77 09E	10 14N	1350	van der Maesen LJG	4814	K
dalzelliana	wild	Kerala, Coimbatore, Idikki Dist, Thekkady	IND	77 12E	9 36N	850	Vivekananthan	46705	CAL
dalzelliana	wild	Madras	IND	76 38E	11 20N		Wight	795	K
dalzelliana	wild		LKA				Gardner	234	K
dalzelliana	wild		LKA	80 50E	7 08N		Shwaits M	1473	P
dalzelliana	wild	McDonald's Valley, below Hakgala, Nuwara Eliya	LKA	80 50E	6 55N	2500	V Rudd, Balakrishman	3167	K
dalzelliana	wild		LKA						K
dalzelliana	wild	Khaw Pok Hill, Lower Thailand	THA	99 45E	7 46N	70	Hauiff	3934	K
exilis	wild	Raichaburi	THA	99 49E	13 32N	200	Kerr AFG	9017	K
exilis	wild	Makam, Chanta Buri	THA	102 12E	12 40N	100	Kerr AFG	9599	K

Species	Status	Locality	Cty	Longitude	Latitude	Alt.m	Collector	Coll. no	Herb.
exilis	wild	Petchaburi	THA	99 57E	13 07N	20	Kerr AFG	11065	K
exilis	wild	Nam chat, Ranong	THA	98 38E	9 58N	<50	Kerr AFG	11725	K
exilis	wild	Rat Buri	THA	99 49E	13 32N	200	Marcan A	1463	BM
exilis	wild	Sam Roi Yawt	THA	99 57E	12 12N	25	Put	2503	K
exilis	wild	Ta Salao Kan Buri	THA	99 12E	14 02N		Put	3061	K
exilis	wild	Khao Sawang Phi Nawng	THA			400	T Smitinand et al.	1374	L
grandiflora	wild	Phnom Penh	CMB	104 55E	11 33N		Harmand		P
grandiflora	wild	Palanampo	THA	100 09E	15 43N	30	Kerr AFG	2152	K
grandiflora	wild	Bangkok	THA	100 31E	13 45N	25	Kerr AFG	4354	K
grandiflora	wild	Bangkok	THA	100 31E	13 45N	25	Marcan A	326	BM
grandiflora	wild	Bak nampo	THA	100 09E	15 43N	30			BM
hirtella	wild	Bhamo, Upper Burma	BUR	97 14E	24 16N	320	Forrest G	9204	K
hirtella	wild	West Central Burma	BUR	94 00E	22 00N	1500	Kingdon-Ward	22733	BM
hirtella	wild	Pegu	BUR	96 30E	17 20N		Kurz S	2546	K
hirtella	wild	Maymyo Plateau, E. of Mandalay	BUR	96 29E	22 01N	1200	Lace JH	6330	K
hirtella	wild	Irrawadi	BUR	95 08E	16 32N			5588H	K
hirtella	wild	Sjemen, Yunnan	CHN	102 37E	24 59N	1700	Henry A	12685	K
hirtella	wild	Pembliangan, East Kalimantan	IDN	117 1E	3 59N		Amjah	840	K
hirtella	wild	Ramghur 1500, Lohardngga, Chda Nagpore	IND	83 35E	23 18N		Clarke CB	20912	BM
hirtella	wild	Linjeban 5000, Sikkim	IND	88 27E	27 27N		Clarke CB	25483	BM
hirtella	wild	Hab. Khassia, Regio Trop, Khasi Hills	IND	91 50E	25 31N	660	Hooker & Thompson	2298	K
hirtella	wild	Hab. Khassia, Regio Trop, Khasi Hills	IND	91 50E	25 31N	660	Hooker & Thompson	1850year	K
hirtella	wild	Upper Assam	IND	95 04E	27 54N		Jeskins & Hooker	1841year	K
hirtella	wild	Caro Hills, Assam	IND	92 19E	24 04N			902	K
hirtella	wild	Simla, Himachal Pradesh	IND	77 10E	31 06N				K
hirtella	wild	Pakson, Bolovens Plateau	LAO	106 14E	15 11N	1200	Poilane E	28392	K

Species	Status	Locality	Cty	Longitude	Latitude	Alt.m	Collector	Coll. no	Herb.
hirtella	wild	Perlis (Flora of Perlis) = Kangar	MYS	110 11E	6 26N			15125	K
hirtella	wild	Bank of Lebir River, Kelantan	MYS	102 13E	5 26N		Ridley HN	Type	K
hirtella	wild	Khaw Khieo National Park, Si Racha Dist., Chonburi Prov.	THA	101 03E	13 16N	350	Maxwell JF	75-1050	L
hirtella	wild	Chiang Dao. Chiang Mai	THA	98 54E	19 21N	425	Maxwell JF	92-823	P
hirtella	wild	Nakhon Sritamarat, Kiriwong, Khao Rawn Nai Hawn	THA	99 58E	8 25N	25	Plemhit	456	L
hirtella	wild	Doi Nang Ka, Chiang Mai	THA	98 29E	18 35N		Put	3367	BM
hirtella	wild	65.5km W of Mae Taen, Chiang Mai	THA	98 36E	19 15N	1365	Tomooka N	96120205	K
hirtella	wild	Montee des piecs du Langbian, Near Marecageup a Dankia	VNM	108 30E	12 05N		Evrard F	392	P
hirtella	wild	Cho-fa (Choo Ha?)	VNM	106 35E	21 24N		H Lecomte, A Finet	495	P
hirtella	wild	Dak Gley pio. du Kontune	VNM	107 45E	15 05N		Poilane M	32975	P
khandalensis	wild	Maharashtra, at Purandhar	IND	73 59E	18 17N	1000	Aagari	32639	CAL
khandalensis	wild	Deccan Hills	IND	75 01E	18 57N	1000	JED	sn	K
khandalensis	wild	Poucan	IND				Stock		K
khandalensis	wild	Salaia dist,M-war	IND	79 55E	26 18N	150	Talbot W A	4467	K
khandalensis	wild	Deccan Hills	IND	78 20E	19 02N				K
minima	wild	Northern, Chili	CHN				Abbe E Licent	859	K
minima	wild	Northern, Chantong	CHN				Abbe E Licent	6353	K
minima	wild	Road side, Ching Shan, Lao Shan, Shantung	CHN	117 30E	36 09N		Chiao C Y	2932	K
minima	wild	Wu Yuang, Anhwei	CHN	117 16E	31 50N		Ching R C	8935	K
minima	wild	Dung Ka to Wen Fa Shi, Hainan	CHN	109 33E	18 59N		Chun NK, CL Tso	43637	K
minima	wild	Che-Foo (Yantai)	CHN	121 24E	37 32N		Hancock W	875 June 29	K
minima	wild	Sai Hang Cheung , near Tung Lei, Kiennan Dist., Kiangsi	CHN				Lau SK	4333	BM
minima	wild	Ta Hian, Hainan	CHN	109 32E	19 03N	600	Linsley Gressit J	824	BM
minima	wild	Kouang fou	CHN				Poli		P
minima	wild	Bak Sa, Plants of Hainan	CHN				SK Lau	25893	KYO
minima	wild	Canton	CHN	113 14E	23 08N		S Theophilus		K

Species	Status	Locality	Cty	Longitude	Latitude	Alt.m	Collector	Coll. no	Herb.
minima	wild	Ye Tau Tang, Hung Mo Shan, Hainan	CHN	109 39E	19 04N		Tsang & Fung	418	K
minima	wild	Loting, Luoding County	CHN	111 33E	22 46N		Tsiang Ying	1169	BM
minima	wild	Tai Chung to Wong Fou Town, Ying Tak, Kwantung	CHN	113 24E	24 10N		Tso CL	22224	BM
minima	wild	Eu Wai Shan, Tai Lung Tung, Tai Shan, Kwantung	CHN	117 04E	36 16N		Tso CL	22476	BM
minima	wild	Kampot	CMB	104 10E	10 36N		Geoffray	82	P
minima	wild	Kg (Kampong) Chhnang	CMB	104 39E	12 15N		Muller	455	P
minima	wild	pres de la chaine des Dangrek province Siem Reap entre Pum Tho May, et Anlong Veng	CMB	103 51E	13 21N		Poilane M	13913	P
minima	wild		HKG	114 10E	22 17N		Hance HF	1317	K
minima	wild	Lantao island,, Tungchung + vic. Shantao	HKG	113 56E	22 15N		YW Taam	1729	KYO
minima	wild	Lo Fan Shan	HKG	114 07E	22 24N				K
minima	wild	Tjivonas, Java	IDN				Boeerlau	244	L
minima	wild	Katambe, valley of Lau Alas, near tributary of Lau Ketambe, ca.35 km NW of Kutatjane (Kutacane), Gunung Leuser Nature Reserve, Atjeh , Sumatra	IDN	97 37E	3 43N		de Wilde & de Wilde	12640	L
minima	wild	Java, Merapi	IDN			833	Junghuhn	145	K
minima	wild	Fort van der Capellen, Sumatra	IDN	100 35E	0 27S		Matthew CG		K
minima	wild	Bangarmassing	IDN	114 35E	3 19S		Motley J	404	K
minima	wild	in Bengalia circa Calcuttam	IND	88 21E	22 34N		Helfer JW	79	BM
minima	wild	Concan, Heb. Concan	IND				Hooker		L
minima	wild	Pakse (Pakxe)	LAO	105 47E	15 05N		Poilane M	28507	P
minima	wild	Savannakhet	LAO	104 45E	16 33N		Poilane M	28033	P
minima	wild	Savannakhet	LAO	104 44E	16 33N		Poilane M	28052	P
minima	wild	Pakson, plateau des Boloveu (Bolovenus Plateau)	LAO	106 14E	15 10N	1200	Poilane M	28394	P
minima	wild	km 52 un peu au dela de Pakson sur le plateau des Boloveu	LAO	106 14E	15 10N	1200	Poilane M	28451	P
minima	wild	Nong Tevada	LAO				Tixier P	15	P

Species	Status	Locality	Cty	Longitude	Latitude	Alt..m	Collector	Coll. no	Herb.
minima	wild	km 12 route de Tha Ngon (Ban Thangon), Vientiane	LAO	102 38E	18 08N		Vidal JE	1942	P
minima	wild	Phou Kaokhouay, Vientiane	LAO	102 46E	18 22N		Vidal JE	5736	P
minima	wild	Tamu Darat, at mile 7 Kota Belud-Ranau Road, N. Borneo	MYS	116 27E	6 15N	70	Damton S	74	BM
minima	wild	Kampong Taalbei, Balabei, Borneo	MYS				Grabowsky		BM
minima	wild	Kata Tongkat, Pahang	MYS	103 28E	3 30N		Evans IHN		K
minima	wild	11 1/4mls. Kuala Trengganu-Besut Road, Kuala Trengganu	MYS	103 08E	5 18N		JSinclair,Kiah bin Saller	40467	K
minima	wild	Coping	MYS	100 55E	4 33N		King	990	K
minima	wild	Batang	MYS				King	1035	K
minima	wild	Kangar, Perlis	MYS	100 12E	6 26N			15726	K
minima	wild	Bued River, Benguet Prov., Luzon	PHL	120 36E	17 35N		Merril ED	4279	K
minima	wild	Prov. of Rizal, Luzon	PHL	121 10E	14 35N		Ramos M	10890	K
minima	wild	Antipolo, Prov. of Rizal, Luzon	PHL	121 15E	14 40N		Ramos M	22259	K
minima	wild	Itogon, Prov. Benguet, Northern Luzon	PHL	120 40E	16 21N		Williams RS	1408	K
minima	wild	Mababrum, Orru, Western Dist.	PNG	142 00E	7 30S		Henry E	49608	K
minima	wild	Pak Thong Chai, 40-50km S of Korat along the highway.	THA	102 01E	14 43N		C Charoenphol et al.	4516	P
minima	wild	Kanchanaburi	THA	98 45E	14 55N	750	CF van Beusekom et al.	3575	K
minima	wild	Phu Rua Nat.Park Summit road c.200m park HQ. Loei Prov	THA	101 19E	17 30N	350	Chantaranothai P et al.	90/479	K
minima	wild	NE: Phu Phan Nat. Park headquarters, ca. 30km SW of Sakonnakhon.	THA	104 00E	17 00N	380	G Murata,C Phengkai	50693	KYO
minima	wild	N.: Thung Salaeng Luang, Nat. Park, 20km Phitsanulok	THA	100 36 38E	16 49N	465	G Murata et al.	38445	KYO
minima	wild	NE. Khonkaen, Phu Khieo Game reserve ca. 80km E of Pheetchabun	THA	101 58E	16 50N	850	G Murata	41618	KYO
minima	wild	NE. Khonkaen, Phu Khieo Game reserve ca. 80km E of Pheetchabun	THA	101 45E	16 28N	875	G Murata	41843	KYO
minima	wild	N: Phitsanuloki Thung Salaeng Luang National Park, about 20km E of Phitsanulok	THA	100 36 38E	16 49N	465	G Murata et al.	38451	KYO

Species	Status	Locality	Cty	Longitude	Latitude	Alt.m	Collector	Coll. no	Herb.
minima	wild	NE: Khonkaen, Phu Khieo, Game Reserve, ca. 80km east of Pheichabun	THA	101 45E	16 28N	875	G Murata et al.	41779	KYO
minima	wild	NE: Phu Phan national park, ca. 30km SW of Sakonnakhon, near Nam Phun Dam	THA	104 00E	17 00N	375	G Murata et al.	50604	KYO
minima	wild	NE: Loei, Nam Nao National Park, road check point	THA	101 23 28E	16 48 49N		G Murata et al.	51649	KYO
minima	wild	NE: Sakonnakhon, Phu Plan National Park	THA	104 00E	17 00N	415	G Murata,C Phengklai	51175	KYO
minima	wild	NE: route 212, Dongman village, Muang, Mukdahan Prov.	THA			220	H Koyama et al.	30870	KYO
minima	wild	Ta Ruang, Chantaburi	THA	100 21E	16 27N	100	Kerr AFG	9723	K
minima	wild	Padang Besar	THA	99 14E	16 45N	50	Kerr AFG	13566	K
minima	wild	Bangkok	THA	100 31E	13 45N	25	Kerr AFG		BM
minima	wild	Bangkok	THA	100 31E	13 45N	25	Marcan A	447	BM
minima	wild	Doi chang Dao, SE. foothills,Ban Yang Toong Bong Forest Sta.	THA	98 54E	19 21N	500	Maxwell JF	89-1410	L
minima	wild	Mae Yom National Park, Dow Boon, Song, Prae	THA	100 12E	18 43N	500	Maxwell JF	91-853	P
minima	wild	Sam-Kae, Phu-Krading	THA	101 48E	16 48N	1000	Native	327	P
minima	wild	Sakonnakhan	THA	104 00E	17 00N		S Mitsuta et al.	50503	KYO
minima	wild	NE: Phu Phen, ca. 30km SW of Sakonnakhon	THA	104 00E	17 00N		S Mitsuta et al.	50487	KYO
minima	wild	NE: Phu Phan National Park, ca. 30km SW of Sakonnakhon	THA	104 00E	17 00N		S Mitsuta et al.	50498	KYO
minima	wild	Kaohsiung Co;. Tengchu ,base of Mt. Tsuyun-shan	TWN			900	H Ohashi et al.	23773A	TUS
minima	wild	Tungkan, Ilan Co., Estuary,	TWN	121 48E	24 46N	5	S F Huang	1936	TUS
minima	wild	Kaohsiung Co., Chisan, Liuguizhuang (Sanhoxi)	TWN	121 30E	22 40N		S Okamoto		TUS
minima	wild	Kaohsiung Co. , Teng-chih, Taiwan	TWN	120 45E	22 59N	1500	T C Huang et al.	14391	TUS
minima	wild	Keelung Co: Pengchiayu	TWN	121 43E	25 07N	75	T C Huang et al.	15769	TUS
minima	wild	Kaohsiung: Chuyunhsan	TWN	120 32E	22 56N	875	T C Huang et al.		TUS
minima	wild	Pingtung co., Laiyu	TWN	120 41 56E	22 31 22N	250	TC Huang et al.	16044	TUS
minima	wild	Pingtung Co., Laiyi	TWN	120 41 56E	22 31 22N	250	TC Huang et al.		TUS
minima	wild	Pingtung Hsien: Wutai Hsiang, to Haocha from old Haocha,	TWN	120 41 31E	22 42 37N	340	Tsui-Ya Liu et al.	762	TUS

Species	Status	Locality	Cty	Longitude	Latitude	Alt.m	Collector	Coll. no	Herb.
minima	wild	Hualien Co:, Takangkou, Shihtiwan	TWN	121 30E	23 30N	50	Y Tateishi et al.	16044	TUS
minima	wild	Pingtung Co. Foot of Mt. Peitawu-shan,	TWN	120 40E	22 36N	800	Y Tateishi et al.	19719	TUS
minima	wild	Kaohsiung Co: Above Meishan, along Southern Cross Road	TWN	120 53E	23 16 N	1200	Y Tateishi et al.	25165	TUS
minima	wild	Dalat, ravin au sud du Langbian palace	VNM	108 26E	11 56N		Evrard F	1785	P
minima	wild	Phanrang, Ca-na Prov.	VNM	108 59E	11 33N	500	Poilane M	9037	P
minima	wild	Phanrang, Ca-na Prov.	VNM	108 59E	11 33N		Poilane M	9196	P
minima	wild	Phanrang, Ca-na Prov.	VNM	108 59E	11 33N	800	Poilane M	9262	P
minima	wild	Phaudaui, Ca(ba)-na Prov.	VNM	107 59E	15 59N		Poilane M	9602	P
minima	wild	montee du col de Blao (Bao Loc) km 98 de al route No.20, Prov. Du Haut Donnai, delegation du Djiring	VNM	107 48E	11 32N	250	Poilane M	19829	P
minima	wild	Kil 119 de la route col No.90, du Haut Danai (Bao Loc Deo)	VNM	107 44E	11 28N	750	Poilane M	21230	P
minima	wild	btw. Dinh Guan and Caocang Pro: de Bienhoa	VNM	107 23E	11 11N		Poilane M	21402	P
minima	wild	Haut Donai, Manif du Bi-Doup Prov	VNM	107 44E	11 28N	2000	Poilane M	30969	P
minima	wild		VNM	108 11E	11 54N		Tixier P	29105802	P
mungo		Mount Abu	IND	72 42E	24 36N	1000	Ali	sn	DUH
mungo		Natrahat,palabu dist,Bihar	IND	84 16E	23 28N	1000	Babu	sn	DUH
mungo		Nandapur, Jeypari, Karapit,Orissa	IND	82 44E	18 34N	600	Babu	sn	DUH
mungo		Tara Devi Hills	IND	77 41E	22 06N	600	Chandel	6010	NBPGR
mungo		Mashobra village,Simla dist,HP	IND	77 14E	31 08N	3500	Chandel	58,7,6	NBPGR
mungo		Bihar,Paranath,Chota Nagpore	IND	74 18E	23 09N	400	Clarke	21391	CAL
mungo		Coimbatore,Bellagi	IND	76 58E	11 00N	1500	Fischer	1347	CAL
mungo		Mahendragiri, Ganjam, AP	IND	85 04E	19 23N	<50	Fischer	s.n.	CAL
mungo		Benghal,Singbhum,Chaibassa	IND	85 48E	22 34N	350	Haines HH	2329	K
mungo		Castle Rock	IND	74 20E	15 25N	600	Meebold	10211	CAL
mungo		Village Sukhijodi, Dharampur,dist Solan, MP	IND	77 07E	30 55N	1000	Nayar	55,48	NBPGR
mungo		Assam near Shillong	IND	91 52E	25 34N	1200	Panigrahi	4489	CAL

Species	Status	Locality	Cty	Longitude	Latitude	Alt.m	Collector	Coll. no	Herb.
mungo		Poona, Mysore, Shimoga,Mattiga nr Yedur	IND	76 28E	13 28N	750	Raghavan	83026	CAL
mungo		Giurnar, Junagadh, Saurashtra	IND	70 27E	21 31N	200	Raizada	21164	FRI
mungo		Dehra Dun,	IND	78 03E	30 20N	600	Raizada	82263	FRI
mungo		Gujarat, Dangs,1km from Sapatura to Ahwa	IND	73 41E	20 45N	350	Remanandau	4619	CAL
mungo		Rajpur, Dehra Dun	IND	78 04E	30 19N	930	Saxena	1321	FRI
mungo		Karwar, Madras	IND	74 52E	23 06N	300	Sedgwick & Bell	6659	K
mungo		Almora dist, in Deorahat graveyard	IND	79 38E	29 36N	2000	Wadhwa	57402	DD
mungo mungo	cult.	Kasyapagma	LKA				Jayasuriya & Adachi	17102	PDA
mungo silvestris	wild	Akyab, Arakan	BUR	92 54E	20 10N		Kurtz S		K
mungo silvestris	wild	Sullia, South Cavara (Chavara), S. India	IND	76 32E	8 58N		Barber CA	2158	K
mungo silvestris	wild	Karwar, Ambo, Tamil Nadu	IND	74 9E	14 49N		Kerr or Kurz	6659	K
mungo silvestris	wild	Bailadila, Bastar	IND	81 56E	19 12N		Mooney HF	1514	K
mungo silvestris	wild	Khandala, Maharashtra	IND	74 01E	18 03N		NBPGR	NI635	BR
mungo silvestris	wild	Mount Abu	IND	72 42E	24 36N		Raizada MB	20596	K
mungo silvestris	wild	Singhur, Bombay	IND	72 52E	19 01N		Talbot WA		K
mungo silvestris	wild	Simla, Himachal Pradesh	IND	77 10E	31 06N		Thomson S		K
nakashimae	wild	Meng Shan, Fei Hsien, Shantung	CHN	117 53E	35 35N		Cheo TY & Yen L	282	BM
nakashimae	wild	Che-Foo (Yantai), Shan Tung	CHN	121 24E	37 32N		Fauvel M		P
nakashimae	wild	Peiitaho Lienpongshan, N. China	CHN				Licent E		BM
nakashimae	wild	Tsing fan	CHN				Limmermann	263	P
nakashimae	wild	Sazikou, Sandong Province	CHN				S Miki		KYO
nakashimae	wild	Hiikawa, Fukuoka-shi, Fukuoka	JPN	130 23 06E	33 32 53N		K Nakajima		TUS
nakashimae	wild	Shikanoshima, Kasuya-gun, Fukuoka	JPN	130 28 06E	33 36 25N		K Nakajima		TUS
nakashimae	wild	near Taiira, Uku-shima, Goto Islands, Nagasaki	JPN	129 08 19E	33 15 35N		Y Tateishi	4586	TUS
nakashimae	wild	Umenoki, Uku-shima, Goto Islands, Nagasaki	JPN	129 07 07E	33 16 44N		Y Tateishi	4657	TUS
nakashimae	wild	Kiba, Uku-shima, Goto Islands, Nagasaki	JPN	129 06 23E	33 16 50N		Y Tateishi	4678	TUS

Species	Status	Locality	Cty	Longitude	Latitude	Alt.m	Collector	Coll. no	Herb.
nakashimae	wild	Uku-shima, Goto Isslands, Nagasaki	JPN	129 06 39E	33 16 20N	100	Y Tateishi, H Hoshi	8777	TUS
nakashimae	wild	Mt. Kwanak, Anyang city, Kyonggi-do	KOR	126 55E	37 23N		G Koidzumi		KYO
nakashimae	wild	Mt. Bungakuza, Jinsen, Kyonggi-do	KOR				H Chang	1012	KYO
nakashimae	wild	Dorkjin, Choalanpuk-do	KOR				H Chang		KYO
nakashimae	wild	Chunchon Dam, Chunchon city, Kangwon-do	KOR	127 51E	37 56N		H Migo	18316	TUS
nakashimae	wild	Masan City, Kyongsangnam-do	KOR	128 35E	35 13N		H Migo	18317	TUS
nakashimae	wild	Glinei, Shinpan-ri, Furin (Ulrung gun?), Kyongsangnam-do	KOR	128 17E	35 18N		Kin Kuaknjyn	1014	TUS
nakashimae	wild	Mt. Bisul, Chilgok-gun, Talsong-gun, Kyongsangpuk-do	KOR	128 33E	36 05N		Oh Soo-Young	326-85-241	TUS
nakashimae	wild	Mt. Taeduk, Taemyong-dong, Namgu, Taegu city, Kyongsangpuk-do	KOR	128 36E	35 52N		Oh Soo-Young	0326-85-98	TUS
nakashimae	wild	Near Ritsuri	KOR	127 57E	37 21N		PHDorsett, WMorse	6282	K
nakashimae	wild	Shinkum-ri, Bongchae-myeon, Kohwung-gun, Choranam-do	KOR	127 19E	34 35N		Ranhodoo student		KYO
nakashimae	wild	Ohi-do, Kunja-dong, Shihung-city, Kyonggi-do	KOR	126 43E	37 24N		TNemoto, BH Choi	2737	TUS
nakashimae	wild	Mt. Scolmsan, Sunam-ri, Dong-myeon, Chonwon-gun, Chungchongnam-do	KOR	127 12E	36 48N		TNemoto, BH Choi	2856	TUS
nakashimae	wild	Mt. Kwanak, Anyang city, Kyonggi-do	KOR	126 55E	37 23N		TNemoto,BH Choi	2770	TUS
nakashimae	wild	Mt. Namsan, Seoul	KOR	127 02E	37 33N		T Uchiyama		TUS
nakashimae	wild	Chinnampo (Namp'o)	PKR	125 24 29E	38 43 58N		Faurie M	444	P
nakashimae	wild	Peitaiho	CHN	119 30E	39 48N		Cowdry NH		K
nepalensis	wild	Selim, Sikkim	IND	88 27E	27 28N	3500	Clark CB	36752B	K
nepalensis	wild	Darjeeling, Sikkim	IND	88 15E	27 01N	1700	Gamble JS	9820	K
nepalensis	wild	Sikkim Himalaya	IND	88 27E	27 28N	330	Hooker	348	K
nepalensis	wild	Sikkim Himalaya	IND	88 27E	27 18N		Hooker	1852	K
nepalensis	wild	Khasi Hills	IND	91 30E	25 30N	1330	Hooker & Thomson		K
nepalensis	wild	Sikkim Himalaya	IND	88 25E	27 42N		Trentler (presented)	97	K
radiata radiata	cult.	118km S of Axum-Gondar, Taccaze R bridge, Begender Prov.	ETH	38 00E	13 23N		de Wilde JJFE	7125	BR

Species	Status	Locality	Cty	Longitude	Latitude	Alt.m	Collector	Coll. no	Herb.
radiata radiata	cult.	Antilles Francaise (Guadeloupe)	GLP	61 43W	16 00N		Sagot	138	P
radiata radiata	cult.	Heijo and Vicinity	PKR	125 44E	39 01N		PHDorsett, WMorse	6204	K
radiata radiata	cult.	Central Rukwa, Kafukola	TZA	31 57E	8 05S		Siame		BM
radiata radiata	cult.	Kimuenza, Kinshasa	ZAF	18 20E	4 25S		van Hove	37	BR
radiata sublobata	wild	New Holland	AUS				Banks & Solander		BM
radiata sublobata	wild	Moreton Dist. Somaset Dam, Queensland	AUS	152 43E	27 34S		Blake ST	17398	K
radiata sublobata	wild	Kimberley Research Station, Kununurra, Western Australia	AUS	128 44E	15 46S		Lazarides M	6778	K
radiata sublobata	wild	Elcho island, Northern Territory	AUS	135 34E	12 02S		Maconochie JR	2157	K
radiata sublobata	wild	near Adelaide R. Township	AUS	138 34E	34 55S		McKean J	B453	K
radiata sublobata	wild	Yorkey's Knob, Cairns, Queensland	AUS	145 43E	16 48S		Mekee HS	9018	K
radiata sublobata	wild	Mareeba, Cook Dist., Queensland	AUS	145 25E	16 59S		Pedley L	2232	K
radiata sublobata	wild	Slade Point, Mackay	AUS	149 10E	21 09S		Lawn RJ	CQ 2225	BM
radiata sublobata	wild	Endeavour R., 500 m upstream from monument, Queensland	AUS	145 29E	15 35S		Lawn RJ	CQ2235	BM
radiata sublobata	wild	Mt. Stuart road, Queensland	AUS	148 40E	23 12S		Remanamdan P	4187	K
radiata sublobata	wild	72 km from Townsville, Ayr road, Queensland	AUS	146 14E	18 54S		Remanamdan P	4208	K
radiata sublobata	wild	South Bay, Bickerton island, Gulf of Australia, Carpentaria	AUS	136 06E	13 45S		Sprecht RL	540	K
radiata sublobata	wild	Ipswich, Moreton Dist., Queensland	AUS	152 45E	27 37S		Wood A		K
radiata sublobata	wild	Tongsa	BHU	90 30E	27 30N			T6	K
radiata sublobata	wild	Maymyo Plateau, E. of Mandalay	BUR	96 28E	22 02N	1200	Lace JH	4248	K
radiata sublobata	wild	Yunnan	CHN	101 10E	24 47N		Delavay M	4561	P
radiata sublobata	wild	Yunnan	CHN	101 10E	24 47N		Delavay M		P
radiata sublobata	wild	Pres Kaetle	CMR				Letouzey R	6548	K
radiata sublobata	wild		CMR				Talbot PA		BM
radiata sublobata	wild	fluvium Tacaze prope Djeladjeranne	ETH				Schimper	1626	BM
radiata sublobata	wild	In valle Mai-Mezanno prope Djeladjeranne	ETH	38 47E	8 59N		Schimper	1718	L
radiata sublobata	wild	Bank of Black Volta near Lawra	GHA	2 56W	10 37N		Hall JB	763	K

Species	Status	Locality	Cty	Longitude	Latitude	Alt.m	Collector	Coll. no	Herb.
radiata sublobata	wild	Wegrand in Makassar, Sulawesi	IDN	119 24E	5 08S		de Froideville CM	437	L
radiata sublobata	wild	Batavia (Jakarta), Java	IDN	106 49E	6 07S		Horsfield T	110	K
radiata sublobata	wild	Sumba	IDN	120 00E	10 00S		Iboet	125	L
radiata sublobata	wild	Besuki, W. Kutoredjo, btw Gradjagan-Pangang Bay, East Java	IDN	114 20E	8 30S		Jacobs M	4989	K
radiata sublobata	wild	Mt. Ndeki, West Flores	IDN	120 01E	8 47S		Kostermans, Wirawan	165	L
radiata sublobata	wild	West Nisar-Sesok, Kl. Soenda Eil, Flores	IDN	120 01E	8 47S	200	Schmutz PE	2153	L
radiata sublobata	wild	Palau Selaru, E. of Adaut, Taniimbar Islands	IDN	131 08E	8 07S		van Borssum Waalkes J	3158	K
radiata sublobata	wild	Kebar valey, ca 100km W of Manokwari Irian Jaya	IDN	133 52E	2 30S	550	van Royen P	5106	L
radiata sublobata	wild	Dindigul, Pulney Hills, Madura Dist	IND	77 58E	10 21N	330	Anglade L	1149	K
radiata sublobata	wild	Dolkhamb forest, Washala Range, Thana Dist., Maharashtra	IND				Billore KV	111833	K
radiata sublobata	wild	Lower Pulneys, Machun, Madras	IND	77 39E	10 12N		Bourne	130	K
radiata sublobata	wild	Pulneys	IND	77 29E	10 14N		Bourne	2004	K
radiata sublobata	wild	Bengal	IND	88 40E	22 53N		Hooker		K
radiata sublobata	wild	Dindigul, Kodaikanal Dist., Tamil Nadu	IND	77 58E	10 21N		KM & KT Matthew		K
radiata sublobata	wild	8km to Mandi, Himachal Pradesh	IND	76 57E	31 38N		van der Maesen LJG	2940	K
radiata sublobata	wild	25km to Kalka, before Dharampur, Himachal Pradesh	IND	77 07E	30 55N	1650	van der Maesen LJG	2956	K
radiata sublobata	wild	44km E of Belgaum, Karnataka	IND	74 52E	15 55N	720	van der Maesen LJG	3321	K
radiata sublobata	wild	44km N of Munnar, Kerala	IND	77 14E	10 24N	880	van der Maesen LJG	3470	K
radiata sublobata	wild	39km to Palani, Melpallam, Tamil Nadu	IND	77 16E	10 21N	1570	van der Maesen LJG	4794	K
radiata sublobata	wild	Netrarhat, near Forest Rest House, Bihar	IND	84 16E	23 28N	1170	van der Maesen LJG	5014	K
radiata sublobata	wild	Canary Hill, Hazaribagh, Bihar	IND	85 22E	24 00N		van der Maesen LJG	5057	K
radiata sublobata	wild	Quarry near Slla	LKA				Simpson ND	9632	K
radiata sublobata	wild	c. 9km NE of Antsalova, Tsingy de Bemaraha (western margin), Ambinda, Ambodiriana, Mahajanga(Majunga) Prov.)	MDG	44 42 11E	18 41 2S	60	DuPuy DJ et al.	M885	K
radiata sublobata	wild		MDG	46 22E	19 26S		Gembloux	NI366	BR
radiata sublobata	wild	Vallee de l'Ifasy en avald' Anaborano (Distr. D'Ambilobe:Nord)	MDG	49 51E	13 04S	125	Humbert & Capuron	25919	K

Species	Status	Locality	Cty	Longitude	Latitude	Alt.m	Collector	Coll. no	Herb.
radiata sublobata	wild	vers Mahaboboka	MDG	47 37E	16 53S		Peltier M	4959	K
radiata sublobata	wild	Forest Research Center, road to Orang Hutan Center, Sandankan, Sabah	MYL	118 06E	5 50N		Stevens PF et al.	711	L
radiata sublobata	wild	Biu Plateau	NGA	12 30E	11 30N	870	Royer HA	B15	BM
radiata sublobata	wild	Wadi Darbaat, 13km NE of Taquah, Dhofar	OMN	54 29E	17 9N	200	Guarino L	53	K
radiata sublobata	wild	2 km N of Dalkhut, Dhofar	OMN	54 00E	17 21N	360	Guarino L	53a	K
radiata sublobata	wild	J.Qara, Spur above W.Hinna, Dhofar	OMN	54 00E	17 21N	620	Miller AG	7730	K
radiata sublobata	wild	Khadrafi, Dhofar	OMN	53 09E	16 42N	600	Smith A Radcliffe	5314	K
radiata sublobata	wild	Maluko and vicinity, Bukidnon, Mindanao	PHL	124 58E	8 24N		Ramos M & G Edano	38444	K
radiata sublobata	wild	Konosia	PNG			20	Carr CE	11194	K
radiata sublobata	wild	Sea Shore, Port Moresby	PNG	147 10E	9 29S		Carr CE	11875	K
radiata sublobata	wild	Mt. Eriama, Central Dist.	PNG	147 10E	9 30S	100	Gillison AN	22190	L
radiata sublobata	wild	near W end of Dobdura air strip, Samboga river, Northern Div.	PNG	148 22E	8 46S	50	Hoogland RD	3795	K
radiata sublobata	wild	near Medino village (N coast of Cape Vogel Peninsula, Milne Bay District	PNG	150 29E	10 24S	25	Hoogland RD	4773	BM
radiata' sublobata	wild		PNG				Koch JWR	687	L
radiata sublobata	wild	Port Moresby	PNG	147 10E	9 29S		McGregor W		K
radiata sublobata	wild	Waigani Reservoir, Port Moresby sub-dist., Central Dist.	PNG	147 10E	9 26S		Pulaford M	298	L
radiata sublobata	wild	Tavai Creek, ca. 43 miles SE of Port Moresby, Central District	PNG	147 26E	9 42S		Pullen R	6921	L
radiata sublobata	wild	Nazareth Mission, Laloki Rv. Pt. Moresby, Central District	PNG	147 11E	9 22S	27	Streimann H, A Kairo		L
radiata sublobata	wild	Nova Guinea neerlandica meridionalis	PNG				Versteegh	1850	L
radiata sublobata	wild	in Lande der Djur ges v G. Schweinfurth	SUD				Auftrage	2409	K
radiata sublobata	wild	Tozi, Blue Nile Prov.	SUD	30 45E	10 50N	50	Jobea	15	K
radiata sublobata	wild	Jabel Abel, near Abu-Naama reserve dam	SUD				Khatib O	13855	K
radiata sublobata	wild	Bangkok	THA	100 31E	13 45N	25	Kerr	9327	K
radiata sublobata	wild	Cental Timur South	TMP	125 10E	9 25S		Walik MS	94	BM

Species	Status	Locality	Cty	Longitude	Latitude	Alt.m	Collector	Coll. no	Herb.
radiata sublobata	wild	Takow	TWN	120 17E	22 37N		Corn G, H Playfair		K
radiata sublobata	wild	Tainan Co., Henglu,	TWN			200	H Ohashi et al.	24066	TUS
radiata sublobata	wild	Taitung Co., Lungchang-Chintsun,	TWN	121 17E	22 57N	50	H Ohashi et al.	24530	TUS
radiata sublobata	wild	Taiwan	TWN	120 51E	23 43N		Henry A	1123	K
radiata sublobata	wild	Hsiaokangshan	TWN	120 19E	22 48N	<200	K C Yang	3484	TUS
radiata sublobata	wild	Wu shantou, Tainan Co., Around the dam	TWN			250	S F Huang	1784	TUS
radiata sublobata	wild	Pingtung, Chingshuiyen	TWN	120 22E	22 32N		TC Huang, M Wu		TUS
radiata sublobata	wild	Pingtung, Chingshuiyen	TWN	120 22E	22 32N		TC Huang, MWu	14803	TUS
radiata sublobata	wild	Tainan Co., Wushantou Reservoir	TWN			150	Y Tateishi	18560	TUS
radiata sublobata	wild	Pingtung Co., Tachi	TWN	120 56E	22 28N	100	Y Tateishi	25291	TUS
radiata sublobata	wild	Tainan, Liuchia, Tutichi	TWN	120 13E	22 54N	250	Y Tateishi,T Kajita	25051	TUS
radiata sublobata	wild	Tainan Co., Wushantou Reservoir, Tutichi,Chienshanlukou	TWN			200	Y Tateishi, Y Endo	20866	TUS
radiata sublobata	wild	Mouth of Kavu river, Sonta plain, Ufipa Dist.	TZA	31 45E	7 39S	780	Richards HM	16010	K
radiata sublobata	wild	Sandy flood plain, Sonta, Rukwa, Mpanda, Dist.	TZA	32 29E	8 05S	780	Richards HM	18285	K
radiata sublobata	wild	Rukwa, Baherine, Central Rukwa, Ufipa Dist	TZA	32 29E	8 05S	780	Richards HM	19229	K
radiata sublobata	wild	Tumba, Rukwa valley	TZA	31 10E	7 07S		Siame	130	K
radiata sublobata	wild	Island nr. Nha Trang, recoltee dans la partie sud de l'ile sol rocheux	VNM	109 10E	12 13N		Poilane M	2955	P
reflexo-pilosa	wild	12 miles NE of Katherine	AUS	132 24E	14 23S		Wilson IB	81	K
reflexo-pilosa	wild	New Holland	AUS				Banks & Solander	65	BM
reflexo-pilosa	wild	Koang-Tcheou Prov. (Zhanjiang)	CHN	110 23E	21 11N		Cavalerie M	3278	P
reflexo-pilosa	wild	Hainan	CHN	109 39E	19 04N		Henry A	8351	K
reflexo-pilosa	wild	Mt. Laufoka	FJI	177 27E	17 37S		Greenwood W	367	K
reflexo-pilosa	wild		FJI	177 58E	17 46S		Seemann B	117	K
reflexo-pilosa	wild	Jampit, Java	IDN	114 08E	8 05S	1300	Backer CA	25064	L
reflexo-pilosa	wild	Java	IDN			1000	Backer CA	37114	L

Species	Status	Locality	Cty	Longitude	Latitude	Alt.m	Collector	Coll. no	Herb.
reflexo-pilosa	wild	roet G Ualintang, Sumatra	IDN			800	Bunnemeyer	4434	L
reflexo-pilosa	wild	Soerabaji (Surabaya), Java	IDN	112 44E	7 14S		Dirgelo JD	823	L
reflexo-pilosa	wild	Loewock, btwn Pakoa & Pinapoeang, Menado, Sulawesi	IDN	122 47E	0 56S		Eyma	3754	K
reflexo-pilosa	wild	Wissel Lake Region, Boebeiro, Irian Jaya	IDN	136 15E	3 55S		Eyma	5095	L
reflexo-pilosa	wild	Timor	IDN	125 06E	9 13S		Forebes HO	3672	BM
reflexo-pilosa	wild	Padang Panjang, Sumatra	IDN	100 25E	0 26S		Matthew CG	5000	K
reflexo-pilosa	wild	Flores, Lesser Sunda Islands	IDN	121 14E	8 47S		Pater JJ	21	L
reflexo-pilosa	wild	Ambon	IDN	128 12E	3 43S		Robinson CB	233	K
reflexo-pilosa	wild	Sungai Klumbang, Korinchi, Sumatra	IDN	102 00E	2 07S	1500	Robinson HC, C Kloss	135	BM
reflexo-pilosa	wild	Korinchi Prov.,, Sungai Kumbang, Sumatra	IDN	101 31E	2 08S		Robinson HC, C Kloss		K
reflexo-pilosa	wild	Sundaran Agong, Korinchi, Sumatra	IDN	101 31E	2 08S		Robinson HC, CKloss	77	BM
reflexo-pilosa	wild	Sungai Kumlang, Korinchi, Sumatra	IDN	101 31E	2 08S		Robinson HC, CKloss	213	BM
reflexo-pilosa	wild	W Shapoctih, NW Seran, Moluccas	IDN	129 00E	3 00S	50	Rutten		L
reflexo-pilosa	wild	Minjambau, Arfak Mountains, Irian Jaya	IDN	133 58E	1 05S	1250	Versteegh	BW 12635	L
reflexo-pilosa	wild	Java	IDN				Zeef	317	L
reflexo-pilosa	wild		IND	88 52E	23 34N		Hooker & Thompson		L
reflexo-pilosa	wild	Loochoo, Keenchan, Okinawa	JPN				ex herb Tashiro		KYO
reflexo-pilosa	wild	Ikema Is., Miyako, Okinawa	JPN	125 14 44E	24 55 21N		G Koidzumi		KYO
reflexo-pilosa	wild	Sate, Kunigami, Kunigami-gun, Okinawa	JPN	128 13 37E	26 47N		G Koidzumi		KYO
reflexo-pilosa	wild	Mt. Katsau-dake, Nago, Kunigami-gun, Okinawa	JPN	127 56 17E	26 37 01N		H Ohashi, Y Tateishi	1023	TUS
reflexo-pilosa	wild	Ada, Kunigami, Okinawa	JPN	128 15 31E	26 45 16N		J Murata	17092	KYO
reflexo-pilosa	wild	Mt. Nekumachiji-dake, Ohgimi, Okinawa	JPN	128 08 25E	26 40 55N		K Shinjo	12063	TUS
reflexo-pilosa	wild	Tomino, Ishigaki Is., Okinawa	JPN	124 11E	24 26N		M Furuse	3769	K
reflexo-pilosa	wild	Tomino, Ishigaki Is., Okinawa	JPN	124 11E	24 26N		M Furuse	4273	K
reflexo-pilosa	wild	Shirahama - Sonai, Iriomote Is., Okinawa	JPN	123 45 13E	24 22 28N		M Furuse	5488	K
reflexo-pilosa	wild	Nr Inutabu, Isen, Tokunoshima, Kagoshima	JPN	128 54 35E	27 42 34N	100	S Hatusima	19441	L

Species	Status	Locality	Cty	Longitude	Latitude	Alt.m	Collector	Coll. no	Herb.
reflexo-pilosa	wild	Nr Nishinakama, Sumio, Amami-Ohshima, Kagoshima	JPN	129 24 34E	28 15 42N	10	S Hatusima	20070	L
reflexo-pilosa	wild	Okinoerabu Is., Kagoshima	JPN	128 36 17E	27 22 30N		S Hatusima,Y Miyagi	39053	TUS
reflexo-pilosa	wild	Asani, Naze, Amami Ohshima Is., Kagoshima	JPN				T Hosoyama		KYO
reflexo-pilosa	wild	Okinawa	JPN				Tashiro		KYO
reflexo-pilosa	wild	Okinawa	JPN				U Faurie	3916	KYO
reflexo-pilosa	wild	Kunigami, Ginama, Okinawa Is., Okinawa	JPN	128 15 28E	26 50 38N		Walker EH	8209	L
reflexo-pilosa	wild	Genka river, Nago, Okinawa	JPN	128 03 36E	26 37 28N	35	Y Tateishi, Yamashiro	45160	TUS
reflexo-pilosa	wild	near seaside, Ada, Higashi, Kunigami-gun, Okinawa	JPN	128 19 13E	26 44 35N		Y Tateishi et al.	43512	TUS
reflexo-pilosa	wild	Southwestern part of Amami Ohshima Is., Kagoshima	JPN	129 14 37E	28 14 50N		YTashiro		KYO
reflexo-pilosa	wild	Yoneyama, Iriomote Is., Okinawa	JPN	123 45E	24 20N	120	Y and T Ankei	750051	KYO
reflexo-pilosa	wild	Son-nua	LAO	104 40E	19 57N		Poilane M	2122	P
reflexo-pilosa	wild	ile Des Pins, Oumagne	NWC	167 28E	22 36S	30	MacKee	23585	K
reflexo-pilosa	wild	Amieu	NWC	165 48E	21 37S	350	MacKee M	12285	K
reflexo-pilosa	wild	Vallee de Houailou	NWC	165 37E	21 17S	300	McKee HS	9880	L
reflexo-pilosa	wild	Camiguin Island, Babunyanes	PHL	121 54E	18 55N		Fenix E	39771	K
reflexo-pilosa	wild	Nr. Tanculan, Bukidnon, Mindanao	PHL	125E	8 00N		Fenix E	2-107	K
reflexo-pilosa	wild	Island of Polillo	PHL	121 56E	14 51N		McGregor RC	2 39	K
reflexo-pilosa	wild	San Antonio, Laguna Prov. Luzon	PHL	121 11E	14 9N		Ramos M	12020	BM
reflexo-pilosa	wild	Mt. Iraya, Batan Island, Batanes Prov.	PHL	121 56E	20 23N		Ramos M	8-373	K
reflexo-pilosa	wild	Siasi, Sulu Archipelago	PHL	120 49E	5 33N		Vidal	2650	K
reflexo-pilosa	wild	Tappen-Biah, G. Wawah	PNG				Aet & Tdjan	40	L
reflexo-pilosa	wild	Ekwap, Wantoat, Kaiapit Subdistrict, Morobe	PNG	146 00E	6 10S	1500	Millar AN	NGF12135	L
reflexo-pilosa	wild	along Bliri River,Karandu Village, Aitape, Sepik Dist.	PNG	141 59E	3 14S	80	Darbyshire, Hoogland	8219	L
reflexo-pilosa	wild	Vicinity of Finschhafen	PNG	147 51E	6 36S		Sawyer FE	246	L
reflexo-pilosa	wild	Butemu, Saidor sub-dist., Madang Dist. T.N.G.	PNG	146 05E	5 50S	1700	Sayers CD	19791	L
reflexo-pilosa	wild	Nova Guinea necrlandica meridionalis(PNG)	PNG				Versteegh	1123	L

Species	Status	Locality	Cty	Longitude	Latitude	Alt.m	Collector	Coll. no	Herb.
reflexo-pilosa	wild	1 mile N Aiome Patrol Post, Atemble track, Lower Ramu-Atitau, Madang	PNG	144 43E	5 09S	100m	Pullen R	1009	L
reflexo-pilosa	wild	Bangkok	THA	100 31E	13 45N	0-50	Kerr AFG	3970	BM
reflexo-pilosa	wild	Vavau, Friendly Island	TON				Barclay GW	3395	BM
reflexo-pilosa	wild	Kaohsiung Co., Shanping	TWN			700	H Ohashi et al.	13422	TUS
reflexo-pilosa	wild	Pingtung Co., Mt. Nanjen-shan	TWN	120 51E	22 05N	425	H Ohashi et al.	14373	TUS
reflexo-pilosa	wild	Miaoli Co., Taian Spa, along Wenshuihsi River	TWN			500	H Ohashi et al.	20767	TUS
reflexo-pilosa	wild	Kkaohsiung Co., Tengchu, base Mt. Tsuyun-shan	TWN			800	H Ohashi et al.	23772	TUS
reflexo-pilosa	wild	Tainan Co., Pinglin, Kuangshan	TWN	120 07E	23 13N	225	H Ohashi et al.	24113	TUS
reflexo-pilosa	wild	Bankin sinp	TWN	120 53E	23 37N	.	Henry A	392	K
reflexo-pilosa	wild		TWN	120 53E	23 37N		Henry A	2020	K
reflexo-pilosa	wild	Taipei Co., Pingling	TWN	121 32E	25N		J M Hu	590	TUS
reflexo-pilosa	wild	Jayi county, Wufeng town, Mt. Ali crossroad	TWN	120 45E	22 44N	1200	J Murata et al.	19091	TUS
reflexo-pilosa	wild	Tanshui	TWN	121 25E	25 11N		Richard Oldman	194	K
reflexo-pilosa	wild	Tanshui	TWN	121 25E	25 11N		Richard Oldman	195	K
reflexo-pilosa	wild	Chiintan, Taipei Co.	TWN			150	SF Huang	1705	TUS
reflexo-pilosa	wild	Antungshan, Hualien Co.	TWN	121 21E	23 18N	400	SF Huang	1708	TUS
reflexo-pilosa	wild	Chiangbin, Taitung Co.,	TWN			500	SF Huang	1722	TUS
reflexo-pilosa	wild	Puli, Mantou Co.	TWN	120 57E	23 58N	250	SF Huang	1937	TUS
reflexo-pilosa	wild	Erohen, Ilan Co.	TWN			25	SF Huang	2109	TUS
reflexo-pilosa	wild	Hualien Co., Meilunbi, Hualien	TWN	121 37 55E	24 00 53N	25	SF Huang et al.	5212	TUS
reflexo-pilosa	wild	Taitung Co., Lanyu, Hongtu-Tienchih	TWN			175	SF Huang, YC Hsu	4723	TUS
reflexo-pilosa	wild	South Cape	TWN	120 47E	21 58N		Schmirer S	361	K
reflexo-pilosa	wild	Kaohsiung Co., Liu-Kuei Tsai-tieh-ku	TWN	120 38E	23 00N	100	TC Huang	14457	TUS
reflexo-pilosa	wild	Taitung Co., Lanzu Isl. Yujen-tsun, seaside	TWN	121 32E	22 02N		T Kajita	826	TUS
reflexo-pilosa	wild	Takao	TWN	120 16E	22 38N		U Faurie	1111	KYO

Species	Status	Locality	Cty	Longitude	Latitude	Alt.m	Collector	Coll. no	Herb.
reflexo-pilosa	wild	Kaohsiung Co, Taoyuan, along Laonung-his River	TWN	120 46E	23 10N	500	Y Tateishi	20196	TUS
reflexo-pilosa	wild	Kaohsiung Co., Tachiuyuan	TWN	120 23E	22 42N	300	Y Tateishi & Y Endo	20087	TUS
reflexo-pilosa	wild	Kaohsiung Co., Lioukue, Tsaitiehku	TWN			500	Y Tateishi & Y Endo	20213	TUS
reflexo-pilosa	wild	Kaohsiung Co., Hsinliao	TWN	120 38E	23 00N	200	Y Tateishi & Y Endo	20224	TUS
reflexo-pilosa	wild	Isl. Lanyu, Hungtou Airport	TWN	121 33E	22 02N	20	Y Tateishi et al.	15413	TUS
reflexo-pilosa	wild	Hualien Co., Zuepei	TWN			180	Y Tateishi et al.	15589	TUS
reflexo-pilosa	wild	Hualien Co., Chimei, Hsiehtenu	TWN	121 26E	23 30N	150	Y Tateishi et al.	15688	TUS
reflexo-pilosa	wild	Hhhualien Co., Takangkou	TWN	121 29E	23 28N	25	Y Tateishi et al.	16048	TUS
reflexo-pilosa	wild	Hualien Co., Tungmen, Wunlan	TWN	121 29E	23 58N	200	Y Tateishi et al.	16226	TUS
reflexo-pilosa	wild	Pingtung Co., Namren-shan	TWN	120 51E	22 05N	400	Y Tateishi et al.	18487	TUS
reflexo-pilosa	wild	Pingtung Co., Foot of Mt. Peitawu-shan, Taiwu,	TWN	120 40E	22 36N	800	Y Tateishi et al.	19720	TUS
reflexo-pilosa	wild	Miaoli Co., Kuantaoshan	TWN	120 49 02E	24 22 10N	450		15452	TUS
reflexo-pilosa	wild	Tonkin	VNM	105 00E	22 00N		Balamsa B	1187	K
reflexo-pilosa	wild	Phuc Nhac	VNM	106 03E	20 09N		Bon M	383	P
reflexo-pilosa	wild		VNM				Bon M	3982	P
reflexo-pilosa	wild	Bas Song Cao, pres Nha Trang	VNM	109 12E	12 14N		Evrard F	681	P
reflexo-pilosa	wild	Cha-pa (Sa Pa)	VNM	103 52E	22 21N		Lecomte H & A Finet	550	P
reflexo-pilosa	wild	coufin sud de la pro: du guang Nam entre les srillags Moi de Moo et Mang	VNM	108 31E	15 07N	1500	Poilane M	31728	P
reflexo-pilosa	wild	S of Luaignan prov. near Moi de Dao Bo village	VNM	106 44E	21 14N	1500	Poilane M	31907	P
reflexo-pilosa	wild	Aneityum (=Anatom)	VUT	169 49E	20 12S		Morrison A		K
reflexo-pilosa	wild	Mt. Erskine, Efate	VUT	168 20E	17 37S		Morrison A		K
reflexo-pilosa	wild	Industrial Forestry Plantation, Shark Bay, Santo Island	VUT				Curry P	650	K
riukiuensis	wild	Oohama, Ishigaki Is., Okinawa	JPN	124 11 48E	24 20 54N		E Takaryo	53	KYO
riukiuensis	wild	Yonaguni Is., Okinawa	JPN	122 58 30E	24 26 40N		G Koidzumi		KYO
riukiuensis	wild	Iriomote Is., Okinawa	JPN	123 50 46E	24 23 46N		G Koidzumi		KYO

Species	Status	Locality	Cty	Longitude	Latitude	Alt.m	Collector	Coll. no	Herb.
riukiuensis	wild	Komi to Funaura, Iriomote Is., Okinawa	JPN	123 52 12E	24 23 22N		G Murata et al.	67562	KYO
riukiuensis	wild	btw. Komi and Ohara, Iriomote Is., Okinawa	JPN	123 54 08E	24 18 16N		G Murata, H Tabata	734	KYO
riukiuensis	wild	Komi-pasture to Komi, Iriomote Is., Okinawa	JPN	123 55 14E	24 20 14N	15	H Okada et al.	307	KYO
riukiuensis	wild	Komi, Iriomote Is., Okinawa	JPN	123 54 10E	24 19 26N		K Shinaguku, Y Miyagi	10258	KYO
riukiuensis	wild	Toyohara, Iriomote Is., Okinawa	JPN	123 50 46E	24 23 46N		K Iwatsuki et al.	308	KYO
riukiuensis	wild	Mt. Omoto, Ishigaki Is., Okinawa	JPN	124 10 56E	24 25 12N		K Iwatsuki, N Fujita	103	KYO
riukiuensis	wild	Kainan, Ishigaki Is., Okinawa	JPN	124 11 36E	24 23 55N		M Furuse	781	K
riukiuensis	wild	Oohama, Ishigaki Is., Okinawa	JPN	124 12E	24 21N		M Furuse	789	K
riukiuensis	wild	Oohama, Ishigaki Is., Okinawa	JPN	124 12E	24 21N		M Furuse	795	K
riukiuensis	wild	Kubura, Yonaguni Is., Okinawa	JPN	122 56 37E	24 26 47N		M Furuse	1273	K
riukiuensis	wild	Kubura, Yonaguni Is., Okinawa	JPN	122 56 37E	24 26 47N		M Furuse	1274	K
riukiuensis	wild	Arakawa, Ishigaki Is., Okinawa	JPN	124 08 24E	24 21 04N		M Furuse	1535	K
riukiuensis	wild	Shirahama to Sonai, Iriomote Is., Okinawa	JPN	123 45E	24 23N		M Furuse	2099	K
riukiuensis	wild	Tonoshiro,Ishigaki Is., Okinawa	JPN	124 10E	24 20N		M Furuse	2604	K
riukiuensis	wild	San-ni-dai, Yonaguni Is., Okinawa	JPN	123 01 51E	24 27 10N		M Furuse	3198	K
riukiuensis	wild	Shirahama to Sonai, Iriomote Is., Okinawa	JPN	128 45 13E	24 22 28N		M Furuse	3274	K
riukiuensis	wild	Tonoshiro, Ishigaki Is., Okinawa	JPN	124 10E	24 20N		M Furuse	3821	K
riukiuensis	wild	Omoto, Ishigaki Is., Okinawa	JPN	124 11E	24 25N		M Furuse	4002	K
riukiuensis	wild	San-ni-dai, Yonaguni Is., Okinawa	JPN	123 01 51E	24 27 10N		M Furuse	4532	K
riukiuensis	wild	Agari-Henna-Zaki, Gusukube, Miyako Is., Okinawa	JPN	125 28E	24 43N		M Furuse	4639	K
riukiuensis	wild	Agari-Henna-Zaki, Gusukube, Miyako Is., Okinawa	JPN	125 28E	24 43N		M Furuse	4848	K
riukiuensis	wild	Agari-Henna-Zaki, Gusukube, Miyako Is., Okinawa	JPN	125 28E	24 43N		M Furuse	4853	K
riukiuensis	wild	Mibaru, Iriomote Is., Okinawa	JPN	123 50 46E	24 23 46N		N Fukuoka, M Ito	364	KYO
riukiuensis	wild	Tindabana, Yonaguni Is., Okinawa	JPN	122 59 30E	24 28 01N		Y Ankei	78536	KYO
riukiuensis	wild	Formosa	TWN					686	KYO
riukiuensis	wild	Pingtung coungry, Chinaluosuei	TWN	120 51E	22 00N		C M Kuo	9548	KYO

Species	Status	Locality	Cty	Longitude	Latitude	Alt.m	Collector	Coll. no	Herb.
riukiuensis	wild	Kelung	TWN	121 44E	25 08N		Faurie U	395	P
riukiuensis	wild	Pingtung-shan, Nanjenkeng	TWN	121 51E	23 05N	100	H Ohashi et al.	14005	TUS
riukiuensis	wild	Pingtung Co., Nanren-shan, Chiopeng	TWN	120 51E	22 05N	50	H Ohashi et al.	14407	TUS
riukiuensis	wild	Pingtung Co; Oluanpi	TWN	120 51E	21 54N	30	H Ohashi et al.	14610	TUS
riukiuensis	wild	Pingtung Co., Chialuosue	TWN	120 37E	22 20N	15	H Ohashi et al.	14614	TUS
riukiuensis	wild	Taitung Co. Fengtian , Haping seaside	TWN	121 19E	23 03N		H Ohashi et al.	24531	TUS
riukiuensis	wild	Pingtung Co: Chialoshui	TWN	120 51E	22 00N	0	Jer-Ming Hu et al.	898	TUS
riukiuensis	wild	Nanfanao, Ilan Co.	TWN	121 51 39E	24 38 48N	10	SF Huang	1951	TUS
riukiuensis	wild	Aoti-Fulung, Taipei Co.	TWN	121 55E	25 03N	10	SF Huang	3078	TUS
riukiuensis	wild	Ilan Co; Nanfangao	TWN	121 51 39E	24 38 48N	25	SF Sellow	5083	TUS
riukiuensis	wild	South Cape, Formosa	TWN	120 49E	21 57N		Schmiere	1273	K
riukiuensis	wild	Sozan, Formosa, Taiwan	TWN				U Faurie		KYO
riukiuensis	wild	Taipei, Santiaochiaoo	TWN	122 00E	25 01N	150	Y Tateishi & T Kajita	24639	TUS
riukiuensis	wild	Taipei Co., Keelung, Isl. Hoping-dao	TWN	121 46E	25 10N		Y Tateishi et al.	17519	TUS
riukiuensis	wild	Pingtung co., Chialuosue	TWN	120 37E	22 20N	10	Y Tateishi et al.	25277	TUS
riukiuensis	wild	Pingtung Co., Taichi	TWN	120 56E	22 28N	100	Y Tateishi et al.	25287	TUS
riukiuensis	wild	Pingtung Co., Sheting Park	TWN	120 49E	21 58N	150	Y Tateishi et al.	25318	TUS
r-p glabra	cult.		PHL				Gembloux	NI532	BR
r-p glabra	cult.	Haiphong	VNM	106 40E	20 51N		Balansa B	1187	P
r-p glabra	cult.	Tu-Phap (Thu Phap)	VNM	105 19E	21 03N		Balansa B	2276	P
r-p glabra	cult.	Nha-trang and vicinity	VNM	109 12E	12 15N		Robinson CB	1364	P
r-p glabra	cult.	Base du Col de Blao	VNM	107 44E	11 28N		Tixier P		P
stipulacea	wild	Java	IDN	110 22E	7 47S		Horsfield T	111	K
stipulacea	wild	G(O?)nstrehen Uraton, Java	IDN				Kooper	3901	L
stipulacea	wild	Kust O.V. Pasar Ikan, Batavia, Java	IDN	106 48E	6 07S		Meer & Hoed	2014	L
stipulacea	wild	Pampanoea, Sulawesi	IDN	102 2E	5 19S		Noerkas	84	L

Species	Status	Locality	Cty	Longitude	Latitude	Alt.m	Collector	Coll. no	Herb.
stipulacea	wild	Batavia (Jakarta)	IDN	106 48E	6 10S		Raap H	420	L
stipulacea	wild	Tjengkareng, Tangerang en Batavia, Java	IDN	106 37E	6 11S		van Ooststroom SJ	13012	L
stipulacea	wild	Java,Tjilintjing nr Dekali	IDN	124 04E	8 30S	350	van Steenis	562	BO
stipulacea	wild	Batavia (Jakarta), Java	IDN	106 49E	6 07S				K
stipulacea	wild	Dindigul, Pulney Hills, Madura Dist	IND	77 58E	10 21N		Anglade L	1150	K
stipulacea	wild	Poona(=Pune)	IND	73 52E	18 32N	700	Bell	6223	K
stipulacea	wild	Bonda N.W.P.	IND				Bell AS	84	K
stipulacea	wild	Biccavol, Godavery Dist., Madras	IND	82 00E	16 57N		Bourne	3196	K
stipulacea	wild	Tapeswaraw, Godavery Dist., Madras	IND				Bourne	3259	K
stipulacea	wild	Khundrum?, Central Prov.	IND				Duttie IJ	8271	K
stipulacea	wild	Bezwada, Kistna Dist., Madras	IND	77 20E	16 24N		Gamble JS	12566	K
stipulacea	wild	Bengal	IND	88 46E	23 19N		Hooker & Thomson	636	K
stipulacea	wild	Calcutta	IND	88 21E	22 34N		Kelpen	82	BR
stipulacea	wild	Chhuikhadan, Chhuikhadan State CP., Sambalpur, MP	IND	80 59E	21 31N		Mooney HF	2348	K
stipulacea	wild	Lalpura forest, Banswara Dist., Rajastan State	IND	74 26E	23 33N	700	Singh V	2979	K
stipulacea	wild	Maisor & Carnatic (tropical), Madras	IND				Thomson S	26	K
stipulacea	wild	M6, ICRISAT Farm, 30km NW of Hyderabad, AP	IND	78 16E	17 33N		van der Maesen LJG	2263	K
stipulacea	wild	ICRISAT Center, Patancheru, 25km NW of Hyderabad, AP	IND	78 21E	17 34N		van der Maesen LJG	4968	K
stipulacea	wild	Nuriger	IND				Wallich		BM
stipulacea	wild	Madras	IND	80 16E	13 04N		Wight	760	K
stipulacea	wild	Ruhuna Nat. Park,Komawa Wewa plot R17, Hambantota, S. Prov.	LKA	81 28E	6 22N	25	Cooray RG	70032512R	K
stipulacea	wild	culvert 39/7 on A12, Anuradhapura Dist.	LKA	80 24E	8 19N	80	RHMaxwell et al.	798	PDA
stipulacea	wild	Marunkan	LKA	80 01E	8 49N		Simpson ND	9350	BM
stipulacea	wild	Btw. Anuradhapura-Galkulama, Anuradhapura, N. Central Pr.	LKA	80 14E	8 12N		VE Rudd	3308	PDA
stipulacea	wild	Environjs de Majunga	MDG	46 19E	15 43S		Bathie P	17250	K

Species	Status	Locality	Cty	Longitude	Latitude	Alt.m	Collector	Coll. no	Herb.
stipulacea	wild		MDG				Bathie P	17618	K
stipulacea	wild	Meranke, Achterweg, Afd. Zuid, N. Guinea	PNG	141 24E	9 30S		Sijde	4007	L
stipulacea	wild	Meranke Dist., Plains west of Meranke	PNG	140 24E	8 30S	5	van Royen P	4935	L
stipulacea	wild		VNM	106 40E	10 45N		Pierre L		P
subramaniana	wild	Cuddafa hills	IND	78 49E	14 28N		Beddom RH	2224	BM
subramaniana	wild	Bellary district	IND	76 55E	15 09N	500	Gamble JS	17775	K
subramaniana	wild	Junnua bridge, Jaunsar, North West of Himalaya	IND	77 18E	31 10N		Gamble JS		K
subramaniana	wild	Bbillary Kerle, Bellary Dist., Madras	IND	76 55E	15 08N	330	Gamble JS		K
subramaniana	wild	Palaman Dist.	IND				Haines	4639	K
subramaniana	wild	Wild in Sal forest, Singhbum	IND				Haines HH	417	K
subramaniana	wild	Kodema forest, Hazari Gagb, Tamil Nadu	IND	85 21E	23 59N		Haines HH	4642	K
subramaniana	wild	between Palampur and Bajnath, W. Himalaya	IND	76 35E	32 04N	1150	Heybroek HM	890	L
subramaniana	wild	Dindigul, Kodaikanal Dist. Palamalai, St. Michael's Estate, behind bungalow, Palni Hills, Tamil Nadu(Madras)	IND	78 00E	11 32N	1250	Matthew KM	51762	K
subramaniana	wild	Jamuniapat, Khudia, Jashpur State	IND	78 49E	29 16N		Mooney HF	1885	K
subramaniana	wild		IND				photo Y Tateishi		TUS
subramaniana	wild	59km to Dalhousie, Punjab	IND	75 33E	32 13N		van der Maesen LJG	2879	K
subramaniana	wild	28km to Dalhousie, Himachal Pradesh	IND	76 07E	32 32N		van der Maesen LJG	2886	K
subramaniana	wild	8km to Mandi, Himachal Pradesh	IND	76 56E	31 42N		van der Maesen LJG	2938	K
subramaniana	wild	Sal forest grassed , Raipur, E. of Dehra Dun	IND	78 04E	30 18N		van der Maesen LJG	2977	K
subramaniana	wild	near Jajal, ca.23km to Tehri from Rishikesh, Uttar Pradesh	IND	78 22E	30 15N		van der Maesen LJG	2998	K
tenuicaulis	wild	Doi Tung, route 1149, 9-10km W of junction, Chiang Rai	THA	99 50E	20 19N	785	Tomooka N	96120503	K
trilobata	wild	Dindigul, Periakulam Dist., Kumbakkarai, Tamil Nadu	IND	77 33E	10 07N	350	Charles M	51368	K
trilobata	wild	Vaudalur, Chinglepet (Chenglpattu), Madras	IND	79 58E	12 41N		Cowne	11081	K
trilobata	wild	Sakesar, Shahpur, Punjab	IND	77 35E	29 22N		Drummond JR	14556	K
trilobata	wild	Godavari Dist., Madras	IND	82 14E	16 57N		Gamble JS	15748	K

Species	Status	Locality	Cty	Longitude	Latitude	Alt.m	Collector	Coll. no	Herb.
trilobata	wild	Concan, Madras	IND				Hooker	6073	K
trilobata	wild	Cape Comorin (Kannyakumari), Kannyakumari,Tamil Nadu	IND	77 33E	8 04N	30	Kramer KU, GB Nair	19250	L
trilobata	wild	Tirudhi, Thuraiyur Dist, Peramangalum	IND	78 36E	11 09N	150	Matthew KM et al.	15708	K
trilobata	wild	Kusumjhori, Dhenkanal State, Orissa	IND	85 35E	20 38N		Mooney HF	2734	K
trilobata	wild	1 km S of Sujangarh, Churu Dist. Rajasthan	IND	75 27E	26 36N	345	Roy GP	6374	K
trilobata	wild	Gangatic Plain, North Western India	IND				Staivak		K
trilobata	wild	near Dehli Ridge	IND	77 12E	28 38N		van der Maesen LJG	2860	K
trilobata	wild	70km N of Munnar, Palni hills, Tamil Nadu	IND	77 04E	10 27N	360	van der Maesen LJG	3454	K
trilobata	wild	Along Road Talaimannar to Mannar, mile stone 159, Vici.	LKA				AG Robyns	7340	PDA
trilobata	wild	Beach behind Yala campsite	LKA	81 31E	6 21N	2	Comanor PL	680	BM
trilobata	wild	Wilpattu Nat. Park, Kollankanatta Beach, plot W22B	LKA	79 51E	8 26N		Mueller-Dombois et al.	69042722	PDA
trilobata	wild	W. of Wilpattu Nat. Park, Pallgaturai, Puttalam, N. W. Prov.	LKA	79 51E	8 24N	1	Davidse G et al.	8210	K
trilobata	wild	Wilpattu Nat Park at Kail Villu plot W9	LKA	79 51E	8 24N	125	Fosberg FR et al.	50972	P
trilobata	wild	Beach, E. Butawa Modera (Ruhuna Yala Nat. Park), S. Prov.	LKA	81 30E	6 20N	2	Fosberg FR	50318	K
trilobata	wild	just north of Arugam Bay, Amparai Dist., Eastern Prov.	LKA	81 50E	6 53N	25	Fosberg FR,MH Scher	53042	K
trilobata	wild	Kari Villu Plot W9 (Wilpattu National Park), Puttalam Dist.	LKA	80 01E	8 24N	125	Fosberg FR et al.	50955	K
trilobata	wild	Nahasi Villa, Southern Ruhuna Nat. Park, Southern Prov.	LKA	81 28E	6 22N		G Davidse	7784	PDA
trilobata	wild	Yala Nat. Park, Uraniya Lagoon, Hambantota Dist., S. Prov.	LKA	81 28E	6 17N	25	Nowicke & Jayasuriya	391	K
trilobata	wild	Pottuvil, ad oram sabulosam oceani	LKA	81 50E	6 53N		L Bernardi	16011	PDA
trilobata	wild	Batticaloa dist., Mankeni (estate of Cadju corp.), Eastern Prov.	LKA	81 29E	8 01N		LH Cramer	4778	PDA
trilobata	wild	short-grass villu in plot, Wilpattu Nat. Park, at Kali'Villu	LKA	79 51E	8 24N		Muller-Dombois et al.	68091107	PDA
trilobata	wild	Opposite Bungalow, Wilpattu Nat. Park at Kali Villu	LKA	79 55E	8 29N		Muller-Dombois et al.	6905112R	PDA
trilobata	wild	In plot W9 on loose sand, Wilpattu Nat. Park, Kali Villu	LKA	79 55E	8 29N		N Wirawan et al.	1007	PDA
trilobata	wild	Roadside, Wilpattu Nat. Park, Mannar-Puttalam Road	LKA	79 54E	8 24N		RG Cooray	70020116R	PDA
trilobata	wild	Puttalam Lagoon Isthmus, ocean shore, track from Daluwa	LKA	79 42E	8 04N	1	RHMaxwell et al.	822	PDA

Species	Status	Locality	Cty	Longitude	Latitude	Alt.m	Collector	Coll. no	Herb.
trilobata	wild	Puttalam Lagoon Isthmus, exposed shore of ocean, End of track from Daluwa	LKA	79 42E	8 04N	1	RHMaxwell et al.	824	PDA
trilobata	wild	Pallagaturai, W. of Wilpattu W. Sanct., Puttalam, N.W. Prov.	LKA	79 51E	8 24N	1	Robyns AG	7336	K
trilobata	wild	Pallagaturai. West of Wilpattu west Sanctuary	LKA	79 50E	8 24N	25	Robyns AG	7337	BR
trilobata	wild	Talaimannar-Mannar, mile 159, near Pesarlai, Mannar, N. Prov.	LKA	79 49E	9 05N	25	Robyns AG	7340	BR
trilobata	wild	Btwn Jaffna and Elephant Pass, Jaffna Dist. N. Prov.	LKA	80 15E	9 38N	25	Rudd VE	3286	K
trilobata	wild	boundary Yala Nat. Park, Amaduwa area nr Brown's Beach Motel, Hambantota Dist	LKA	81 26E	6 17N		S H Sohmer et al.	8994	PDA
trilobata	wild	2-3 miles up road to Wilpattu Nat. Park from highway A12, near Maragahawara, roadside	LKA	80 07E	8 14N	80	T Koyama et al.	16005	PDA
trilobata	wild	Batticaloa, Batticaloa Dist., Eastern Prov.	LKA	81 42E	7 43N	25	Townsend CC	73/274	K
trilobata	wild	10m from shore in shelter of thicket (protected from goats), btwn Jaffna causeway and Kayts, Jaffna, N. Prov.	LKA	79 56E	9 39N		V E Rudd	3268	PDA
trilobata	wild	Near Kankesanturai, Jaffna District, Northern Prov.	LKA	80 02E	9 49N		V E Rudd	3280	PDA
trilobata	wild	N of Trincomalee, Alos Garden	LKA	81 13E	8 35N	1	Veldkamp JF	7810	L
trilobata	wild	Keeli-Kudah, Batticaloa Dist. Eastern Prov.	LKA	81 34E	7 43N	25	Waas S	2119	K
trilobata	wild	80 1E, Huludu Island, Maldives	MDV	73 14E	0 36S		Adams CD et al.	14943	K
trilobata	wild	Wadi Zabid	YEM	43 19E	14 12N		McKilligan SN		BM
trilobata	wild	Sautada				500	Rothe SP	6229	K
trinervia	wild		BUR				Heyne	5589	K
trinervia	wild	Moheli (=Nzwani)	COM	43 43E	12 19S		Schlieben HJ	11255	K
trinervia	wild		COM				Bosser J	18079	K
trinervia	wild	Talus depiste des camions a 300m au Sud de la Scierie de Nioumbaajou	COM	43 18E	11 46S	550	Floret JJ	641	K
trinervia	wild	Bandung, Java	IDN	107 36E	6 54S		Backer CA	34695	K
trinervia	wild	Batavia, Bogor, Java	IDN	106 47E	6 35S		Bakh van der Brinh	1087	K
trinervia	wild	Malino, S.W.Sulawesi	IDN	119 50E	5 15S		Bunnemeyer	10868	L

Species	Status	Locality	Cty	Longitude	Latitude	Alt.m	Collector	Coll.no	Herb.
trinervia	wild	Saumlaki to Vlilj, Tanimber Isl. (Timor Laut) P. Jamdena, Moluccas	IDN	131 18E	7 58S		Buwalda P	4020	L
trinervia	wild	Buibenzorg, Kotta Basu, Java	IDN	106 47E	6 35S		Dihm	243	L
trinervia	wild	Sumatra	IDN				Forbes HO	2007	BM
trinervia	wild	Timor	IDN	125 06E	9 13S		Forbes HO	3951	BM
trinervia	wild	Wegbern, Omgeving Watampone, Sulawesi	IDN	120 20E	4 32S		Froideville	292	L
trinervia	wild	Buitenzorg (=Bogor), Bototolis (=Batutulis), Java	IDN	106 46E	6 37S		Hallier		K
trinervia	wild	Wamena, Garten des Cottage Hotels, Zentralgebirge, Baliem-Tal	IDN	138 44E	3 55S		Hiepko P et al.	631	K
trinervia	wild	Soeka-radja, Bogor, Java	IDN	108 36E	6 56S		Hochreutiner	1698	L
trinervia	wild	Java	IDN				Horsfield T	113	K
trinervia	wild	Ranre Darongan, Helling Smerve, Java	IDN				Kleinhoonk	346	L
trinervia	wild	Atambua, Nourd Belu, Timor	IDN	124 53E	9 06S		Kooy CW	788	L
trinervia	wild	Central Java	IDN	110 04E	7 28S		Lorzing JA	205	L
trinervia	wild	Cibodas, near the garden, Java	IDN	106 48E	6 45S		Nitta A	15044	L
trinervia	wild	Kota botoe, bij Biuteurarg, Java	IDN				Raap H	94	L
trinervia	wild	Tanahsareal (Jakarta Raya), Java	IDN	106 48E	6 09S		Raap H	469	L
trinervia	wild	Regio calida, Prov. Batavia, Ad ripas, Java	IDN	106 48E	6 10S	250	Schiffner V	2071	L
trinervia	wild	Depak (Depok), Bogor, Java	IDN	106 49E	6 23S		Soegandiredja	252	L
trinervia	wild	Batavia (Jakarta), Bnisenrong, Java	IDN	106 48E	6 10S		van Brink B	1007	L
trinervia	wild	Tjibodas (G.Gede), Java	IDN			1350	van Ooststroom SJ	13083	L
trinervia	wild	Tjibodas (G.Gede), Java	IDN				van Ooststroom SJ	13133	L
trinervia	wild	Tjibodas (G.Gede), Java	IDN				van Ooststroom SJ	13679	L
trinervia	wild	Mgeing Tjibedes (Gade), Java	IDN	106 43E	6 50S		van Ooststroom SJ	13934	L
trinervia	wild	Preanger, Tjidadap (Cidadap), by Tjibeber (Cibeber), Java	IDN	107 08E	7 01S		Winckel WT	977	L
trinervia	wild	Bandoeng, Bruakland, Java	IDN	107 36E	6 54S			264	L
trinervia	wild	Java	IDN						L

Species	Status	Locality	Cty	Longitude	Latitude	Alt.m	Collector	Coll. no	Herb.
trinervia	wild	Pulney Hills	IND	77 35E	10 19N		Beddom RH	2228	BM
trinervia	wild	Vilpatti, Pulneys, Madras	IND	77 58E	10 05N		Bourne	1088	K
trinervia	wild	Sembaganur, Pulneys, Madras	IND	77 29E	10 14N		Bourne	7006	K
trinervia	wild	Courtallam, Madras	IND	77 17E	8 55N		Bourne		K
trinervia	wild	Bombay	IND	72 49E	18 59N		Dalzell		K
trinervia	wild	Trovancine, Madras	IND				Gamble JS	14816	K
trinervia	wild	Nilgiris, Madras	IND			1000	Gamble JS	15669	K
trinervia	wild	13km E of Sultan's Battery, Calicut (Dist.), Kerala	IND	75 54E	11 18N		van der Maesen LJG	2665	K
trinervia	wild	Mile post 20 Passeluwa, Kandy, Central Prov.	LKA	80 38E	7 06N	1500	Jayasuriya M et al.	997	K
trinervia	wild	Collines au sud-ouest de Kandy	LKA	80 36E	7 16N		Poli		P
trinervia	wild	McDonald's Valley, Hakagara, Nuwara Eriya Dist.	LKA	80 47E	6 58N	1000	Rudd V, Balakrishnam	3171	K
trinervia	wild	btwn Bandarawela-Haputale, marker 9/9, Badura, Uva Prov.	LKA	80 58E	6 48N	1500	Rudd V, Balakrishnam	3200	K
trinervia	wild		LKA				Thwaites	1476	P
trinervia	wild	Anjuan Johanna	MDG				Blackburn J		K
trinervia	wild	Nanisana, Antananarivo	MDG	47 31E	18 55S		Jacqeline, M Peltier	3813	K
trinervia	wild	Nosi-be(=Nosy-be)	MDG	48 15E	13 20S		Hildebrandt JM	3401	BM
trinervia	wild	Gunong Beremban, Kamp. Balukar, Cameron Highland, Pahang	MYS	101 27E	4 28N		Lewis GP	223	K
trinervia	wild	Waggan Kinnalun	PHL			1000	Conklin HC et al.	2549	K
trinervia	wild	Motupore Island, Bootless Inlet, Central Prov.	PNG	147 17E	9 31S	35	Frodin DG, G Leach	5258	L
trinervia	wild	Mlingano Sisal Research Station, Ngomeni	TZA	38 54E	5 09S		Osborne JF	L27	K
trinervia bourneae	wild	Poombarai riverside, Pulneys, Madras	IND	77 25E	10 12N		Bourne	1291	K
trinervia bourneae	wild	Kodai Kanal, Ghat, Pulneys, Madras	IND	77 25E	10 12N		Bourne	2005	K
trinervia bourneae	wild	12km from Kodai Kanal, Tamil Nadu	IND	77 25E	10 12N	1530	van der Maesen LJG	3574	K
trinervia bourneae	wild	Pulney Mountains, Madras	IND	77 25E	10 12N		Wight	793	K
umbellata	wild		HKG	114 10E	22 17N		Shiu Ying Hu	10375	K
umbellata	cult.	Amanatun, Timor	IDN	124 59E	9 16S	750	Kooy CW	3	L

Species	Status	Locality	Cty	Longitude	Latitude	Alt.m	Collector	Coll. no	Herb.
umbellata	wild	Utakwa River to Mt. Carstensz, Irian Jaya	IDN				Kloss CB		BM
umbellata	cult.	Kasyapagma	LKA				M Jayasuriya et al.		PDA
umbellata	cult.	Lipshe	NEP	83 26E	28 35N	1850	M Mikage et al.	9682402	TUS
umbellata	wild	Pok chong, Pakchong	THA	101 25E	14 42N	300	Marcan A	1563	K
umbellata	cult.	Pingtung Co., Santimen	TWN			150	H Ohashi et al.	20150	TUS
umbellata	cult.	Shueli, Nantou Co.	TWN			200	SF Huang	1935	TUS
umbellata	cult.	Hsinwulu, Taitung Co.	TWN			250	SF Huang	3210	TUS
umbellata	cult.	Taitung Co., Tongho	TWN	121 17E	22 58N		Y Tateishi et al.	25350	TUS
umbellata	wild	Noh Pong Shan, Taam-chau Dist. Hainan	CHN	109 21E	19 45N		Tsang, Wai-Tak	16370	K
umbellata	wild	Meyz, Yunnan	CHN	102 37E	24 59N	1500		11227	K
umbellata	wild	Utakwa River to Mt. Carstensz, Irian Jaya	IDN				Kloss CB		BM
umbellata	wild	Buton (Butung) Is., Baubau, Oha, Wakunti forest, Sulawesi	IDN	122 35E	5 28S	200	Widjaja EA	548	L
umbellata	wild		IND	89 52E	24 34N		Shibpur Exp Farm		L
umbellata	wild		IND						K
umbellata	wild	Layang Layang Jahor, Johore	MYS	103 28E	1 48N		Rosenquist EA		L
umbellata	wild	Bangkok	THA	100 31E	13 45N	25	Kerr AFG	9355	K
umbellata	wild	Huay Ban Kau, Kanchanaburi	THA	98 35E	14 55N	750	van Beusekom et al.	3627	K

REFERENCES

Ahn, C.S. and R.W. Hartman. 1978. Interspecific hybridization among four species of the genus *Vigna* Savi. Pages 240-246 *in* AVRDC ed. Proc. 1st Intl. Mungbean Symp. AVRDC, Taiwan.

Arora, R.K. and E.R. Nayar. 1984. Wild relatives of crop plants in India. NBPGR Sci. Monograph no. 7.

Arumuganathan, K. and E. D. Earle. 1991. Nuclear DNA content of some important plant species. Plant Mol. Biol. Reporter 9:208-218.

AVRDC (Asian Vegetables Research and Development Center). 1988. Mungbean: Proceedings of the Second International Symposium. AVRDC, Shanhua, Tainan, Taiwan.

Babu, C.R., S.K. Sharma and B.M. Johri. 1985. Leguminosae-Papilionoideae: Tribe – *Phaseoleae*. Bull. Bot. Surv. India 27(1-4):1-28.

Baker, J.G. 1876. Leguminosae. Pages 200-207 *in* J.D. Hooker (ed.) The flora of British India 2, (*Phaseolus* and *Vigna* 200-207), Kent Baker.

Baldwin, B.G., D.J. Crawford, J. Francisco-Ortega, S.-C. Kim, T. Sang and T.F. Stuessy. 1998. Molecular phylogenetic insights on the origin and evolution of oceanic island plants. Pages 410-441 *in* P. S. Soltis, D. E. Soltis and J. J. Doyle (eds.), Molecular systematics of plants, II. DNA sequencing. Kluwer Academic Publishers, New York, USA.

Baudet, J.C. 1974. Signification taxonomique des caractères blastogèniques dans la tribudes Papilionaceae-*Phaseoleae*. Bull. Jard. Bot. Nat. Belg. 44:259-293.

Baudoin J.P. and R. Maréchal. 1988. Taxonomy and Evolution of the Genus *Vigna*. Pages 2-12 *in* S. Shanmugasundaram. and B.T. McLean (Eds.) Mungbean: Proceedings of the Second International Symposium. AVRDC, Shanhua, Tainan, Taiwan.

Beebe, S., O. Toro Ch., A.V. Gonzalez, M.I. Chacon, D.G.Debouck, 1997. Wild-weed-crop complexes of common bean (*Phaseolus vulgaris* L., Fabaceae) in the Andes of Peru and Colombia, and their implications for conservation and breeding. Genet. Res. Crop Evol. 44:73-91.

Bentham, G. 1837. Commentationes de leguminosarum generibus. 78 pp. Wien.

Bentham, G. 1865. Leguminosae. *In* G. Bentham and J.D. Hooker, Genera plantarum 1: 434-600.

Birren, B., E.D.Green, S. Klapholz, R.M. Myers and J. Roskams (eds.). 1997. Genome analysis: A laboratory manual. Vol. 1. Analyzing DNA. Cold Spring Harbor Laboratory Press.

Blackhurst, H.T. and J. Creighton. 1980. Cowpea. Pages 327-337 *in* Hybridization of Crop Plants. American-Society of Agronomy Madison, WI, USA.

Borlaug, N. 1973. Building a protein revolution on grain legumes. Pages 7-11 *in* Max Milner (ed.) Nutritional improvement of food legumes by breeding. Protein Advisory Group of the United Nations.

Boutin, S.R., N.D.Young, T.C.Olson, Z.-H. Yu, R.C.Shoemaker and C.E.Vallejos. 1995. Genome conservation among three legume genera detected with DNA markers. Genome 38:928-937.

Brown, A.H.D. 1995. The core collection at the crossroads. Pages 3-19 *in* Core Collections of Plant Genetic Resources T. Hodgkin, A.H.D. Brown, Th.J.L. van Hintum and E.A.V. Morales (eds.). Wiley, Chichester.

Brown, A.H.D., J.P. Grace and S.S. Speer. 1987. Designation of a 'core' collection of perennial *Glycine*. Soybean Genetics Newsletter 14:59-70.

Bujang, I., Y. Egawa, N. Tomooka, S.G. Tan, H.A. Abu Bakar and S. Anthonysamy. 1994. Exploration, collection and electrophoretic variants of wild *Vigna* (subgenus *Ceratotropis*) in Peninsular Malaysia. Page 69-71 *in* M.K. Vidyadaran and S.C. Quah (Eds.) Proc. 3rd Symp. Appl. Biol., Malaysia.

Bunting, A.H. 1960. Some reflection on the ecology of weeds. Pages 11-26 *in* The biology of weeds, J.L.Harper (ed.). Blackwell Scientific Publishers, Oxford.

Chaitieng, B., A. Kaga, O.K. Han, X.W. Wang, S. Wongkaew, P. Laosuwan, N. Tomooka and D.A. Vaughan. 2002. Mapping a new source of resistance to powdery mildew (*Erisiphe polygoni* DC.) in mungbean [*Vigna radiata* (L.) Wilczek]. Plant Breeding (accepted)

Chandel, K.P.S., R.N. Lester and R.J. Starling. 1984. The wild ancestors of urd and mung beans (*Vigna mungo*(L.) Hepper and *V. radiata* (L.) Wilczek. Bot. J. Linn. Soc. 89:85-96.

Chang, T.T. 1983. The origins and early cultures of the cereal grains and food legumes. Pages 65-94 *in* The Origins of Chinese Civilization. D.N. Keightley (ed.) Univ. of California Press, Berkeley.

Chappill, J.A. 1995. Cladistic analysis of the Leguminosae: development of an explicit hypothesis. Pages 1-9 *in* Advances in Legume Systematics. Part 7. Phylogeny. M.D. Crisp and J.J. Doyle (eds.). Royal Botanic Gardens, Kew. U.K.

Chen, H.K., M.C. Mok, S. Shanmugasundaram, and D.W.S. Mok. 1989. Interspecific hybridization between *Vigna radiata*(L.) Wilczek and *V. glabrescens*. Theor. Appl. Genet. 78:641-647.

Chen, N.C., L.R. Baker and S. Honma. 1983. Interspecific crossability among four species of *Vigna* food legumes. Euphytica 32:925-937.

Convention on Biological Diversity. 1992. Convention on Biological Diversity:Text and Annexes. pp. 1-34. Secretariat of the Convention on Biological Diversity, Montreal.

Credland, P.F. 1990. Biotype variation and host change in bruchids: Cause and effects in the evolution of bruchid pests. Pages 271-287 *in* K. Fujii., A.M.R. Gatehouse., C.D. Johnson., R. Mitchel and T. Yoshida (eds.), Bruchids and Legumes: Economics, Ecology and Coevolution. Kluwer Academic Publishers, Dordrecht, Netherlands.

Dana, S. and P.G. Karmakar. 1990. Species relationships in *Vigna* subgenus *Ceratotropis* and its implications in breeding. Plant Breeding Reviews 8:19-42.

Date, R.A. 1995. Collecting *Rhizobium*, *Frankia* and mycorrhizal fungi. Pages 551-560 *in* L. Guarino, V. Ramanatha Rao and R. Reid (eds.) Collecting Plant Genetic Diversity: Technical Guidelines. CAB International, Wallingford, U.K.

Date, R.A. and J. Halliday. 1987. Collecting, isolation, cultivation and maintenance of rhizobia. Pages 1-27 *in* G.H. Elkan (ed.) Symbiotic Nitrogen Fixation Technology. Marcel Dekker, New York.

Debouck, D. 2000. Biodiversity, ecology and genetic resources of *Phaseolus* beans – Seven answered and unanswered questions. Pages 95-123 *in* 7th MAFF International Workshop on Genetic Resources. Part 1. Wild legumes. NIAR, Tsukuba, Japan.

De Candolle, A.P. 1825. Leguminosae, Pages 93-524 *in* Prodromus Systematis Naturalis Regni Vegetabilis 2, (*Phaseolus* 390-396), Paris.

Dela Vina, A.C. and N. Tomooka. 1994. Genetic diversity in mungbean [*Vigna radiata*(L.) Wilczek] based on two enzyme systems. Philipp. J. Crop Sci. 19:1-9.

De Wet, J.M.J. and J.R. Harlan. 1975. Weeds and domesticates: Evolution in man-made habitat. Econ. Bot.

29:99-107.

Doebley, J. and A. Stec. 1993. Inheritance of the morphological differences between maize and teosinte: comparison of results for two F_2 populations. Genetics 134:559-570.

Doi, K., A. Kaga, N. Tomooka and D.A. Vaughan. 2002. Molecular phylogeny of genus *Vigna* subgenus *Ceratotropis* based on rDNA ITS and *atpB-rbcL* intergenic spacer region of cpDNA sequences. Genetica 114:129-145.

Doyle, J.J. 1995. DNA data and legume phylogeny: progress report. Pages 11-30 *in* Crisp, M. and J.J. Doyle (eds.) Advances in Legume Systematics. Part 7. Phylogeny. Royal Botanic Gardens, Kew. U.K.

Duke, J.A. 1981. Handbook of Legumes of World Economic Importance. Plenum Press, New York.

Egawa, Y. 1988. Phylogenetic differentiation between three Asian *Vigna* species, *V. radiata, V. mungo* and *V. umbellata*. Bull. Natl. Inst. Agrobiol. Resour. 4:189-200.

Egawa, Y., M. Nakagahra and G.C.J. Fernandez. 1988. Cytogenetical analysis of tetraploid *Vigna glabrescens* by interspecific hybridization involving diploid Asian *Vigna* species. Pages 200-204 *in* Mungbean. Proceedings of the second international symposium. AVRDC, Shanhua, Taiwan.

Egawa, Y., M. Nakagawara and G.C.J. Fernandez. 1990a. Cross compatibility and cytogeneical relationships among Asian *Vigna* species. Pages 201-208 in K. Fujii, A.M.R. Gatehouse, C.D. Johnson., R. Mitchel and T. Yoshida (eds.), Bruchids and Legumes: Economics, Ecology and Coevolution. Kluwer Academic Publishers.

Egawa, Y., D. Siriwardhane, K. Yagasaki, H. Hayashi, M. Takamatsu, M. Saito, Y. Nomura, Toshiya Okabe, F. Idezawa, S. Miyazaki. 1990b. Collection of millets and grain legumes in Shimoina district of Nagano Prefecture, 1989. Annual Report on exploration and introduction of plant genetic resources (NIAR, Tsukuba, Japan). 6:1-22.

Egawa, Y. and N. Tomooka. 1994. Phylogenetic differentiation of *Vigna* species in Asia. Pages 112-120 *in* Plant Genetic Resources Management in the Tropics. JIRCAS International Symposium Series no 2. JIRCAS, MAFF, Tsukuba, Japan.

Egawa, Y., I.B. Bujang, S. Chotechuen, N. Tomooka and Y. Tateishi. 1996. Phylogenetic differentiation of tetraploid *Vigna* species, *V. glabrescens* and *V. reflexo-pilosa*. JIRCAS Journal, 3:49-58.

Epino, P.B. and B. Morallo-Rejesus. 1982. Mechanism of resistance in mungbean [*Vigna radiata* (L.) Wilczek] to *Callosobruchus chinensis* (L.). Philipp. Ent. 5:447-462.

FAO UN 1998. The State of Food and Agriculture. 371 pages. Rome, Italy.

Fatokun, C.A., D.I. Menacio-Hautea, D. Danesh and N.D. Young. 1992. Evidence of orthologous seed weight genes in cowpea and mung bean based on RFLP mapping. Genetics 132:841-846.

Fatokun, C.A., D. Danesh, N.D. Young and E.L. Stewart. 1993. Molecular taxonomic relationships in the genus *Vigna* based on RFLP analysis. Theor. Appl. Genet. 86:97-104.

Fernandez, G.C.J. and S. Shanmugasundaram. 1988. The AVRDC mungbean improvement program: Past, present and future. Pages 58-70 *in* Mungbean, Proceedings of 2nd International Symposium. AVRDC, Shanhua, Taiwan.

Fery, R.L., 1980. Genetic of *Vigna*. Hort. Rev. 2:311-394.

Fujii, K. and S. Miyazaki. 1987. Infestation resistance of wild legumes (*Vigna sublobata*) to azuki bean weevil, *Callosobruchus chinensis* (L.) (Coleoptera: Bruchidae) and its relationship with cytogenetic classification. Appl. Ent. Zool. 22:319-322.

Fujii, K., M. Ishimoto and K. Kitamura. 1989. Patterns of resistance to bean weevils (Bruchidae) in *Vigna radiata-mungo-sublobata* complex inform the breeding of new resistant varieties. Appl. Ent. Zool. 24:126-132.

Guarino, L., V. Ramanatha Rao and R. Reid (eds.). 1995. Collecting Plant Genetic Diversity. Technical Guidelines. CAB International, Cambridge, U.K.

Guarino, L., A. Jarvis, R.J. Hijmans and N. Maxted. 2002. Geographic information systems (GIS) and the conservation and use of plant genetic resources. Pages 387- 404 *in* J.M.M. Engels, V. Ramanatha Rao, A.H.D. Brown and M.T. Jackson (eds.). Managing Plant Genetic Diversity. IPGRI.

Hanelt, P. (ed.) 2001. Mansfeld's Encyclopedia of Agricultural and Horticultural Crops. Springer, Berlin.

Harlan, J.R. 1965. The possible role of weed races in the evolution of cultivated plants. Euphytica 14:173-176.

Harlan, J.R. and J.M.J. De Wet. 1965. Some thoughts about weeds. Econ. Bot. 19:16-24.

Harlan, J.R. 1967. A wild wheat harvest in Turkey. Archaeology 20:197-201.

Harlan, J.R. 1992. Crops and Man. Second Edition. American Society of Agronomy, Madison, Wisconsin, USA. 284 pages.

Harlan, J.R. and J.M.J. De Wet. 1971. Towards a rational classification of cultivated plants. Taxon 20:509-517.

Herklots, G.A.C. 1972. Vegetables in South-East Asia. London: George Allen and Unwin.

Heywood, V.H. (ed.) 1978. Flowering plants of the world. Oxford University Press, U.K. 335 pages.

Heywood, V.H. and S.R.Chant. 1982. Popular encyclopedia of plants. Cambridge University Press, U.K. 367 pages.

Imamura, K., 1996. Jomon and Yayoi: the transition to agriculture in Japanese prehistory. Pages 442-464 *in* D.R. Harris (ed.) The origin and spread of agriculture and pastoralism in Eurasia. University College, London Press.

Innan, H., R. Terauchi, G. Kahl and F. Tajima. 1999. A method for estimating nucleotide diversity from AFLP data. Genetics 151:1157-1164.

IBPGR (International Board for Plant Genetic Resources regional committee for Southeast Asia). 1980. Descriptors for Mung Bean. IBPGR, Rome.

Jaaska, V. and V. Jaaska. 1990. Isozyme variation in Asian beans. Bot. Acta 103:281-290.

Jain, H.K. and K.L. Mehra. 1980. Evolution, adaptation, relationships and uses of the species of *Vigna* cultivated in India. Pages 459-468 *in* Summerfield, R.J. and A.H. Bunting (eds.) Advances in legume science. Volume 1 of the Proceedings of the International Legume Conference, Kew, U.K.

Jaiwal, P.K., R. Kumari, S. Ignacimuthu, I. Potrykus, C. Sautter. 2001. *Agrobacterium tumefaciens* - mediated genetic transformation of mungbean (*Vigna radiata* L. Wilczek) - a recalcitrant grain legume. Plant Science 161:239-247.

Jepson, P., J.K. Jarvie, K. MacKinnin and K.A. Monk. 2001. The end of Indonesia's lowland forest? Science 292:859-861.

Kaga, A. 1996. Construction and application of linkage maps for azuki bean (*Vigna angularis*). PhD. Thesis Kobe University. 210 pages.

Kaga, A., M. Ohnishi, T. Ishii and O. Kamijima. 1996a. A genetic linkage map of azuki bean constructed with molecular and morphological markers using an interspecific population (*Vigna angularis* x *V. nakashimae*). Theor. Appl. Genet. 93:658-663.

Kaga, A., N. Tomooka, Y. Egawa, K. Hosaka and O. Kamijima.1996b. Species relationships in the subgenus

Ceratotropis (genus *Vigna*) as revealed by RAPD analysis. Euphytica 88:17-24.

Kaga, A. and M. Ishimoto. 1998. Genetic localization of a bruchid resistance gene and its relationship to insecticidal cyclopeptide alkaloids, the vignatic acids; in mungbean (*Vigna radiata* L. Wilczek). Mol. Gen. Genet. 258:378-384.

Kaga, A., T. Ishii, K. Tsukimoto, E. Tokoro and O. Kamijima. 2000a. Comparative molecular mapping in *Ceratotropis* species using an interspecific cross between azuki bean (*Vigna angularis*) and rice bean (*V. umbellata*). Theor. Appl. Genet. 100:207-213.

Kaga, A., M.S. Yoon, N. Tomooka and D.A. Vaughan 2000b. Collaborative research on the *Vigna* species in East Asia. 1. Collecting mission on the islands of southern Okinawa prefecture, Japan. 21st - 29th October 1999. Report to East Asia Genetic Resources Coordinators Meeting and IPGRI pages 2-25.

Kajiwara, H. and N. Tomooka. 1998. Comparative analysis of genus *Vigna* seeds using antiserum against a synthesized multiple antigenic peptide. Electrophoresis 19:3110-3113.

Kashiwaba, K., N. Tomooka, A. Kaga, O.K.Han, D.A.Vaughan. 2002. Characterization of resistance to three bruchid species (*Callosobruchus* spp., Coleoptera, Bruchidae) in cultivated rice bean, [*Vigna umbellata* (Thunb.) Ohwi & Ohashi]. J. Econ. Ento. (submitted)

Kobayashi, N., R. Ikeda, and D.A. Vaughan. 1993. Resistance to rice tungro viruses in wild species of rice (*Oryza* spp.). Jpn. J. Breed. 43:247-255.

Kobayashi, T., N. Shimada, N.Q. Thang and L.T. Tung. 1994. Exploration and collection of grain legume germplasm in Vietnam. Annual report on exploration and introduction of plant genetic resources (NIAR, Tsukuba, Japan). 10:141-169.

Koinange, E.M.K., S.P. Singh and P. Gepts. 1996. Genetic control of the domestication syndrome in common-bean. Crop Sci. 36:1037-1045.

Konarev, A.V., N. Tomooka and D.A. Vaughan. 2002. Proteinase inhibitor polymorphism in the genus *Vigna* subgenus *Ceratotropis* and its biosystematic implications. Euphytica 123:165-177.

Lackey, J.A. 1981. Tribe 10. PHASEOLEAE DC. (1825). Pages 301-327 *in* R.M. Polhill and P.H. Raven (eds.) Advances in Legume Systematics Part 1. Royal Botanic Gardens, Kew, U.K.

Ladizinsky, G. 1975. Collection of wild cereals in the upper Jordan Valley. Econ. Bot. 29:264-267.

Ladizinsky, G. 1987. Pulse domestication before cultivation. Econ. Bot. 41:60-65.

Lambrides, C.J., R.J. Lawn, I.D. Godwin, J. Manners and B.C. Imrie. 2000. Two genetic linkage maps of mungbean using RFLP and RAPD markers. Aust. J. Agric. Res. 51(4):415-425.

Lawn, R.J. 1995. The Asiatic *Vigna* species. Pages 321-326 *in* J. Smartt and N.W. Simmonds (eds.), The Evolution of Crop Plants, pp. 321-326. 2nd edition. Longman, Harlow, U.K.

Lawn, R.J. and A. Cottrell. 1988. Wild mungbean and its relatives in Australia. Biologist 35:267-273.

Li, C.-D., C.A. Fatokun, B. Ubi, B.B. Singh and G.J. Scoles. 2001. Determining genetic similarities and relationships among cowpea breeding lines and cultivars by microsatellite markers. Crop Sci. 41:189-197.

Lock, J.M. and J. Heald. 1994. Legumes of Indo-China: A check list. Royal Botanic Gardens, Kew. 164 pages.

Lumpkin, T.A. and D.C. McClary. 1994. Azuki bean : Botany, production and uses. CAB International, Wallingford, U.K. 268 pages.

Maekawa, F. 1955. Topo-morphological and taxonomical studies in *Phaseoliae*, Leguminosae. Jap. J. Bot. 15: 103-116.

Maekawa, F. and T. Shidei. 1974. Geographical background to Japan's flora and vegetation. Pages 1-31 in M. Numata (ed.) The Flora and Vegetation of Japan. Kodansha Ltd., Tokyo.

Maréchal, R., J.M. Mascherpa and F. Stainier. 1978. Etude taxonomique d'un groupe complexe d'espécies des genres *Phaseolus* et *Vigna* (Papilionaceae) sur la base de données morphologiques et polliniques, traitées par l'anayse informatique. Boissiera 28:1-273.

Maréchal, R., J.M. Mascherpa and F. Stainier. 1981. Taxonometric study of the *Phaseolus-Vigna* complex and related genera. Pages 329-335 *in* Advances in Legume Systematics. R.M. Polhill and P.H. Raven (eds.) Royal Botanic Gardens, Kew. U.K.

Maughan, P.J., M.A. Saghai Maroof and G.R. Buss. 1996. Molecular-marker analysis of seed weight: genome locations, gene action, and evidence for orthologous evolution among three legume species. Theor. Appl. Genet. 93:574-579.

Maxted, N. 1995. An ecogeographic study of *Vicia* subgenus *Vicia*. Systematic and Ecogeographic Studies on Crop Genepools. 8. International Plant Genetic Resources Institute, Rome, Italy.

Maxted, N., B.V. Ford–Lloyd and J.G. Hawkes (eds.) 1997. Plant Genetic Conservation: The *in situ* approach. Chapman and Hall. 444 pages.

Maxted, N., L. Guarino, L. Myer and E.A. Chiwona. 2002. Towards a methodology for on-farm conservation of plant genetic resources. Genet. Res. Crop Evol. 49:31-46.

Maxwell, R.H. 1991. Tribe *Phaseoleae*. Pages 236-381 *in* M.D. Dassanayake and F.R. Fosberg (eds.) A Revised Handbook to the Flora of Ceylon. A.A. Balkema, Rotterdam.

Meadows, R.H. 1996. The origin and spread of agriculture and pastoralism in northwestern South Asia. Pages 390-412 *in* D.R. Harris (ed.) The origins and spread of agriculture and pastoralism in Eurasia. UCL Press London.

Menacio-Hautea, D., L. Kumar, D. Danesh and N.D.Young. 1992. A genome map for mungbean [*Vigna radiata* (L.) Wilczek] based on DNA genetic markers (2n=2x=22). Pages 6.259-6.261 *in* J.S. O'Brien (ed.) Genetic Maps 1992. A compilation of linkage and restriction maps of genetically studied organisms. Cold Spring Harbor, New York.

Menacio-Hautea, D., C.A. Fatokun, L. Kumar, D. Danesh and N.D. Young. 1993. Comparative genome analysis of mungbean (*Vigna radiata* L. Wilczek) and cowpea (*V. unguiculata* L. Walpers) using RFLP mapping data. Theor. Appl. Genet. 86:797-810.

Miller, A.G. and J.A. Nyberg. 1995. Collecting herbarium specimens. Pages 561-573 *in* L. Guarino, V. Ramanatha Rao and R. Reid (eds.) Collecting Plant Genetic Diversity. Technical Guidelines. CAB International, Wallingford, U.K.

Mimura, M., K. Yasuda and H. Yamaguchi. 2000. RAPD variation in wild, weedy and cultivated azuki beans in Asia. Genet. Res. Crop Evol. 47:603-610.

Miyazaki, S. 1982. Classification and phylogenetic relationships of the *Vigna radiata-mungo-sublobata* complex. Bull. Natl. Inst. Agr. Sci. Ser. D 33:1-61. (in Japanese)

Morell, V. 2002. China's Hengduan Mountains. National Geographic. April. 98.-112.

Murata, K., S. Shirai, M. Hara, K. Chiba, O. Adachi, N. Shimada, M. Fujita, S. Iida and Y. Shinada. 1995. A report on the collection of genetic resources in azuki bean and the evaluation of their characteristics, including field surveys in Nepal and Bhutan. Hokkaido Agricultural Experiment Station, Memuro, Hokkaido. pp.1-22. (in Japanese)

Murray, M.G., J.D. Palmer, R.E. Cuellar, W.F. Thompson. 1979. Deoxyribonucleic acid sequence organization in the mungbean genome. Biochemistry 18:5259-5264.

Myers, N., R.A. Mittermeier, C.G. Mittermeier, G.A.B. da Fonseca and J. Kent. 2000. Biodiversity hotspots for conservation priorities. Nature 403:853-858.

Nagamine, T. and H. Takeya. 1999. The descriptors for evaluation in plant genetic resources. Vol. 1. NIAR, Ministry of Agriculture, Forestry and Fisheries, Japan.

Neupane, R.K. 1999. Wild relatives of grain legume crops in Nepal. Pages 102-108 *in* National Conference on wild relatives of cultivated plants in Nepal. The Green Energy Mission, Nepal.

Ohwi, J. 1953. (English edition published in 1965) Flora of Japan. Meyer, F.G. and E.H. Walker (eds.) Smithsonian Inst., Washington DC, USA.

Ohwi, J. and H. Ohashi. 1969. Azuki beans of Asia. Jap. J. Bot. 44:29-31.

Oka, H. I. and H. Morishima, 1971. The dynamics of plant domestication: Cultivation experiments with *Oryza perennis* and its hybrid with *O. sativa*. Evol. 25:356-364.

Orf, J. and T. Hymowitz. 1979. Inheritance of the absence of the Kunitz trypsin inhibitor in seed protein of soybeans. Crop Sci. 19:107-109.

Piper, C.V. 1926. Studies in American *Phaseoleae,* Contr. U. S. Nat. Herb. 22, 663-701.

Piper, C.V. and W.J. Morse. 1914. Five oriental species of beans, Bull. U.S.D.A. 119, 1-32.

Potokina, E., D.A. Vaughan, E.E. Eggi and N. Tomooka. 2000. Population diversity of the *Vicia sativa* agg. (*Fabaceae*) in the flora of the former USSR deduced from RAPD and seed protein analyses. Genet. Res. Crop Evol. 47:171-183.

Purseglove. J.W. 1974. Tropical crops. Longman, London, UK. 719 pages.

Quattrocchi, U., 2000. CRC World Dictionary of Plant Names: Common names, scientific names, eponyms, synonyms and etymology. Vol 4. CRC press, Boca Raton.

Rachie, K.O. and L.M. Roberts. 1974. Grain legumes in the lowland tropics. Adv. Agron. 26:1-132.

Reddy, K. S., S.E. Pawar and C.R. Bhatia. 1987. Screening for powdery mildew (*Erysiphae polygoni* DC.) resistance in mungbean (*Vigna radiata* (L.) Wilczek) using excised leaves. Proc. Indian Acad. Sci. (Plant Sci.) 97:365-369.

Roxburgh, W. 1874. *Phaseolus*. Pages 287-300 in Carey (ed.) vol. 3 of Flora Indica.

Ryan, C.A. 1990. Protease inhibitors in plants: genes for improving defenses against insects and pathogens. Ann. Rev. Phytopath. 28:425-449.

Sacks, F.M. (1977) A literature review of *Phaseolus angularis* – the adzuki bean. Econ. Bot. 31:9-15.

Schulze-Lefert, P. and J. Vogel. 2000. Closing the ranks to attack by powdery mildew. Trends in Plant Science 5:343-348.

Sehgal, V.K. and R. Ujagir. 1988. Insect pests and pest management of mungbean in India. Pages 315-328 *in* Asian Vegetables Research and Development Center. Mungbean: Proceedings of the Second International Symposium, AVRDC, Shanhua, Tainan.

Singh, B.B., S.R. Singh and O. Adjadi. 1985. Bruchid resistance in cowpea. Crop Sci. 25:736-739.

Singh, H.B., B.S.Joshi, K.P.S.Chandel, K.C. Pant and R.K.Saxena. 1974. Genetic diversity in some Asiatic *Phaseolus* species and its conservation. Indian J. Genet. 34A:52-57.

Sinha, S.S. and H. Roy. 1979. Cytological studies in the genus *Phaseolus*. 1. Mitotic analysis in fourteen species. Cytologia 44:191-199.

Skerman, P.J., D.G. Cameron and F. Riveros. 1988. Tropical forage legumes. Food and Agriculture Organization of the United Nations. Rome. 692 pages.

Smartt, J. 1990. Grain legumes: Evolution and genetic resources. Cambridge University Press. Cambridge, U. K. 379 pages.

Takeya, M. and N. Tomooka. 1997. The illustrated legume genetic resources database on the world wide web. Misc. Publ. Inst. Agrobiol. Resour. 11:1-93.

Tateishi, Y. 1983. Leguminosae collected in the Arun valley, East Nepal. Pages 131-146 in M. Numata (ed.), Structure and dynamics of vegetation in Eastern Nepal, pp. 131-146. Chiba.

Tateishi, Y. 1984. Contributions to the genus *Vigna* (Leguminosae) in Taiwan. Sci. Rep. Tohoku Univ. 4th ser. (Biology) 38:335-350.

Tateishi, Y. 1985. A revision of the Azuki bean group, the subgenus *Ceratotropis* of the genus *Vigna* (Leguminosae). Ph. D. Dissertation, Tohoku University, Japan. 292 pages.

Tateishi, Y. 1996. Systematics of the species of *Vigna* subgenus *Ceratotropis*. Pages 9-24 in P. Srinives, C. Kitbamroong and S. Miyazaki (eds.) Mungbean germplasm: Collection, evaluation and utilization for breeding program. JIRCAS, Japan.

Tateishi, Y. and H. Ohashi. 1990. Systematics of the Azuki bean group in the genus *Vigna*. pp. 189-199. In Fujii, K. A.M.R. Gatehouse., C.D. Johnson., R. Mitchel and T. Yoshida (eds.). Bruchids and Legumes: Economics, Ecology and Coevolution. Kluwer Academic Publishers, Netherlands.

Tateishi, Y. and N. Maxted. 2002. New species and combinations in *Vigna* subgenus *Ceratotropis* (Piper) Verdcourt (Leguminosae, Phaseoleae). Kew Bull. 57(3): 625-633.

Thuan, N.V. 1979. Flore du Cambodge du Laos et du Vietnam 17. Legumineuses-Papilionoidees, Phaseolees.

Tomooka, N., C. Lairungreang, P. Nakeeraks, Y. Egawa and C. Thavarasook. 1991. Mungbean and the genetic resources, the subgenus *Ceratotropis*. Tropical Agriculture Research Center, Tsukuba, Japan. 67 pages.

Tomooka, N., C. Lairungreang, P. Nakeeraks, Y. Egawa and C. Thavarasook. 1992a. Center of genetic diversity and dissemination pathways in mung bean deduced from seed protein electrophoresis. Theor. Appl. Genetics. 83:289-293.

Tomooka, N., Ithnin Bin Bujang, S.A. Samm and Y. Egawa. 1993. Exploration and collection of wild *Ceratotropis* species in Peninsular Malaysia. Annual Report of Exploration and Introduction of Plant Genetic Resources (NIAR, Tsukuba, Japan) 9:127-142.

Tomooka, N., H. Nakayama, K. Yamada and A. Sugimoto. 1994. Exploration for collecting landraces of cultivated crops in Tanegashima and Yakushima islands, Kagoshima prefecture. Annual report of exploration and inroduction of plant genetic resources (NIAR, Tsukuba, Japan).10:15-24. (In Japanese with English summary)

Tomooka, N., C. Lairungreang and Y. Egawa. 1996. Taxonomic position of wild *Vigna* species collected in Thailand based on RAPD analysis. Pages 31-40 in P. Srinives, C. Kitbamroong and S. Miyazaki (eds.) Mungbean germplasm: Collection, evaluation and utilization for breeding program. JIRCAS, Tsukuba, Japan.

Tomooka, N., S. Chotechuen, N. Boonkerd, B. Taengsan, S. Nuplean, D.A. Vaughan, Y. Egawa, T. Yokoyama and Y. Tateishi. 1997. Collection of seed samples and nodule samples from wild subgenus *Ceratotropis* species (genus *Vigna*) in Central and Northern Thailand. Annual Report of Exploration and Introduction of Plant Genetic Resources (NIAR, Tsukuba, Japan) 13:189-206. (in Japanese with English summary)

Tomooka, N., D. Vaughan, R. Q. Xu and K. Doi. 1999. Wild relatives of crops conservation in Japan with a focus on *Vigna* spp. Introduction. Annual Report on Exploration and Introduction of Plant Genetic

Resources (NIAR, Tsukuba, Japan) 14:45-61.

Tomooka, N., Y. Egawa and A. Kaga. 2000a. Biosystematics and genetic resources of the genus *Vigna* subgenus *Ceratotropis*. Pages 37-62 *in* Proceedings of the 7th MAFF International Workshop on Genetic Resources. Wild Legumes. NIAR, Tsukuba, Japan.

Tomooka, N., K. Kashiwaba, D.A. Vaughan, M. Ishimoto and Y. Egawa. 2000b. The effectiveness of evaluating wild species: searching for sources of resistance to bruchid beetles in the genus *Vigna* subgenus *Ceratotropis*. Euphytica 115:27-41.

Tomooka, N., A.S.U. Liyanage and J. Takahashi. 2000c. Field survey of the *Vigna* subgenus *Ceratotropis* species in Sri Lanka. Annual Report on Exploration and Introduction of Plant Genetic Resources (NIAR, Tsukuba, Japan) 16:187-196. (in Japanese with English summary)

Tomooka, N., Y. Egawa, Y. Tateishi, T. Yamashiro and D.A. Vaughan 2000d. Wild relatives of crops conservation in Japan with a focus on *Vigna* species. 1. Collecting mission on Okinawa, Ishigaki and Iriomote. 22nd -26th Feb. 1999. Annual Report on Exploration and Introduction of Plant Genetic Resources.(NIAR, Tsukuba, Japan) 16: 39-49.

Tomooka, N., P. Srinives, D. Boonmalison, S. Chotechuen, B. Taengsan, P. Ornnaichart and Y. Egawa. 2000e. Field survey of high temperature tolerant Asian *Vigna* species in Thailand. Annual Report on Exploration and Introduction of Plant Genetic Resources (NIAR, Tsukuba, Japan) 16:171-186. (in Japanese with English summary)

Tomooka, N., D.A.Vaughan, R.Q. Xu, K. Kashiwaba and A. Kaga. 2001a. Japanese native *Vigna* genetic resources. Jpn. Agric. Res. Quart. 35:1-9.

Tomooka, N., A. Kaga, D.A. Vaughan, N. Kobayashi, T. Yoshida, T. Nobori, T. Komatsuzaki, M. Akiba, T. Omizu, T. Taguchi and B. Pickersgill. 2001b. Monitoring and collecting of the azuki bean complex (*Vigna angularis*) genepool in Tottori, Okayama, Ibaraki and Tochigi Prefectures, Japan, 2000. Annual Report on Exploration and Introduction of Plant Genetic Resources (NIAR, Tsukuba, Japan) 17:17-32. (in Japanese with English summary)

Tomooka, N., N. Maxted, C. Thavarasook and A.H.M. Jayasuriya. 2002a. Two new species, new species combinations and sectional designations in *Vigna* subgenus *Ceratotropis* (Piper) Verdcourt (*Leguminosae, Phaseoleae*). Kew Bull. 57(3): 613-624.

Tomooka, N., M.S. Yoon, K. Doi, A. Kaga and D.A. Vaughan. 2002b. AFLP analysis of diploid species in the genus *Vigna* subgenus *Ceratotropis*. Genet. Res. Crop Evol. 49(5):521-530.

Vaillancourt, R.E. and N.F. Weeden. 1993. Chloroplast DNA phylogeny of Old World *Vigna* (Leguminosae). Syst. Bot. 18:642-651.

van der Maesen, L.J.G. and S. Somaatmadja (eds.). 1989. Plant Genetic Resources of South-East Asia no. 1 Pulses. Pudoc Wageningen 105 pages.

Vanderborght, T. n.d. Wild *Phaseoleae - Phaseolinae* collection. IPGRI base collection for wild *Phaseolus* and *Vigna* species. (unpublished paper from National Botanic Garden of Belgium, Meise).

Vaughan, D.A., N. Tomooka, R.Q. Xu, A. Konarev, K. Doi, K. Kashiwaba and A. Kaga. 2000. The *Vigna angularis* complex in Japan. Pages 159-176 *in* Wild legumes. 7th MAFF international Workshop on Genetic Resources. NIAR, Tsukuba, Japan.

Vaughan, D.A. and A. Kaga. 2000. Wild relatives of crops conservation in Japan with a focus on *Vigna* species. 2. Monitoring and collecting mission in Tottori and Okayama prefectures, Japan. 22-24th September 1999. Annual Report on Exploration and Introduction of Plant Genetic Resources. (NIAR,

Tsukuba, Japan) 16:51-57.

Vavilov, N., 1922. The law of homologous series in the case of variation. J. Genet. 12:47-89.

Vavilov, N.I. 1926. Studies on the origin of cultivated plants. Inst. Appl. Bot. Plant Breed., Leningrad.

Verdcourt, B., 1970. Studies in the *Leguminosae-Papilionoideae* for the 'Flora of Tropical East Africa', IV. Kew Bull. 24:507-569.

Vitousek, P.M., L.L. Loope and H. Adsersen (eds.). 1995. Islands: biological diversity and ecosystem function. Springer-Verlag, Berlin, Germany.

Wang, X.W., A. Kaga, N. Tomooka and D.A.Vaughan. 2002. Development and characterization of SSR markers in azuki bean [*Vigna angularis* (Willd.) Ohwi & Ohashi]. In process

Ward, G.C. 2002. India's Western Ghats. National Geographic January 90-109.

Wiersema, J.H. and LB. León. 1999. World economic plants. A standard reference. CRC Press, New York.

Xiong, L., K. Liu, X. Dai, C. Xu and Q. Zhang. 1999. Identification of genetic factors controlling domestication-related traits of rice using an F2 population of a cross between *Oryza sativa* and *O. rufipogon*. Theor. Appl. Genet. 98:243-251.

Xu, R.Q. N. Tomooka and D.A. Vaughan. 2000a. AFLP markers for characterizing the azuki bean complex. Crop Sci. 40:808-815.

Xu, R.Q., N. Tomooka, D.A. Vaughan and K. Doi. 2000b. The *Vigna angularis* complex: Genetic variation and relationships revealed by RAPD analysis, and their implications for *in situ* conservation and domestication. Genet. Res. Crop Evol.47:123-134.

Yamada, T., M. Teraishi, K. Hattori and M. Ishimoto. 2001. Transformation of azuki bean by *Agrobacterium tumefaciens*. Plant Cell, Tissue and Organ Culture 64: 47-54.

Yamaguchi, H. 1992. Wild and weed azuki beans in Japan. Econ. Bot. 46:384-394.

Yamaguchi, H. and Y. Nikuma. 1996. Biometric analysis on the classification of weed, wild and cultivated azuki beans. Weed Res. (Japan) 41:55-62.

Yasuda, K. and H. Yamaguchi. 1998a. Life history of wild and weed azuki beans under different weeding conditions. J. Weed Sci. Tech. 43:114-121.

Yasuda, K. and H. Yamaguchi. 1998b. A gathering experiment concerning the early stage of domestication in azuki bean. Noukou no Gijutsu to Bunka 21:137-155. (in Japanese)

Yee, E., K.K. Kidwell, G.R. Sills and T.A. Lumpkin. 1999. Diversity among selected *Vigna angularis* (Azuki) accessions on the basis of RAPD and AFLP markers. Crop Sci. 39:268-275.

Yokoyama, T., S. Ando, N. Tomooka, D.A. Vaughan and K. Tsuchiya. 1999. The common nodulation genes of the genus *Bradyrhizobium* isolated from soybean (*Glycine max*) compared with those isolated from wild *Vigna* species in Thailand. Pages 45-56 *in* Proceedings of International Conference on Asian Network on Microbial research. Chiang Mai, Thailand Nov. 29th -1st Dec. 1999.

Yoon, M.S., A. Kaga, N. Tomooka and D.A. Vaughan. 2000. Analysis of genetic diversity in the *Vigna minima* complex and related species in East Asia. J. Plant Res. 113:375-386.

Yoshizaki, M. 1995. Appearance of the cultivated plants in ancient Japan. Archaeology Quarterly 50:18-24. (in Japanese)

Yoshizaki, M. 1997. Jomon agriculture: Retrieval of evidence. The Quaternary Research 36:343-346. (in Japanese with English summary)

Young, N.D., L. Kumar, D. Menacio-Hautea, D. Danesh, N.S. Talekar, S. Shanmugasundarum and D.-H. Kim. 1992. RFLP mapping of a major bruchid resistance gene in mungbean (*Vigna radiata*, L. Wilczek).

Theor. Appl. Genet. 84:839-844.

Young, N.D., D. Danesh, D. Menancio-Hautea and L. Kumar. 1993. Mapping oligogenic resistance to powdery mildew in mungbean with RFLP's. Theor. Appl. Genet. 87:243-249.

Zeven, A.C. and De Wet, J.M.J. 1982. Dictionary of cultivated plants and their regions of diversity. Pudoc, Wageningen. 263 pp.

Zink, D., K. Schumann and W. Nagl. 1994. Restriction length polymorphisms of the phytohemagglutnin genes in *Phaseolus* and *Vigna* (Leguminosae). Pl. Syst. Evol. 191:131-146.

Zohary, D. 1989. Pulse domestication and cereal domestication: How different are they? Econ. Bot. 43:31-34.

Zong, X.X., A. Kaga, N. Tomooka, X.W. Wang and D.A. Vaughan. 2002. Genetic diversity of the azuki bean complex using AFLP markers. (in process)

INDEX TO SCIENTIFIC NAMES

Page numbers in bold italics refer to plates. Page numbers in italics refer to species description.

INDEX

Page numbers in bold italics refer to plates

The manufacturer's authorised representative in the EU is Springer
Nature Customer Service Centre GmbH, Europaplatz 3, 69115 Heidelberg,
Germany. If you have any concerns regarding our products, please
contact ProductSafety@springernature.com

Printed and bound by CPI Group (UK) Ltd, Croydon, CR0 4YY
23/04/2026
02095585-0005